HUME AND THE HEROIC PORTRAIT

HUME AND THE HEROIC PORTRAIT

Studies in Eighteenth-Century Imagery

EDGAR WIND

EDITED BY JAYNIE ANDERSON

CLARENDON PRESS · OXFORD
1986

Oxford University Press, Walton Street, Oxford OX2 6DP
Oxford New York Toronto
Delhi Bombay Calcutta Madras Karachi
Kuala Lumpur Singapore Hong Kong Tokyo
Nairobi Dar es Salaam Cape Town
Melbourne Auckland
and associated companies in
Beirut Berlin Ibadan Nicosia

Oxford is a trade mark of Oxford University Press

Published in the United States
by Oxford University Press, New York

British Library Cataloguing in Publication Data
Wind, Edgar
Hume and the heroic portrait: studies in 18th-century imagery.
1. Painting, English 2. Painting, Modern—17th–18th centuries—England
I. Title. II. Anderson, Jaynie
759.2 ND466
ISBN 0-19-817371-7

Set by Eta Services, Beccles, Suffolk
Printed in Great Britain
at the University Printing House, Oxford
by David Stanford
Printer to the University

Editor's Preface

THE title 'Hume and the Heroic Portrait' was chosen by Edgar Wind many years ago for a projected translation of his paper 'Humanitätsidee und heroisiertes Porträt in der englischen Kultur des 18. Jahrhunderts', the principal study in the present collection of his essays on eighteenth-century imagery—a companion volume to *The Eloquence of Symbols.* Wind delivered this paper on English portraiture in 1931, at the Warburg Library in Hamburg, as one of a series of lectures on 'England und die Antike' given by leading German and English scholars. The Hume essay marked a change of direction in his own development, a move away from the aesthetic and philosophical problems that preoccupied him during his early years to the concrete analysis of particular works of art. Wind was the first scholar of the original Warburg group to make a study of English art, and his philosophical approach to portrait painting was unique not only for the period under consideration, but also for the study of this genre in other contexts.

Eighteenth-century studies continued to occupy Wind during his London years, from 1933 to 1939. They were then gradually driven aside by an absorbing interest in Renaissance imagery, which had begun under the influence of Warburg and was to become the dominant theme of his later research. The interconnection of Renaissance and eighteenth-century ideas and images is concisely demonstrated in his short note on Reynolds's parody of *The School of Athens*, where Bellori's important observation about subject-matter in Raphael's painting is illustrated in Reynolds's caricature of English visitors to Rome.

For many of these essays, Wind left annotated offprints, additional notes, and references to further sources and bibliography. For example, to his brief article on 'The Mænad under the Cross', and Reynolds's observations on the subject, he added many years later Goethe's reflections on the same Warburgian theme, and these have been incorporated in an additional note. When additions to the notes are my own, they are indicated by 'see now' or by the use of square brackets.

The translation of the essay on Hume was shared by Mr T. J. Reed, Professor P. F. Ganz, Dr Christian Gooden, Dr Lotte Labowsky, and Mrs Joyce Reid. I am grateful to Dr P. M. S. Hacker, Professor Raymond Klibansky, and Mrs Olive Sayce for enlightenment on many difficult questions in this study. Professor Klibansky has also contributed to the essay 'Shaftesbury as a Patron of Art' an important note on Shaftesbury's copies of Horace and Marcus Aurelius.

While editing these papers, I have frequently consulted Mr and Mrs D. M. Davin, Dr

John Gage, Professor Austin Gill, Professor Hugh Lloyd-Jones, Professor Charles Mitchell, and my husband, Dr Richard Pau, and have greatly benefited from their advice. For the solution of particular problems, I am indebted to Dr Robert Beddard, Dr David Blayney Brown, Mr Howard Colvin, the late Mr John Cooper, Professor Gerhard Croll, Miss Elizabeth Einberg, Dr A. Levi Fox, Mr Patrick Gardiner, Dr Kenneth Garlick, Professor Francis Haskell, Mr Rupert Hodge, Mr John Ingamells, Mrs Patricia Jaffé, Mr Ian Lowe, Monsieur Bruno Neveu, the late Professor P. H. Nidditch, Professor John North, Mr Peter Parsons, Dr Nicholas Penny, Lord Quinton, Mr Albi Rosenthal, Mr Bede Rundle, Dr Daniel Waley, Sir Ellis Waterhouse, and Dr Bryan Wilson.

In the later stages of the work on this volume, I have received invaluable assistance from Mrs Reid, who applied her critical acumen to every aspect of the text. But as with *The Eloquence of Symbols*, by far my greatest debt is to Mrs Margaret Wind. She has not only advised and encouraged me throughout; she has also undertaken a great deal of the editorial work.

Permission to reprint the essays has been kindly given by the editors of the periodicals concerned, notably the *Journal of the Warburg and Courtauld Institutes*. Lastly, these volumes and their editor owe much to the patience and dedication of the staff of the Clarendon Press.

JAYNIE ANDERSON

Wolfson College, Oxford

Contents

The Delegates of the Oxford University Press are grateful to Miss Louise Crane and the Josephine B. Crane Foundation for a generous grant towards the costs of publication

List of Illustrations

Grateful acknowledgement is made to museums, libraries, and private collectors for permission to reproduce works of art in their possession. Where no source is given in brackets the photograph has been supplied by the owners.

Illustrations within the Text

I · Hume and the Heroic Portrait

Introduction
The Interaction of Philosophy and Art in the Eighteenth Century

IN this paper I shall attempt to relate English portrait painting of the eighteenth century to the philosophical views of the time on the nature and dignity of man. This is not to say that I shall attempt with the hindsight of *Geistesgeschichte* to establish some link that was hidden from contemporaries. On the contrary, I shall seek through the study of documentary evidence—both texts and pictures—to rediscover a relation which Englishmen of the eighteenth century were well aware of, but which has become obscured by the passage of time. It is well known to historians of art that Reynolds and Gainsborough were rivals engaged in both a private and a professional feud, and that their patrons regarded them as champions of two opposing styles.[1] It is equally well known to historians of philosophy that the sceptical views of David Hume set off a philosophical controversy in which Samuel Johnson, James Beattie, and others passionately opposed Hume's view of human nature.[2] What has been forgotten is that these artistic and philosophical debates were but phases of one and the same conflict between two parties, each of which united artists and thinkers in a common cause. This came to be forgotten not least because an over-timid division of labour in modern historical studies has led scholars to examine artistic and philosophical documents in isolation from each other, instead of studying them in their interaction.

I am deeply indebted to a number of English collectors and museums. In the Witt Library I was able to peruse material that gave direction to my research, and it was through the kind offices of Sir Robert and Lady Witt that I was able to visit some of the most important private collections. K. T. Parker gave me valuable assistance both in the Print Room of the British Museum and far beyond it, and on specific matters C. H. Collins-Baker of the National Gallery and Henry M. Hake of the National Portrait Gallery offered most useful advice. In matters of theatrical history I have consulted the Burney Collection in the British Museum, the Enthoven Collection in the Victoria and Albert Museum, and the Collection of the Corporation of Stratford-upon-Avon. A visit to the collection of the Garrick Club proved especially valuable, owing to the illuminating comments of Dr Bruce Richmond who kindly showed me round.

I should like particularly to thank those collectors who have allowed me to see and describe or to reproduce pictures in their possession, among whom I mention here only Lady Leconfield, Lord Lee of Fareham, Lord Sackville, Sir Philip Sassoon, and Oscar Ashcroft, Esq.

[1] See, for example, W. T. Whitley, *Artists and their Friends in England 1700–1799*, 2 vols. (1928).

[2] See R. Metz, 'Bibliographie der Hume-Literatur', *Literarische Berichte aus dem Gebiete der Philosophie*, XV–XVI (1928), pp. 39–50; T. E. Jessop, *A Bibliography of David Hume and of Scottish Philosophy from Francis Hutcheson to Lord Balfour* (1938); and now R. Hall, *Fifty Years of Hume Scholarship. A Bibliographical Guide* (1978).

'Humanitätsidee und heroisiertes Porträt in der englischen Kultur des 18. Jahrhunderts', in *England und die Antike*, *Vorträge der Bibliothek Warburg 1930–1931* (1932), pp. 156–229; an expanded version of a lecture delivered at the Bibliothek Warburg, Hamburg. Further material is here added from Wind's papers.

Even the most elementary statements on art by Reynolds or Gainsborough contain philosophical assertions which only need to be made explicit in order to show that the painter has a philosophical bent or even consciously adheres to a philosophical school. When Gainsborough, for example, voices a mistrust of all intellectual pronouncements, indeed of anything with the least pretence to intellectual sophistication, and when he condemns cultural refinement acquired through the reading of books, preferring to rely on the natural instinct of the painter, his beliefs have an affinity with the philosophical tenets of the sceptics. For they aim to show, with subtle arguments, that reason is unreliable in all things, and that one must depend instead on natural feeling as the ultimate, indeed the only meaningful, arbiter of what is just and true.[3] And conversely, the almost Philistine earnestness with which Reynolds in his very activity as a painter proclaims his adherence to a philosophical system displays a faith in the power and efficacy of reason such as can only be possessed by a man who, quite untroubled by the misgivings of the sceptics, seeks to develop and perfect his artistic technique through constant reference to reason, who will not regard the success of a work as secure unless it can be justified on an intellectual plane. In his youth, Reynolds was introduced to Platonic philosophy by a clergyman, the Revd Zachariah Mudge of Exeter, and the 'dispositions to generalize and to view things in the abstract' remained with him.[4] In the literary Club which he founded, as he said, to hear Johnson talk, and at whose meetings such men as Burke and Goldsmith were also heard, it was customary to discuss artistic questions in the closest possible connection with those of moral philosophy.[5]

On the other side, too, there is evidence of the interaction between artistic and philosophical movements. When Hume attacked that view of human dignity which sees an exalted existence as the only one fitting for man, and when he rejected the absolute urge to perform heroic deeds as a senseless and presumptuous attempt to exceed the limits imposed upon man by Nature, he came into conflict not only with moralists like Johnson and philosophers like Beattie, but also with painters like Reynolds. One negative outcome was that Hume did not have his portrait painted by Reynolds.[6] More significantly, when Reynolds

[3] Gainsborough's scepticism is vividly shown in the sarcasm and humour of his letters to William Jackson of Exeter, which are preserved in the Royal Academy and have been published by Whitley, *Thomas Gainsborough* (1915), pp. 379 ff., albeit with 'slight expurgations, for Gainsborough's intimate letters sometimes contain coarse expressions'; see also *The Letters of Thomas Gainsborough*, ed. M. Woodall (1963), nos. 48–62 (hereafter in this edn.).

[4] A letter from Edmund Burke to Edmond Malone, 4 May 1797, tells of Reynolds's mentor the Revd Mr Mudge, who disposed him to 'philosophize in the spirit of the Platonists'. It was published in part by Malone in the introduction to his 2nd edn. of *The Works of Sir Joshua Reynolds*, I (1798), pp. xxxii ff., then in full by C. R. Leslie and T. Taylor, *Life and Times of Sir Joshua Reynolds*, II (1865), pp. 637 ff., together with other notes by Burke on Reynolds. See also F. W. Hilles, *The Literary Career of Sir Joshua Reynolds* (1936), p. 7.

[5] Boswell's *Life of Johnson*, ed. G. B. Hill and L. F. Powell, 6 vols. (1934–50), remains the most important source of information on these meetings.

[6] Although an occasional picture, such as the one sold at Christie's in 1859, has been alleged to be a portrait of Hume by Reynolds (A. Graves and W. V. Cronin, *A History of the Works of Sir Joshua Reynolds P.R.A.*, II, 1899, p. 494), there is, in no instance, any proof of authenticity, despite the wealth of source material. [For Hume's diffident reaction to his portrait appearing as an engraved frontispiece to his *History of England*, see *The Letters of David Hume*, ed. J. Y. T. Greig (1932), I, no. 193; and II, nos. 354 and 411. Nevertheless, Hume was eventually persuaded to sit for a drawing by John Donaldson, which was then engraved by Ravenet and prefixed to the 1768 edn. of the *Essays and Treatises* and the 1770 edn. of the *History*.]

had the opportunity to paint Hume's adversary Beattie, he did not fail to include Hume in the allegorical trappings of Beattie's portrait as the defeated and castigated thinker.

Hume's own choice of portraitist emphasizes the contrast and gives artistic significance to the philosophical dispute. For the most important likeness we have of Hume comes from the brush of his friend Allan Ramsay—the very painter who, until Gainsborough moved to London, was Reynolds's greatest rival, painting in the fluent, gallant manner Gainsborough was later to perfect.[7]

These examples, cited here to indicate the method I intend to follow, show how the artist can enter with his work into a philosophical debate, and support in his painting those who appear to confirm his artistic maxims. But they also show that the philosopher whom he is opposing is, for his part, not restricted to expressing his views on man in his philosophical treatises, but can also express them indirectly—and artistically—in his choice of portraitist.[8] Primarily, of course, pictures are there to be seen and philosophical propositions to be understood. But while remaining fully aware of the difference between the two, we may nevertheless look for philosophical implications in a picture and artistic implications in a philosophical proposition, thus allowing each type of document to shed light on the other. In the eighteenth century, most English artists felt impelled to clarify things rationally; almost without exception they were *raisonneurs*. Conversely, practically all philosophers of the Enlightenment were convinced of the importance of a 'just taste' and sought to show this in their choice of subject-matter, in the arrangement of their arguments, and in the care they devoted to their literary style.

Portraiture shows this give and take between artists and philosophers especially clearly. Attached to the painting of a portrait is a social situation, in which the artist has to come to terms with an attitude, that of his sitter, an attitude that will often be supported by philosophical views, and which, should the sitter happen to be a professional philosopher, will result in the artist producing—or being obliged to produce—an argument in paint. Then there is the correspondence between the objectives of the painter and of the philosopher, the one aiming at the representation of individual man, the other enquiring into the nature of man, and both working in an intellectual climate which fostered speculation on the aesthetic and moral qualities of 'the perfect man'. As a result,

[7] A letter from Horace Walpole to Sir David Dalrymple, 25 February 1759, contains a report about Reynolds and Ramsay, which could, if the date were put forward some twenty years, refer just as well to Gainsborough as to Ramsay without any disturbance of the meaning: 'He [Mr. Ramsay] and Mr. Reynolds are our favourite painters and two of the very best we ever had. . . . Mr. Reynolds and Mr. Ramsay can scarce be rivals, their manners are so different. The former is bold and has a kind of tempestuous colouring, yet with dignity and grace; the latter is all delicacy. Mr. Reynolds seldom succeeds in women; Mr. Ramsay is formed to paint them.' (*Horace Walpole's Correspondence*, ed. W. S. Lewis, xv, 1952, p. 47.) [A letter from Ramsay to Mrs Montagu, 11 September 1760 (M. Allentuck, 'David Hume and Allan Ramsay', *Studies in Scottish Literature*, IX, 1972, pp. 265 f.), reveals that it was 'by much drinking with David Hume and his associates', that 'I have learnt to be very historical; and am nightly confirmed in the belief that it is much easier to tell the *How* than the *Why* of any thing; and that it is moreover better suited to the state of man; who, we are all satisfied, from self-examination, is any thing rather than a rational animal'.]

[8] [Comments on the unsatisfactory nature of historical studies which treat developments in parallel instead of in their interaction have been omitted here. See Wind's introduction to *Kulturwissenschaftliche Bibliographie zum Nachleben der Antike*, I (1934), pp. vii ff.]

the historian is able to use the works of art of the period to illustrate the controversy about the idea of Man and to find the conflict of views concerning human nature reflected in the pictorial representations of particular human beings.

In what follows, I shall take Hume's moral philosophy as a starting-point to describe the polemics between the protagonists of a sceptical and of a heroic view of life. I shall then try to demonstrate the precise point at which this essentially ethical controversy suddenly changed into one about the theory of art. Finally, I shall show, by means of concrete examples, what effect this conflict between different conceptions of art actually had on the practice of portrait painting of the time.

I *Critique of Human Nature*
The Ideal of Natural Sentiment—David Hume

In his essay 'Of the Dignity or Meanness of Human Nature',[9] Hume reflects on man's habit of considering himself in two ways. He compares himself with the beasts and discovers that he possesses powers which place him above them. He can plan for the future, he can investigate remote events; he can foresee, remember; he can even form ideas about things far beyond anything experienced in nature. Surveying these powers, he is seized by immense pride, he feels singled out by God, believes himself divinely inspired. He is in a state of enthusiasm. But man is also capable of the reverse sentiments. He conceives of an absolutely perfect being and, comparing himself with this being, he sinks to the level of the beasts. He feels like a worm crawling on the ground; he is filled with an immense fear of a being so greatly superior to himself; and to ensure the benevolence of this supernatural power, he offers it gifts and vows, and attempts to discover its will and intention from mysterious portents. In short, he is in a state of superstition.

Fear, which is the root cause of superstition, and self-confident pride, which is the basis of enthusiasm, are both born of imagination. Whether man is by nature admirable or contemptible, nearer to the deity or to the worm, is a question of comparison. The answer will depend upon one's vantage-point. Only imagination prevents us from seeing that the demigod and the worm may still be substantially the same man. The situation is the same when we marvel at a large dog or a small horse: imagination makes us forget that both, mathematically speaking, may well be the same size. We are similarly deceived by pride or by humility, whether we feel ourselves elevated above common nature or annihilated by fear.

For these excesses of the human mind there is, according to Hume, only one remedy: a sceptical critique of the imagination. The enthusiast must learn from the sceptic that to believe he can aspire to supernatural heights with impunity is to ignore the limits and the

[9] *Essays Moral, Philosophical, and Literary* (1753–6), xi (Oxford, 1963), pp. 81 ff. See also 'Of Superstition and Enthusiasm', *Essays*, x, pp. 75 ff.; 'Of Pride and Humility', in *A Treatise of Human Nature* (1739), Bk. II, part i, now in the edn. of L. A. Selby-Bigge and P. H. Nidditch (1978), pp. 275 ff.; and the corresponding passages in *A Dissertation on the Passions* (1757), Sect. II, in *The Philosophical Works of David Hume*, ed. T. H. Green and T. H. Grose, IV (1875), pp. 144 ff.

weakness of human understanding. And the superstitious man must recognize, his fear duly moderated by scepticism, that the absolutely perfect being whose omnipotence he feared is an idea of his own creation which cannot be entertained without contradictions. Enthusiasm and superstition are dissolved by scepticism and give way to a sense of man's limitations.

This gentle moderation is what Hume calls 'natural'. For him, those excesses of immoderate pride or humility prove themselves unnatural by the mere fact that they cannot long be sustained. In an exceptional moment a man can be so filled with enthusiasm that he feels close to the divine; but very soon he will sink back into the sobering activity of everyday life. In an exceptional moment he may be so filled with humility before God that he feels like a worm; yet very soon his natural impulses will release him from this sense of annihilation. It is, of course, possible to prolong these fleeting states of excess, ostensibly at least, by artificial means. The enthusiast tries to keep his raptures alive by artificial practices; the superstitious man constantly tries to re-awaken his feeling of fear. And yet, the very act of artificially prolonging these emotions beyond the exceptional moment inevitably destroys their authenticity, and makes hypocrites of those who profess them.

Two nations, according to Hume, have sometimes misused the powers of intellect to stir up enthusiasm and superstition artificially—the ancient Greeks and the French.[10] Two men, in particular, seem to him, by the artificiality of their way of life, the very symbols of 'extravagant philosophy': Diogenes and Pascal. It is striking that these two men upheld with equal energy opposite answers to every vital question. Diogenes prided himself on being self-sufficient; Pascal took great pains to foster his feeling of utter dependence. Diogenes affected indifference to all human claims to consideration and railed uncouthly at his friends; Pascal nursed every compassionate impulse so anxiously that he claimed to love even his enemies. Diogenes indulged in the most beastly pleasures, even in public; Pascal denied himself the most innocent pleasures, even in private.[11] These two men differed in every single trait, and yet they have one characteristic in common: an artificiality of conduct which seemed to them the only way of sustaining their 'extravagant philosophy'.

For Hume, these 'artificial lives' spell the breakdown of all morality. Where such artificiality prevails, special ways of life develop which can never be commonly accepted. If this artificiality were general, we should have to despair of any understanding between the different peoples of the world. The theory that there are no generally valid moral laws, that each nation has particular customs which cannot be compared with those of another, would be valid if the artificiality of the way of life in all countries were raised to a principle. But in reality it cannot gain acceptance as a principle, for 'nature is always too strong for principle'.[12] Wherever artificial practices are found to persist, they serve natural needs common to all men, even if they gratify them in different ways. These fundamental needs, which give artificial customs their significance, are easily agreed upon. But where artificial

[10] 'A Dialogue', see Hume's *Enquiries concerning Human Understanding and concerning the Principles of Morals* (1748 and 1751), ed. Selby-Bigge and Nidditch (1975, reprinted from the posthumous edn. of 1777), pp. 333 ff.

[11] Ibid., pp. 342 f.

[12] *An Enquiry concerning Human Understanding*, Sect. XII, part ii, § 128 (Selby-Bigge and Nidditch, p. 160).

rules are made which run counter to these needs, they cannot survive and are breached by life itself. Again and again, the natural feeling they seek to violate proves strong enough to prevail against them. The sceptic can thus confidently rely on nature in his battle against enthusiasm and superstition. However much the rules of conduct based upon them contradict one another, and make men doubt the existence of such things as universal moral laws, the sceptic can only rejoice. For by appealing to this doubt, and exposing the absurdity of these artificial systems, he lays bare natural feeling as the root of moral conduct.

The positive ethical theory which Hume builds on this foundation is surprisingly simple; for all its individual rules can be reduced in the last analysis to the one maxim, that personal merit consists in nothing other than possessing qualities 'useful or agreeable to ourselves or others'.[13] At first glance this may appear trivial. But if we see Hume's proposition alongside the theological and metaphysical systems against which it was directed, we can fully understand the revolutionary effect it had. For Hume was, in fact, demanding that the sole criterion of the value of all actions should be whether they are useful or agreeable to anyone; if they do not fulfil this requirement, they must be condemned as worthless. That meant rejecting any theological and supernatural foundation for morality; and countless actions which theologians traditionally praise as virtuous—acts of asceticism, self-denial, or self-torment—are dismissed as meaningless, precisely because nobody under any circumstances can feel them to be useful or agreeable. Indeed, Hume points out that too much preoccupation with the state of one's soul tends to destroy the social virtues which affect natural feeling so agreeably: 'The steady attention alone to so important an interest as that of eternal salvation is apt to extinguish the benevolent affections, and beget a narrow, contracted selfishness. And when such a temper is encouraged, it easily eludes all the general precepts of charity and benevolence.'[14]

Once a sense of what is 'useful' or 'agreeable' is accepted as the criterion of personal merit—as is, in fact, the case with all those who abandon themselves to their natural feeling—merit is transformed from a hard-won right based on exceptional achievements into a directly perceivable, natural quality which appeals to the beholder's feelings in the same way as physical beauty. Accordingly, Hume's ideal of a meritorious life has nothing in common with the accepted view that it demands some heroic excellence; for precisely the kind of forced endeavour to go beyond what Nature has granted destroys those endearing characteristics which make a man useful or agreeable to himself or others.[15]

[13] *An Enquiry concerning the Principles of Morals*, Sect. IX, part i, § 219, also § 217 (Selby-Bigge and Nidditch, pp. 270 and 268).

[14] *Dialogues concerning Natural Religion* (1779), Part XII, ed. N. Kemp Smith (1947), p. 222.

[15] We must be careful not to interpret this doctrine as the kind of utilitarianism that Bentham and Mill were to develop from it. Hume's ethics are far from envisaging a felicific calculus, just as the psychological doctrine of the association of ideas is free from any intention to turn such laws of association into a rigid system, a 'chemistry of conceptions'. This simplification and the shallowness it entails could only occur when the tension between natural and supernatural modes of explanation, which is the nerve of Hume's philosophy, was so far relaxed that the empiricists were no longer obliged to join battle with theology and could feel wholly at home in their observations and calculations. Hume's own attitude, in contrast, can be understood only from his dual concern—with the struggle against a still powerful theological and metaphysical dogma, which threatened to obscure from view the natural behaviour he was interested in; and with the observation of that natural behaviour itself, which despite the threat from theology kept breaking through in daily life.

Scepticism and Humanity—The Critique of Heroism

One has to ask how Hume's rejection of all artificiality in ethics and his justification of natural feeling can be reconciled with his epistemological theory, where his one aim seems to be to invalidate with the most sophisticated arguments the common-sense view of the world. In his ethics, while applying his critique of the imagination to the extremes of enthusiasm and superstition, Hume still gave natural feeling its proper place; but in his theory of knowledge imagination is not only exposed in its excesses, but constantly shown to be playing its tricks in the most simple acts of the understanding. When we say 'there is a dog barking outside', the mere fact that we ascribe the noise to a dog is ultimately attributable to our faculty of imagination; indeed, even the location of it 'outside' is an act of imagination—for the noise is, of course, 'here'. Thus in our daily life our judgements always go beyond what is immediately present to our senses; we yield ourselves to the play of imagination. And was it not the aim of Hume's epistemology to show how treacherous this is? Was he not bent on proving that inferences from the past to the future, from cause to effect—indispensable operations in everyday life—are neither logical nor reliable? Was not the sceptic out to expose the whole substantiality of things as illusory, precisely because on examination they turn out to be products of the imagination?

We can confidently answer 'no' to this question. It is certainly true that in his critique of substance and causality Hume tried to refute any claim that these terms could be meaningful in a strictly logical sense. He did not, however, do so in order to deny that their use as such was justified, but on the contrary to give it a firmer foundation. If, Hume argued, we were dependent on our understanding when using these concepts, then the notorious unreliability of our understanding and the uncertainty abstract reasoning is prone to would make us constantly vulnerable. But Nature has served us better than that. Instead of entrusting such vital beliefs as the expectation that the future will conform to the past, or that similar causes will produce similar effects, to an understanding that can always err, she has assigned them to our instinct. The understanding must not presume to replace this instinct, but must scrutinize its own performance, so that anything which might disturb the functions of natural beliefs will be dissolved by scepticism.

The theory that scepticism performs only a regulative function, that it is not an end in itself but merely exerts a practical influence as 'mitigated scepticism', is constantly emphasized in Hume's writings—in the *Treatise*, the *Enquiries*, and the *Essays*, and not least in *The History of England*. It emerges most clearly perhaps in the *Dialogues concerning Natural Religion*.[16] We shall see—says Cleanthes, challenging Philo the Sceptic—how deep your scepticism goes, whether after our conversation you leave by the window or the door, whether you really doubt that your body has weight or can be injured by its fall, which is what your fallible senses and even more fallible experience teach you. At bottom one need not take you Sceptics seriously, since for you there are only two possible ways to behave:

[16] Part I (Kemp Smith, pp. 130 ff.).

either you live according to your lights, in which case you will all perish of suicide by neglect; or you will belie your teaching in your behaviour, in which case you are no better than the Stoic whom you despise and ridicule. For how does he differ from you? He sits in his study imagining the acutest pain and the most calamitous event, tries to conjure up exactly how these torments and dangers will feel, and mentally schools himself to endure them. But when he emerges into real life, all these chimeras disappear and 'the philosopher sinks by degrees into the *plebeian*'. To which Philo the Sceptic replies: I allow your comparison with the Stoics. We devise in the study subtle arguments to show that our natural belief is vain and ill-founded; yet in our lives we behave like other men. But this shows precisely the depth of our scepticism; for when the time comes, we shall abandon even our doubts, because we realize that life is stronger than logical principles. Indeed, this is the value of our scepticism—that anything which can resist it must be acknowledged as willed by Nature. However, holding these views, we do not renounce all influence on life, and in this, despite our disagreements, we feel a kinship with the Stoics we oppose: for when by long practice a man has mentally schooled himself to endure pain, as the Stoics do, then though it is true that, when he suffers real pain, he will certainly not fail to feel it, he will nevertheless react differently from those who are not even familiar with it in thought. He will not rise like a hero above human weakness, but he will be somewhat more heroic than other men who surrender to their weakness completely. And Hume notes that it is the Stoics who have given us particular examples of this virtue:

> Vain Wisdom all and false Philosophy:
> Yet with a pleasing sorcerie could charm
> Pain, for a while, or anguish, and excite
> Fallacious Hope, or arm th' obdurate breast
> With stubborn Patience, as with triple steel.[17]

Hume argues that the same 'imperceptible influence' of his way of thinking on his behaviour must be granted to the Sceptic. For if a man has disciplined himself to doubt, then it is true that he will not be able to maintain the same degree of doubt when dealing with real life as in his study, but even so he will not surrender as readily as other men to views which are pressed upon him. Schooled in doubt as he is, he will confront them with more circumspection. Here Scepticism triumphs over Stoicism, because the mind has not been trained, after all, for superhuman acts, heroic deeds, and other feats of presumption, but in the art of discretion, of assenting to a judgement only in so far as seems justifiable; and this practice can only give rise in man to moderate sentiments. Here the Sceptic can develop two virtues to which the Stoic in his enthusiasm will never attain: tolerance and humanity.

This ideal of humanity and tolerance—the belief that the value of abstract reasoning lies in restraining men in their actions, that strength refined by sceptical scruple puts aside

[17] *Paradise Lost*, II, ll. 565–9, quoted by Hume, *Dialogues concerning Natural Religion*, Part I (Kemp Smith, p. 134).

fanaticism, to be transfigured by grace and gentleness, which are the true marks of naturalness—not only guided Hume's writings on social theory and the philosophy of religion, but also determined his historical judgement. In his *History of England*, every character and every action is measured against the ideal of naturalness, and every deed that springs from an excess of imagination is relentlessly condemned. His judgement on the Earl of Strafford is a typical example. When Strafford shortly before his execution writes his famous letter to Charles I, begging him not to intercede on his behalf but to allow him to be sacrificed, because the people would only be further excited by resistance and might even threaten the crown itself, Hume describes this in great detail, simply and factually, as if to arouse admiration in his readers. But he adds the significant remark: 'We might ascribe this step to a noble effort of disinterestedness, not unworthy of the great mind of Strafford, if the measure which he advised had not been, in the event, as pernicious to his master, as it was immediately fatal to himself.'[18] Hume could not rid himself of the suspicion that at the last moment Strafford let himself be carried away by enthusiasm, that the prospect of playing the martyr appealed more to him than the certainly less glorious but more difficult task of meeting a critical situation with political astuteness. Hume thought that the outcome proved Strafford wrong. As a statesman he should have known that the people, having once tasted blood, would not stop even at the king, and should therefore have tried by all means at his disposal to prevent his own execution instead of furthering it, even at the risk of being later reproached with having tried to save his own skin.

In his description of military scenes, Hume expressed the same mistrust of all actions which have an appearance of heroics and in the popular patriotic sense are a 'good show'. He describes the naval battle at Brest in which Sir Edward Howard died, again in full detail.[19] He tells how the admiral was the first to leap boldly on to the French ship, how the cable was cut and he stood alone on the enemy's deck, how nevertheless he did not stop fighting, and laid about him so gallantly that the French had no alternative but to kill him. Yet Hume once again reverses the sense of his account by pointing to the consequences of this heroism: the English fleet, disheartened by the loss of their commander, had to withdraw. He refers in a footnote to Howard's view that only reckless valour could succeed in naval warfare, but adds that even so some discretion, as the result showed, is advisable.

Hume always felt this faint suspicion when an action commended as heroic had resulted from an excess of imagination. This has a counterpart in the particularly indulgent way he treated all sins born of indolence. The Marquis of Newcastle, who in the midst of war chose a poet rather than a warrior for his adjutant, was so outraged by the energy with which Prince Rupert followed up an advantage that he left the king's service. Hume excused this one great lapse by pointing to the habitual gentleness of his disposition.[20] And the life of inactivity which Richard Cromwell chose after the death of his father is actually praised as the expression of a truly mature mind.[21] Indeed, the portrayal of Charles II, in which

[18] *The History of England* (1792–3), VI, ch. liv, pp. 414 ff.

[19] Ibid., III, ch. xxvii, p. 431.

[20] Ibid., VII, ch. lvii, pp. 11 ff.

[21] Ibid., ch. lxii, pp. 297 f.

Hume emphasizes the king's unreliability and does not acquit him of having almost betrayed and destroyed his own country, ends with the forgiving words: 'Yet may all these enormities, if fairly and candidly examined, be imputed, in a great measure, to the indolence of his temper: A fault which, however unfortunate in a monarch, it is impossible for us to regard with great severity.'[22]

Hume tried to order his own life according to this ideal of gentle moderation, and even in the manner of his death he seemed intent on giving decisive proof that he was able to practise what he preached. When he fell victim to an incurable, but in its effects mild, disease, he spent the last months of his life in such a happy and serene frame of mind that in writing the autobiographical essay *My Own Life* he was able to speak jestingly of himself in the past tense: 'I am, or rather was (for that is the style, I must now use in speaking of myself, which emboldens me the more to speak my sentiments); I was, I say, a man of mild disposition, of command of temper, of an open, social, and cheerful humour, capable of attachment, but little susceptible of enmity, and of great moderation in all my passions.'[23] And having described himself clearly and indulgently as if writing of a stranger or even a historical personality, he ends with the words: 'I cannot say, there is no vanity in making this funeral oration of myself; but I hope it is not a misplaced one; and this is a matter of fact which is easily cleared and ascertained.'

Defence of Heroism by Common Sense
Samuel Johnson and James Beattie

Hume could hardly have expected Samuel Johnson's circle to appreciate this sort of self-evaluation; and the attempt to meet death with the same equanimity as an everyday event could for them be only blasphemy. 'He had a vanity in being thought easy', Johnson told Boswell when the latter was outraged that a man could cling to his unbelief even in the face of death. 'And you are to consider, that upon his own principle of annihilation he had no motive to speak the truth.'[24] With grandiose intolerance and inconsistency—for his words are surely contradictory—Johnson saw Hume's attitude as a great lie. For was it not more likely, he asked, that a man would deny his true feelings, than that he should not feel fear when confronted with the totally unknown condition Providence has in store for him? To Johnson death had always been something dreadful, and profound gloom overcame him whenever the subject was even touched on in the conversations Boswell records. His melancholy was a burden which he could ease only by meticulous religious exercises, the *Prayers and Meditations* in which he daily rendered account to himself of his frailties and sin. But strangely enough, this anxiety about the state of his soul did not make him a narrow egotist, which is what Hume thought must result if a man brooded too much on religious

[22] *The History of England*, VIII, ch. lxix, p. 212.

[23] *My Own Life*, 18 April 1776 (see *Dialogues concerning Natural Religion*, Kemp Smith, pp. 239 f.).

[24] Boswell's *Life of Johnson*, 16 September 1777 (Hill and Powell, III, p. 153).

fears. If a man's style can fairly be taken as an expression of his true self, then it was precisely Johnson's transcendental leanings that gave him the noble style which he could turn with such lightning effect to parody.

To support his ethical theory, Hume was wont to invoke a principle of physics: 'The force of the greatest gravity, say the philosophers, is infinitely small, in comparison to that of the least impulse; yet it is certain that the smallest gravity will, in the end, prevail above a great impulse; because no strokes or blows can be repeated with such constancy as attraction and gravitation.' Hence man should never try to overcome by force those imperceptible but ever-present inclinations which move him in accordance with his natural inertia; on the contrary, he must recognize their power as 'natural', and yield to them with good grace: 'It is certain, from experience, that the smallest grain of natural honesty and benevolence has more effect on men's conduct, than the most pompous views suggested by theological theories and systems. A man's natural inclination works incessantly upon him; for it is for ever present to the mind; and mingles itself with every view and consideration: Whereas religious motives, where they act at all, operate only by starts and bounds; and it is scarcely possible for them to become altogether habitual to the mind.'[25] It was precisely this natural inertia, these imperceptible influences, that Johnson never ceased to fight against, and though the struggle was often a matter of 'starts and bounds', he was still able to turn his inner tension not only into noble eloquence, but also into wit.

Admittedly he could never, even in later years, enjoy a serenity like Hume's. He saw his whole life as a desperate struggle of the spirit to free itself from the shackles of the flesh; nor did he at the end, like Hume, surrender to death easily, even light-heartedly, but died struggling, fighting for clarity of mind by refusing all palliatives: 'Then, I will take no more physick, not even my opiates; for I have prayed that I may render up my soul to GOD unclouded.'[26] Even in moments of lucidity he could not reconcile himself to his fate, and the words he wrote on his deathbed movingly express the torment that kept him on the rack: 'the serenity that is not felt, it can be no virtue to feign.'[27] The 'Roman' nobility of his death is echoed in the laments of his friends: '. . . all who saw him, beheld and acknowledged the *invictum animum Catonis*' (—'*atrocem* animum Catonis', commented Malone).[28]

It was left to a far lesser mind than Johnson's to defend against Hume the idea of heroism that animated him. James Beattie enjoyed Johnson's goodwill because he shared his convictions. But the way he tried to defend them has brought him into disrepute with philosophers, as his arguments were mere declamation. Hume tried to reveal through analysis what underlies our beliefs; Beattie appealed to the beliefs themselves and considered any questioning of them wicked in principle: 'Conviction, and steadiness of principle, is that which gives dignity, uniformity, and spirit to human conduct. . . . It constitutes, as it were, the vital stamina of a great and manly character; whereas scepticism betrays a sickly understanding and a levity of mind, from which nothing can be expected

[25] *Dialogues concerning Natural Religion*, Part XII (Kemp Smith, p. 221).

[26] Boswell, December 1784 (Hill and Powell, IV, p. 415).

[27] Ibid., p. 395.

[28] Ibid., November 1784, p. 374; see Horace, *Odes*, II, i, l. 24.

but inconsistence and folly.'[29] Beattie does not even accept that the practice of scepticism may have a moderating influence, by bringing home to us the limits of our faculties. Are the so-called 'sanguinary principles of bigotry and enthusiasm' really worse, he asks, than the 'milky ones of scepticism and moderation'?[30] True, if all strong convictions are removed, the evil effects of fanaticism will be rooted out too. But will a doctor, merely to prevent us from getting a fever, bless us with consumption? And reaching the climax of his Philippic against Hume's analysis of perceptions and feelings, he exclaims:

> Alas! What is become of the magnificence of external nature, and the wonders of intellectual energy, the immortal beauties of truth and virtue, and the triumphs of good conscience! Where now the warmth of benevolence, the fire of generosity, the exultations of hope, the tranquil ecstasy of devotion, and the pang of sympathetic delight! . . . Body and spirit are annihilated; and there remains nothing (for we must again descend into the jargon of metaphysic) but a vast collection, bundle, mass or heap, of impressions and ideas.[31]

Against all these dangers Beattie's magic panacea is 'common sense'. Common sense gives us certainty about all the things Hume dared to question. Although Beattie more than once tries to define it, essentially he uses the term emotively, and it never acquires a clear meaning. What does he not derive from it! The existence of the external world, the reality of the soul, the continuity of personal identity—and, a mere step further, the certain existence of God. That is the strange paradox which links this philosophical school with the scepticism it combated: whereas Hume used highly subtle, almost abstruse arguments to secure the claims of natural feeling, Beattie tried to validate supernatural sanctions from the simple testimony of common sense. But what we find paradoxical here is all part of the same conflict of attitudes which marks the life and thought of that age, and the paradox is also expressed in its art. For Sir Joshua Reynolds, the painter who strove for heroic and sublime effects, was more robust and earthy in his choice of artistic means than Gainsborough, whose ideal was naturalness, but who was in fact the more sophisticated in his use of painterly nuance.

II *Aesthetic Norms*
Simplicity and Refinement—Hume's Concept of 'la belle Nature'

In the aesthetic writing of the age, the champions of a lofty heroic ideal and those who defended a reliance on natural feeling refined by scepticism joined battle over a specific aspect of artistic technique: the right to use heightened forms which stimulate the imagination immoderately. In his essay 'Of Simplicity and Refinement in Writing', Hume begins by warning in his judicious way against easy solutions; for writers, and artists in general, are by their nature in a dilemma.[32] On the one hand, to be artistic at all, literature

[29] *An Essay on the Nature and Immutability of Truth in Opposition to Sophistry and Scepticism* (1777), pp. 127 f.

[30] Ibid., p. 125.

[31] Ibid., pp. 229 f.

[32] *Essays*, xx (Oxford 1963 edn., pp. 196 ff.).

has to transcend ordinary speech, especially the coarse talk of peasants or provincials. A certain refinement and heightening of tone is therefore a condition of art. But it may lead on to affection and baroque excess. For this reason what is expressed in a refined form must at the same time be expressed with simplicity, and the writer must find a balance between the extremes of triviality and extravagance. But although Hume appears to reject both extremes equally, there is a great difference in the way he views the dangers inherent in each of these faults. Excessive refinement is in his eyes the real enemy, since it appeals to an extravagant imagination; whereas he considers excessive simplicity, if not as beautiful, at least as harmless. His reason is twofold: first, any search for artistic form is bound to lead to refinement, the impulse towards unaffected simplicity never being as strong as that towards affectation; secondly, lack of sophistication in art is only a negative fault—it is boring, whereas over-refinement is positively harmful—it is mannered and insidious. Thus, although Hume aims at finding a middle way and rejects both extremes, he nevertheless judges artlessness with goodwill and indulgence. He demands Nature—admittedly, refined Nature: *la belle Nature* as he calls it.

Enthusiasm and Metaphor
Edmund Burke and Oliver Goldsmith

This aesthetic idea of nature bears the marks of the unresolved dilemma from which Hume started: the artist, in order to be an artist at all, must offer more than nature and yet, to be a true artist, must not deviate from nature. Thus, although Hume refrained from being too explicit, a definite ideal of beauty emerges from his disquisition. Against it may be set what Edmund Burke—Johnson's and Reynolds's friend—wrote in his essay on the psychological origins of the Sublime and Beautiful. Throughout this work Burke shows much more interest in the sublime—indeed, he admits that his purpose is not merely to analyse, but to edify.[33] And this very sense of overwhelming power, of the supernatural, which comes over us in the face of the sublime is, Burke claims, more conducive to spiritual edification than the harmonious feeling of pleasure that beauty gives us. The unknown and the obscure are essential elements of the sublime; any thing which we can grasp fully and wholly perceive sinks to the level of the commonplace and destroys our feeling of exaltation. Hence clarity is the enemy of enthusiasm: 'A clear idea is therefore another name for a little idea.'[34]

It was Oliver Goldsmith who tried to analyse the literary techniques most suited to

[33] *A Philosophical Enquiry into the Origin of our Ideas of the Sublime and Beautiful*, 2nd edn. (1759), I, xix: 'The elevation of the mind ought to be the principal end of all our studies, which if they do not in some measure effect, they are of very little service to us.' Cf. Hume: 'The *imagination* of man is naturally sublime, delighted with whatever is remote and extraordinary, and running, without control, into the most distant parts of space and time in order to avoid the objects, which custom has rendered too familiar to it. A correct *Judgement* observes a contrary method, and avoiding all distant and high enquiries, confines itself to common life, and to such subjects as fall under daily practice and experience. . . .' (*An Enquiry concerning Human Understanding*, Sect. XII, part iii, § 130; Selby-Bigge and Nidditch, p. 162.)

[34] Burke, op. cit., II, iii and iv.

express enthusiasm. His essays 'On Metaphors' and 'On Hyperbole'[35] describe the way language 'becomes figurative' whenever a supernatural emotion seizes the mind. Metaphors, and especially hyperbole, are linguistic symptoms of enthusiasm, and the poet, if he wishes to excite the emotions of his audience, must use these figures. They at once uplift and cloud the mind, and he must use them with moderation. The theory of metaphor as Goldsmith develops it is opposed to metaphorical excess. Repeated use of the same type of metaphor—and he quotes Tamerlane, in Nicholas Rowe's tragedy, who abuses the Turkish emperor Bajazet with a string of nautical conceits—may produce boredom, but mixed metaphors reveal a muddled imagination and should be resisted. That is why he thought—and here he was in accord with Hume[36]—that the eighteenth century's enthusiasm for Shakespeare was somewhat questionable.

Eighteenth-century interpretations of Shakespeare show each party trying to protect 'their' Shakespeare from the distortions of opponents. Both sides had to 'alter' him—one side to make the wonderful naturalness of his 'most beautiful' scenes stretch to those passages where he lapsed into a false and dated rhetoric, or had been careless; the other to purge from his 'best' grand heroic scenes the traces of vulgarity which betrayed the untutored taste of a rude age. This contrast runs through criticism and editions, stage productions and individual actors' interpretations. In grotesque form it even extends to Bardolatry. The vivid colour-scheme in which Shakespeare's old tomb at Stratford had been restored in 1748 offended the sensibilities of his idealist admirers, and Malone, to save the honour of the Swan of Avon, had it whitewashed out of hand—an interference which at once provoked a jibe from his adversaries at his editorial methods:

> Stranger to whom this monument is shown
> Invoke the poet's curse upon Malone,
> Whose meddling zeal his barbarous taste betrays
> And daubs his tomb-stone as he mars his plays.[37]

Battle was joined not only over Shakespeare, but also over the interpretation of ancient and modern poets. In *The Vicar of Wakefield* there is a scene in which the Vicar's son and daughter have a veritable *querelle des anciens et des modernes* about whether Gay or Ovid is the greater poet.[38] 'I never sit thus,' says Sophia, 'but I think of the two lovers, so sweetly described by Mr. Gay, who were struck dead in each other's arms. There is something so

[35] [These two essays were first published in *The British Magazine*, July 1761–January 1763, as part of a series 'On the Study of the Belles Lettres'. They were first attributed to Goldsmith in 1801 (*Miscellaneous Works*, IV) and then later republished among his essays, cf. *The Works of Oliver Goldsmith*, ed. J. M. W. Gibbs, I (1884), pp. 361 ff. and 378 ff. In 1924, C. Tupper ('Essays Erroneously Attributed to Goldsmith', *Publications of the Modern Language Association*, XXXIX, 1924, pp. 325 ff.) proved that the 'Belles Lettres' series was demonstrably not by Goldsmith, a view which has been supported by other writers. Cf. R. S. Crane, *New Essays by Oliver Goldsmith* (1927), pp. xiv f. and xix, and the *Collected Works of Oliver Goldsmith*, ed. A. Friedman, III (1966), pp. XIV and 88 f.]

[36] *The History of England*, VI, Appendix to the Reign of James I, pp. 191 f.

[37] [The verse was written in 1810 as an entry in the Stratford Visitors' Book (Shakespeare Birthplace Trust Records Office, ER 1/24/1, p. 167), and was later quoted in *The Gentleman's Magazine*, LXXXV (1815), p. 390.]

[38] *The Vicar of Wakefield*, viii (Friedman, IV, pp. 46 ff.).

pathetic in the description, that I have read it an hundred times with new rapture.' Her brother protests that 'the finest strokes in that description are much below those in the Acis and Galatea of Ovid. The Roman poet understands the use of *contrast* better, and upon that figure artfully managed all strength in the pathetic depends'. Both are interrupted by Mr Burchell, who is Goldsmith's mouthpiece:

> It is remarkable, that both the poets you mention have equally contributed to introduce a false taste in their respective countries, by loading all their lines with epithet. Men of little genius found them most easily imitated in their defects, and English poetry, like that in the latter empire of Rome, is nothing at present but a combination of luxuriant images, without plot or connexion; a string of epithets that improve the sound, without carrying on the sense.

And to show what kind of art he himself champions, he recites a popular ballad:

> Turn, gentle hermit of the dale,
> And guide my lonely way. . . .

Goldsmith never repudiated this taste for the natural style, even during his friendship with Reynolds; as we shall see, he acted as mediator in the quarrel between Hume and Beattie, sharply criticizing Reynolds for showing his partisanship even in his art.

So far we have traced the battle between enthusiasm and scepticism only in literature. We must now consider what technical means the two parties had at their disposal in the realm of painting. In poetry, the issue was metaphor, and the use of exaggerated images whose obscurity was found stimulating and as such was extolled by one side and rejected, or at least restricted, by the other. But what were the corresponding methods in painting?

Goldsmith himself answered this. In *The Vicar of Wakefield*, there is a caricature of a certain extravagant style of painting which corresponds to the exaggerated use of metaphor in poetry—the so-called 'historical family piece'.[39] The Vicar's family decide to be painted in order to outdo their neighbours, who have had their portraits done by an itinerant artist. These seven people had had themselves painted holding seven oranges, 'a thing quite out of taste . . . no composition in the world'. The Vicar's family decide to be represented in a large portrait in the historical style. But they cannot agree on a theme ('artificiality', just as Hume said, leads to contrariness). So it is settled that each shall choose a pose to his taste. The Vicar's wife appears as Venus, with the two children as Cupids; the Vicar kneels before her as a prelate of dignity and hands her his books. One daughter chooses to be an Amazon and the other a shepherdess. And when the picture is almost finished, the Squire, who is paying court to the Amazon daughter, turns up and insists on being painted at her feet in the figure of Alexander the Great. Significantly, the picture has a major function in the story: the Squire's pose tells both the neighbours and the family that he is courting the Vicar's daughter. The picture is immediately given a social interpretation. At the same time it is a symbol of social pride, for it turns out to be far too big for the Vicar's house. The painter had

[39] *The Vicar of Wakefield*, xvi (Friedman, pp. 82 f.).

put up his easel in the kitchen and no one had given a thought to where the picture was actually going to be hung. When it is finished it will not go through the door and so has to remain in the kitchen, much to the amusement of the neighbours, who admire it as one admires those large ships which have inexplicably been got into small bottles.

III *Theory of Style and the Concept of Portraiture 'Great Style' and 'Simple Style'—Reynolds and Gainsborough*

The 'historical' or 'allegorical' portrait that Goldsmith caricatures was a legacy of the late Dutch school, which found increasing favour among eighteenth-century English artists in their confrontation with the French style. It is a starting-point from which to approach the contrast between Reynolds and Gainsborough. For in the whole of Gainsborough's work there is not a single portrait of this kind, whereas it was to be of fundamental importance for the way in which Reynolds constructed his portraits. Not that Reynolds continually painted portraits of this type. ('Damn the man, how various he is!', Gainsborough is said to have exclaimed on looking at Reynolds's portraits, surprised by the ease with which he moved from the heroic to the pleasing and achieved the most contrasting effects.[40]) Yet the heroic portrait takes pride of place in his work. In his theory of art, Reynolds established a hierarchy of genres and gave the 'grand style' an importance which he either quite denied the others or at least granted them only in so far as they had some indirect connection with the superior genre. Similarly in his practice, for all the ease with which he adapted himself to the most varied subjects, the 'grand style' had such central importance that even his work in 'lesser' styles was influenced by it. 'In painting portraits', Burke wrote of him, 'he appeared not to be raised upon that platform, but to descend to it from a higher sphere.' For 'he communicated to that description of the art, in which English artists are the most engaged, a variety, a fancy, and a dignity derived from the higher branches . . .' .[41] Reynolds himself singled out portraiture among the lesser genres as the one with which the 'grand style' could most easily and without debasing itself be combined: 'It may be asserted, that the great style is always more or less contaminated by any meaner mixture. But it happens in a few instances, that the lower may be improved by borrowing from the grand. Thus if a portrait-painter is desirous to raise and improve his subject, he has no other means than by approaching it to the general idea.'[42]

This approach to a general idea was Reynolds's guiding principle in his portraits wherever he sought to dignify the sitter by historical or mythological allusion. Gainsborough detested this attempt to elevate the portrait above the natural limits of its genre and proudly declared that his was the 'simple Portrait way'.[43] If he ever did seek inspiration

[40] Leslie and Taylor, op. cit., II, p. 83.

[41] Burke's obituary notice of Reynolds (ibid., pp. 628 ff.).

[42] *Discourses on Art*, ed. R. R. Wark (1975), IV, ll. 459–63 (hereafter in Wark's edn.).

[43] Gainsborough to David Garrick, 22 August 1768, *The Letters of Thomas Gainsborough*, no. 28.

from another kind of subject in composing his portraits, it was never from historical, much less literary, themes. What did sometimes serve as a mode of approach to his portrait subjects was a landscape—a genre which, given his view of nature, Gainsborough often worked in for its own sake, but which is almost as rare in Reynolds as mythological subjects are in Gainsborough.[44]

This disagreement over their conception of art was matched by a conflict in their lives which caused a great deal of offence on both sides and ran the full gamut from simple professional rivalry to personal chicanery. It erupted into open warfare when Gainsborough, outraged by the bureaucratic attitude of the Exhibition Committee which would not hang his pictures at the correct height, resigned from the Academy and swore 'by God' never to let a picture of his be seen there again. The conflict ended only when Gainsborough, as he lay dying, called Reynolds to him to show him his last works and to make his peace with him in the name of Van Dyck. Reynolds then undertook to deliver the great oration in the Academy, the Fourteenth Discourse, a memorial address on his rival. But however much he tried to praise Gainsborough's singular genius, he was too prudent, perhaps also genuinely too imbued with his own didactic mission, not to mingle his praise with reservations, and he warned his pupils against following Gainsborough's example. He spoke of him as one of the great exceptions who achieved greatness in a way all his own, to follow which must needs create confusion. He recognized Gainsborough's limitations in that he saw the world only 'with the eye of a painter',[45] and not at the same time with the eye of a poet; and although he gave Gainsborough credit for accepting the consequences of this limitation by not attempting the heroic style, it was a very doubtful compliment to praise the grace of his figures as being of a kind more frequently to be found 'in cottages than in courts'.[46] In all he says the reproach rings clear that Gainsborough lacked the imagination which enables the painter to rise above his particular subject and awaken in the spectator the memory of an Idea.

Gainsborough's tendency to cling to the material world and Reynolds's desire to free himself from it are also revealed in their evaluation of their own work. For Reynolds, of course, 'invention' was so pre-eminent a part of artistic achievement that he had nearly all his pictures, including portraits, engraved immediately on completion, so that the pictorial idea could be preserved and reproduced long after the painting itself had darkened and decayed. Gainsborough, by contrast, never bothered to have his pictures engraved. Instead he cultivated drawing, a form of graphic art in which individual performance was the *raison d'être*. The only plan he ever appears to have had for reproducing his works concerned some landscape drawings he wanted to publish as a set of etchings. Yet even here he did not leave the execution, as Reynolds did, to another hand, but worked the plates himself, boldly

[44] In Gainsborough's only mythological picture, the *Diana and Actaeon* (Her Majesty the Queen, Buckingham Palace), the figures are set naturally in an idyllic landscape; whereas one of Reynolds's rare landscapes, *The Thames from Richmond Hill*, is motivated by, and linked to, personal documentary interest—a view from his house (Tate Gallery).

[45] Discourse XIV, ll. 203–5.

[46] Ibid., ll. 233–4.

combining the most varied techniques in an attempt to preserve something of the individuality and freshness of line that marked the original drawings.[47] This love of drawing for its own sake was foreign and offensive to Reynolds. He warned his pupils never to forget that the brush, not the pencil, was their proper tool;[48] the limited number and quality of his own drawings has indeed surprised connoisseurs.[49]

In other ways, too, the attitudes to life and art of these two painters were so opposed that Gainsborough must have seemed the perfect example of how not to meet the demands Reynolds made of his pupils. Reynolds taught that a painter should take unremitting pains over the conception of his pictures, that he must work at it and not be easily satisfied if an important pictorial idea should be vouchsafed him. Yet Gainsborough, as Reynolds recounted, was in the habit of going out into the streets and fields, gathering up grasses, pieces of moss, all sorts of animals, and often cottage children and gypsies to bring back to his studio—where he used the animals and children as models, and assembled small landscapes on his table from the grasses and moss, adding various household articles, pieces of wood, fragments of looking-glass, and so on, all of which he then copied. This natural, playful way of handling subjects was completely alien to Reynolds's idea of the dignity of the painter as an artist. Gainsborough seemed determined to keep painting on the humble level of a craft, whereas Reynolds was striving to ensure it a rightful place as one of the liberal arts. Reynolds preaches again and again that the artist must educate himself. He must travel to Italy, in order to know the world and, above all, the works of the Old Masters. And he must model his style on theirs. For this is the paradox of originality: that only imitation can develop the power of original invention. Not until he had copied the Old Masters could a painter really judge whether his style was original or not: those who claimed to be original but were not acquainted with the classical models were in danger of taking for new effects long since invented by others.[50] Here, too, Gainsborough exemplified the opposite attitude: he had found his original style without ever leaving England, and when he did copy, it was the style, not of the Roman school, which Reynolds revered as the greatest example, but of the Flemish and Dutch schools.[51] Reynolds taught that the painter should be well-read, and it was one of his main aims when President of the Academy to create Professorships of Ancient Literature and Ancient History. But Gainsborough raged against this view of the learned artist. The cultured gentleman was in his eyes the real enemy of the artist's craft. His view is made plain in the contemptuous cursings in his letters to Jackson: '. . . you are dayly throwing your gift away upon *Gentlemen* & only studying how you shall become the *Gentleman* too, now damn Gentlemen, there is not such a set of Enemies to a real artist in the world as they are, if not kept at a proper distance . . . I should not think of

[47] Cf. M. J. Friedländer, 'Gainsborough als Radierer', *Kunst und Künstler*, XIV (1916), pp. 15 ff.; see now also J. Hayes, *Gainsborough as Printmaker* (1971).

[48] Discourse II, ll. 292–3.

[49] Cf. K. T. Parker, *Old Master Drawings*, V (1930), pp. 75 f.; also Friedländer, loc. cit.

[50] Discourse II, ll. 88–9.

[51] Besides the well-known copies and paraphrases of figure paintings by Van Dyck and the reminiscences of Ruisdael's landscapes, it is worth mentioning a copy at Hampton Court Palace after a *Portrait of a Rabbi* by Rembrandt (now in the National Gallery, London).

my pretending to reproach you who are a regular fellow. If I meant anything (which God knows if I did) it was this, that many a *real Genius* is lost in the fictitious Character of the Gentleman. . . .' Gentlemen 'make no part of the Artist'.[52]

The Eclectic Ideal

The many-sided culture of the artist which Gainsborough here rejects and despises was for Reynolds a *sine qua non*, for it was a necessary counterpart to his artistic ideal of eclecticism. Even the copying of the Old Masters was, in his conception, not a mechanical process but a means of education. He warned against mere thoughtless, mechanical copying. The student was to enter into the spirit of the original, and so to assimilate something of the greatness of its style. This was the real value of copying—that through the mechanical process one would simply be forced into the spirit and style of the original. He recommended to his pupils that when executing a composition they had invented themselves they should take as many parts as possible from Michelangelo, and then fill out the rest on their own. He maintained that by this initial copying they would assimilate Michelangelo's style so thoroughly that the remaining parts would inevitably become Michelangelesque too. He further recommended painting 'companion pieces' to the masterpieces they were emulating, so as to reveal the disparity still to be made good between intention and accomplishment. Finally, he proposed giving classical models a new meaning—'to change the purpose of the figures without changing the attitude'. This method of adapting old motifs seemed to him so important for the development of the artist that he anxiously tried to prove that the Old Masters had themselves used it. Had not Tintoretto transformed Michelangelo's Samson into a Jupiter by giving him an eagle and putting a thunderbolt in his right hand instead of a jawbone? Had not Titian introduced God the Father from the Sistine Ceiling into the *Battle of Cadore* and transformed him into a general falling from his horse?—'A real judge who should look at this picture, would immediately pronounce the attitude of that figure to be in a greater style than any other figure of the composition.'[53]

His contemporaries were well aware how frequently quotations of this kind were to be found in Reynolds's work; Horace Walpole tried to justify the ways of Reynolds to men of culture:

Sir Joshua has been accused of plagiarism for having borrowed attitudes from ancient masters. Not only candour but criticism must deny the *force* of that charge. When a single posture is imitated from a historic picture and applied to a portrait in a different dress and with new attributes, this is not plagiarism, but quotation: and a quotation from a great author, with a novel application of the sense, has always been allowed to be an instance of parts and taste; and may have more merit than the original. When the sons of Jacob imposed on their father with a false coat of Joseph, saying: 'Know now whether this be thy son's coat or not?', they only asked a deceitful question—but that interrogation became wit, when Richard I on the pope reclaiming a bishop whom the king had taken

[52] *The Letters of Thomas Gainsborough*, nos. 50 and 51; also Whitley, *Thomas Gainsborough*, pp. 380 f.

[53] Discourse xv, ll. 465–78. See also Wind, 'The Mænad under the Cross', below, pp. 74 ff.

prisoner in battle, sent him the prelate's coat of mail, and in the words of the Scripture asked his holiness whether THAT was the coat of his son, or not?[54]

This technique of quotation which puts an old idea into a new context, achieving an effect of surprise and therefore of 'wit', occurs particularly often in Reynolds's portraits. Thus in the portrait of Mrs Hartley with her child, he painted the actress as a Bacchante but gave her the attitude of the Madonna Doni (Figs. 5–6). In his portrait of the Duchess of Marlborough, he evoked a lunette from Michelangelo's Sistine Ceiling (Figs. 1–2). Likewise Mrs Bouverie's attitude is reminiscent of the Libyan Sibyl, even though she holds a child instead of a book. In the portrait of Mrs Siddons (Figs. 46–8), the posture of her head and bent arm are obviously modelled on Michelangelo's Isaiah, her outstretched arm on Daniel, while the two attendant figures behind the throne not only suggest the general arrangement of the Sistine Prophets and Sibyls but one of them is an almost exact copy of the Genius who mourns to the left of Jeremiah.

But perhaps the most telling example of how strictly Reynolds followed his own precepts is in a pencil sketch which he made in Italy, and which bears the inscription in his own hand: *the lower part / Julio Romano / in Te. / the upper part / invention* (Fig. 4). Since on the same sheet there is a sketch of *A Canon at Mantua*, it is natural to look for the source among Giulio Romano's frescoes in the Palazzo del Te; and in fact the lower part corresponds precisely to Giulio Romano's Persephone, from whom Psyche receives the vase of ointment (Fig. 3). Instead of the vase, Reynolds has placed a child in the woman's arms, and turned the inclined but rigid pose of the head into a gesture of tenderness. Here we can see how he first attempts by literal copying to imbue himself with Giulio Romano's style, and then goes on to free variation within the same design. Admittedly, the heroic attitude in the source seems to have been so successfully transformed into a graceful one in the freely invented part that we are reminded of Reynolds's assertion that the painter must even be able at times to parody his model. But if we remember that this was only a preliminary sketch for a work representing the 'Mother and Child' theme, perhaps even in portrait form, then undoubtedly the allusion to Persephone enhances the figure and raises it to heroic dignity. There are always two points of comparison involved in these borrowings. If we look at the Reynolds figure as a replica of Giulio Romano's Persephone, then it appears stylistically diminished. On the other hand, if we look at it as 'Mother and Child', then the attitude of the woman and the drapery of her dress elevate her to heroic dignity by evoking the original.

The Artificiality of the 'Grand Style' and its Relation to the Art of Acting

This method of an artificially heightened style recalls the 'artificial life' as described and opposed by Hume. There, too, man sets his sights on something beyond his natural

[54] *Anecdotes of Painting in England*, ed. R. N. Wornum, I (1862), p. xvii note 2, from Horace Walpole's advertisement prefixed to the fourth volume of the original edition (1771). See also below, p. 69.

measure, and so falls into enthusiasm. We saw how enthusiasm cannot long be sustained without sinking back to the level of natural behaviour, how the enthusiast has to keep stirring it up in himself by artificial practices. In exactly the same way, Reynolds taught that the elevated Michelangelesque style could not be learnt once and for all, but had to be constantly re-attained by discipline, that is, by copying. The comparison is supported by Reynolds's own testimony: 'It must be remembered, that as this great style itself is artificial in the highest degree, it presupposes in the spectator, a cultivated and prepared artificial state of mind.'[55] He quotes with approval James Harris, who, in his battle against the view that '*fine things should descend to us*', saw all schooling in taste as aspiring upwards and to this end urged himself and others '*even to feign a relish*, *till we find a relish come; and feel*, *that what began in fiction*, *terminates in reality*' (Reynolds's italics).[56]

The feigning of noble sentiments which are not really felt was here recommended as a method of self-education, as the appearance was supposed to engender the reality. But for the enemies of the heroic outlook this connection of hypocrisy and enthusiasm debased enthusiasm itself. In his critical account of the Puritan religious spirit, Hume admits explicitly that pure hypocrisy without any touch of real enthusiasm is impossible, but it is likewise impossible for pure enthusiasm to last long without a touch of hypocrisy:

> So congenial to the human mind are religious sentiments, that it is impossible to counterfeit long these holy fervours, without feeling some share of the assumed warmth. And, on the other hand, so precarious and temporary, from the frailty of human nature, is the operation of these spiritual views, that the religious ecstasies, if constantly employed, must often be counterfeit, and must be warped by those more familiar motives of interest and ambition, which insensibly gain upon the mind.[57]

This positive and negative judgement of enthusiasm is based on an observation which was being developed in the psychological theories of the day and found practical realization in contemporary acting techniques: 'I have often observed,' writes Burke, 'that on mimicking the looks and gestures of angry, or placid, or frighted, or daring men, I have involuntarily found my mind turned to that passion whose appearance I endeavoured to imitate; nay, I am convinced it is hard to avoid it, though one strove to separate the passion from its correspondent gestures.'[58] For the eighteenth-century's declamatory style of acting it was axiomatic that the actor can best train himself to represent an emotion by first

[55] Discourse XV, ll. 394–6.

[56] See James Harris, 'Rules Defended', in *Philological Inquiries* (1781), II, xii (*The Works of James Harris*, 1841, p. 453): 'And here, if we pursue some author of high rank, we perceive we do not instantly relish him, let us not be disheartened; let us even feign a relish, till we find a relish come. A morsel perhaps pleases us; let us cherish it; another morsel strikes us; let us cherish this also. Let us thus proceed, and steadily persevere, till we find we can relish, not morsels but wholes; and feel that what began in fiction, terminates in reality. . . . The film being in this manner removed, we shall discover beauties which we never imagined; and condemn for puerilities, what we once foolishly admired. . . . This speculation applies not to literature only: it applies to music, to painting, and, as they are all congenial, to all the liberal arts. We should in each of them endeavour to investigate what is best, (if I may so express myself,) *there* to fix our abode.'

To 'feign a relish' is a basic motif of idealist aesthetics, cf. Schiller, *Über die ästhetische Erziehung des Menschen* (1795), ix: '. . . die Wahrheit lebt in der Täuschung fort, und aus dem Nachbilde wird das Urbild wieder hergestellt werden.' See also Wind, *The Eloquence of Symbols* (1983), p. 13.

[57] *The History of England*, VI, ch. lv, p. 583 note.

[58] *Of the Sublime and Beautiful*, IV, iv.

assuming its corresponding outward appearance—a point also made by Lessing in *Die Theatralische Bibliothek*.[59] This is the exact counterpart of Reynolds's doctrine that what begins as a mere mechanical exercise gradually imposes upon the imitator the excellencies of his model.

IV *Types of Portrait*

Clearly, then, for moral and aesthetic reasons, a painter's attitude to the art of acting will decisively affect his pictorial representations of man. An exponent of the artificially elevated style will favour theatrical gesture because he sees the exceptional moment in a person's life as decisive; an exponent of the natural style will reject theatrical gesture because he does not want to estrange man from his everyday existence. Thus the artistic controversy between Reynolds and Gainsborough will be most evident in their portraits of actors, for here the very nature of the model forced the painter to confront the problem of allegorically heightened expression. But the same issue will also affect the social characterization of subjects in portraits of men in various professions, of soldiers, statesmen, or scholars, and of women in different ranks of society; not least, it will be reflected in self-portraits. It can be discovered, and even most sharply seen, where one would least expect it, in the portraits of children, and it is with them that I shall begin.

Portraits of Children

For all their elegant manner, Gainsborough's portraits of children always leave the child undisturbed in its natural surroundings. For example, in a double portrait of his daughters the two girls are chasing a butterfly (Fig. 7), and elsewhere he shows a girl tying a bow or arranging another's hair. They are wearing their everyday dress and the general impression is lyrical. Any attempts to heighten the effect, if made at all, tend towards emphasizing the natural, easy, occasionally even gypsy-like appearance of the figures.[60] Gainsborough never imposes theatrical action or a rigid statuesque pose on these subjects.

In contrast, if we look at a picture like *Master Crewe* by Reynolds (Fig. 9), the child is deliberately shown in a pose that transcends his life as a child. He plays the part of Henry VIII and is given the same pose as the king in Holbein's portrait. The spectator is meant to savour the disparity between the imposing original and the childlike ingenuousness of the imitation. Although the child parodies the king's bearing, he is also, through being the instrument of such a witty paraphrase, raised above his natural status.[61]

[59] Lessing, 'Auszug aus dem *Schauspieler* des Herrn Remond von Sainte Albine', *Die Theatralische Bibliothek* (1754), I, iv (*Sämtliche Schriften*, ed. K. Lachmann and F. Muncker, VI, 1890, p. 152).

[60] See, for example, *Miss Linley and her Brother* (Sterling and Francine Clark Art Institute, Williamstown, Mass.).

[61] The contemporary theatre was familiar with corresponding effects. See Garrick's *Lilliput*, where much of the comedy is drawn from the fact that adult roles are travestied by children; also Lessing's tragedy *Philotas*, where the child's desire to be an adult becomes a tragic-heroic motif.

The particular pathos of these child portraits, and the fact that they seriously aim at heroic effect, appears most clearly when the child is portrayed as a classical hero, not merely in the costume of a national one. In *Master Cox as Hannibal* (Fig. 11), the child who looks out at us, all pensiveness and premonition, plays the role of the young Hannibal, forced by his uncle to swear eternal enmity against the Romans (Polybius, III, xi, 5). His right hand rests on his sword, and behind him smoke rises from the libation on the altar. The typical response such a picture elicits is again an awareness of tension, not in this case a tension between the maturity of the original subject, whom the child is to emulate, and his own childishness (for here the heroic prototype is a child), but rather the presentiment of things to come which fills the young Hannibal's mind and, in the imagination of the spectator, is transferred from the historical exemplar to the actual boy depicted in his role. In other words, the tension is due to the visible manipulation of the temporal sequence in the picture.

Shaftesbury's theory of the pregnant moment offers a clue to the interpretation of such pictures.[62] For Shaftesbury the most favourable moment to be represented in a picture is, precisely, one which points back into the past or forward into the future. The artist should record or prophesy; the means for both functions are there in his art. He can unite several moments in one because the mind moves more quickly than the body. The mind can be filled with a new emotion while the body is still showing traces of the preceding one. Thus a painter can show a woman 'newly transported with joy at the sight' of one 'who the moment before had been lamented as . . . deceas'd or lost'.[63] Here the body lags behind the mind and still shows the last emotion although a new one has replaced it, but the mind can also prefigure the future, leaping ahead of the body with its premonitions. In these cases events are anticipated which are not represented in the picture at all. As a particularly striking means of conveying such anticipations, Shaftesbury recommends (and in this he differs from Lessing) the use of emblems. For example, a club or a lion's skin placed in the young Hercules' cradle will signify the deeds he is to accomplish later. Similarly, in our picture the young Hannibal is given a sword: the oath points forward to the heroic deeds that lie in the future.

The heroic effect obtained by giving the child a prophetic awareness of his future is not, of course, confined to portraiture and certainly did not originate there. It stems rather from mythological and religious genre scenes in which many eighteenth-century artists, and

[62] 'A Notion of the Historical Draught or Tablature of the Judgment of Hercules' (1713), I, 10; included in the 2nd edn. of Shaftesbury's *Characteristicks*, III (1714), vii. See also Lessing, *Laokoon* (1766), I, iii and xvi (Lachmann and Muncker, IX, 1893, pp. 19 ff. and 95). On the relation between Shaftesbury and Reynolds, see Karl Heinrich von Stein, *Die Entstehung der neueren Ästhetik* (1886), pp. 183 f.

[63] '. . . That notwithstanding the Ascendency or Reign of the principal and immediate Passion, the Artist has power to leave still in his Subject the Tracks or Footsteps of its Predecessor: so as to let us behold not only a rising Passion together with a declining one; but, what is more, a strong and determinate Passion, with its contrary already discharg'd and banish'd. As for instance, when the plain Tracks of Tears new fallen, with other fresh tokens of Mourning and Dejection, remain still in a Person newly transported with Joy at the sight of a Relation or Friend, who the moment before had been lamented as one deceas'd or lost.' (Loc. cit.)

especially Reynolds himself, depicted the theme of the hero's boyhood.[64] The view of art shared by Reynolds and his circle made such a subject attractive to them, for they accepted Burke's doctrine that any thing which can be clearly grasped in its totality never impresses as truly grand, and that a certain obscurity is a necessary condition of the sublime.[65] Hence a fully-grown hero represented as a complete, well-rounded personality will always seem banal, never heroic; the mystery of heroism can be represented only by showing the hero *in statu nascendi*, as a promise, but not yet a fulfilment. Under the influence of these ideas, Reynolds frequently portrayed the heroic quality of Greek mythological figures and Old Testament prophets by representing them as children: *The Infant Jupiter*, *The Infant Bacchus*, *The Infant Mercury*, *The Infant Hercules*, or *The Infant Samuel*, and indeed *The Infant Moses*. It is a common element in nearly all these pictures that the representation of childhood has the effect either of a parody or of excessive pathos, or both, because the mission of the man is anticipated in the expression of the small boy. The fair-haired infant who plays Jupiter (Fig. 10), and is cared for by the goat Amalthea and accompanied by the eagle, holds the thunderbolts clasped in his hand, and his expression, for all its childlike quality, is that of the Thunderer. Mercury appears as a precocious little thief; Hercules is shown couched on a lion's skin, strangling serpents in his cradle, and surrounded by heroes, seers, and gods who excitedly watch the proceedings. This last picture, painted for Catherine II, was meant to allude to 'the great difficulties which the Empress of Russia had to encounter in the civilization of her empire, arising from the rude state in which she found it'.[66] The idea for the picture was Reynolds's own, the commission being, as *The St. James's Chronicle* somewhat maliciously reported, 'unlimited in Respect to Size, Subject, and Price'. A rather doubtful anecdote has it that Burke was responsible for the Hercules fable. At all events, contemporaries sensed that the bombastic style was not unconnected with the painter's literary and philosophical advisers: 'As the national Honour is in some Respect concerned in this Production, we wish the Imagination of the Artist had been formed on Greek Literature, and not on the false Splendour of Dr. Johnson.'[67]

[64] It would be interesting to find out whether an influence from classical antiquity is at work here, for works on divination show that representations of children in heroic roles already existed, as in Cicero's *De divinatione*, I, xxxvi, 79. This passage, emended by Winckelmann in *Geschichte der Kunst des Alterthums* (1764), II, ii, p. 343, was well known in the eighteenth century.

[65] See above, p. 13.

[66] James Northcote, *The Life of Sir Joshua Reynolds*, 2nd edn. rev. (1819), II, p. 215. See J. S. G. Simmons, 'Samuel Johnson "On the Banks of the Neva": A Note on a Picture by Reynolds in the Hermitage', in *Johnson, Boswell and their Circle. Essays presented to Lawrence Fitzroy Powell* (1965), pp. 208 ff.

[67] *The St. James's Chronicle*, 29 April 1788, apropos of the exhibition in the Academy. *The Infant Jupiter* was destroyed by fire in 1816 in Belvoir Castle and has come down to us only in an engraving. Sotheby's *Poetical Epistle to Sir George Beaumont* (Graves and Cronin, op. cit., III, pp. 1164 f.) includes a poetic transposition of Reynolds's heroic infant portraits:

> Where boundless genius brooding o'er the whole,
> Stamps e'en on babes sublimity of soul;
> Whether, while terror crowns Jove's infant brow,
> Before the godhead awed Olympus bow,
> Or while from heav'n celestial grace descends,
> Meek on his knees the infant Samuel bends,
> Lifts his clasp'd hands, and as he glows in prayer,
> Fixes in awful trance his eyes on air.

Since the description of Samuel corresponds exactly with the picture, it is worth considering whether the description of Jupiter should not also be taken literally. The composition would then be only a part of the original picture, and we should have to imagine Jupiter, like Hercules in Reynolds's Leningrad picture, or Romney's *Infant Shakespeare* (Fig. 14), surrounded by many figures offering homage. The Hercules picture also exists in a mezzotint by C. H. Hodges, which reproduces only the part with the child on the lion's skin in his cradle.

There is another group of paintings in which the contrast between infancy and maturity is used as a means of raising the subject to heroic stature. These works are not at all intended as portraits of children, but as depictions of a legendary childhood, freely invented, *ex post facto*, as it were, in celebration of a famous man.

In Romney's *Infant Shakespeare* (Fig. 14), the baby is attended by the Tragic and Comic Muses: Thalia has given the child her flute to play with, Melpomene has placed her cup of poison in front of him. On the side of the Tragic Muse, which is painted in dark colours, the darker passions are symbolized, while the figures gathered round the Comic Muse, symbols of gentle feelings, are illumined by a shaft of light. Linking the two groups, there hovers a benevolent figure shining with more than earthly brilliance—Nature. The legend of the poet's infancy has here become a prophetic fairy-tale: while Nature watches over the child, the Muses nourish him with the contrary passions his poetry will express.[68]

But strangest of all is the transference of these child-heroics to the painter's contemporaries—that is, neither figures of a mythological past, nor heroes of national history or literature, but men whom the painter actually knew, in all their idiosyncratic individuality. The child that gazes with brooding profundity and unspeakable sadness at his little crossed legs, pressing the fingers of one hand with the other, is an idealized portrait of Samuel Johnson (Figs. 12–13). The near-grotesque impact of the monstrous figure, the incessant moralist, the scholar burdened by encyclopaedic knowledge, the sage (so Boswell called him—Reynolds had him sit as model for Teiresias), plagued and disfigured by physical suffering yet possessed by his imagination, is rendered here, true to the type of the legendary infant-hero: the grandeur is not shown in fully developed form, but only prefigured in embryo. The wit and charm of the pictorial conception, which is again meant to be both solemn and parodistic, demands that the spectator be willing and able to savour fully the contrast between the real and the infant Johnson, just as the Lilliputian Henry VIII presupposes that the spectator is familiar with Holbein's picture and will appreciate the disparity between the child and the original.[69]

Looking at these pictures, we can understand what Reynolds meant when he said that the artist's task was not only to please the senses, but also to stir the imagination. The intention

[68] The painting in the collection of the Shakespeare Memorial Theatre, Stratford-upon-Avon, was engraved for the title-page of Boydell's *Graphic Illustrations of the Dramatic Works of Shakespeare* (1802). Variants of a related subject by Romney, *Nature unveiling the Infant Shakespeare, attended by the Tragic and Comic Muses*, also engraved for Boydell, are at Petworth House, Sussex, and the Walker Fine Art Gallery, Liverpool. [On Romney's variations on the theme of natural genius in the form of a new-born child, see now D. Piper, *The Image of the Poet. British Poets and their Portraits* (1982), pp. 107 f.]

[69] [In his study 'Portraits of the Author: Lifetime Likenesses of Samuel Johnson' (*English Portraits of the Seventeenth and Eighteenth Centuries*, Papers read at a Clark Library Seminar, UCLA, 1974, pp. 47–88), H. W. Liebert questions the identification of this infant as Johnson. Four years after Reynolds's death, the painting was sold at auction with the title 'Study of a Naked Boy'; subsequently, when in the collection of the Marquess of Lansdowne, it was called 'Infancy'. As late as 1844, Mrs Anna Brownell Jameson, in her guide to the public collections in and near London, remarked that 'Sir Joshua said he intended to represent Dr. Johnson when a year old' (cf. Liebert, op. cit., pp. 58 f.). Although Liebert casts doubt on this, Mrs Jameson's judgements have often been vindicated by later research (for example, her account of the Hampton Court collection); see J. Shearman, *The Early Italian Pictures in the Collection of Her Majesty the Queen* (1983), passim.]

is to carry the beholder beyond what is shown in the picture, to stimulate ideas in his mind, and it was precisely because Gainsborough's child portraits never did point to the realm of ideas in this way that Reynolds, even while recognizing their qualities as paintings, always regarded them as works of an inferior order.

Portraits of Young Girls

In Reynolds's portraits of young girls we can observe the same principle at work, albeit modified by the nature of the subject. Once again a group of portraits is marked by that interplay between two stages in the subject's life, in which the condition of the grown-up person is projected back into the mind of the child whose pose and expression prophetically, as it were, hint at her future situation. In *The Age of Innocence*, the emphasis on purity has the effect of attributing to the child a self-conscious awareness of maidenhood threatened and preserved. *The Strawberry Girl*, for which the model was Reynolds's niece Theophila Palmer, is characterized by a posture expressing a particular solicitude, like that of a little old lady, which again can be understood only as anticipating a future stage in the girl's development. If one fails to see the picture in this light, choosing to interpret the coyness or the solicitude as qualities of the actual child, they seem false and precocious and offend the naturalistic eye. But there can be no doubt that Reynolds's contemporaries, trained as they were in an eclectic tradition, could read such pictures on several levels. For them the precocity and maidenliness were reminders or anticipations of a stage of life not actually shown in the picture, and hence only indirectly attributed to the child.

But the line between this type of metaphorical portraiture and genre painting proper is not a clear one; there is often doubt whether the picture is indeed a portrait of a child with genre features, or a real genre painting for which that particular child served as model. Yet the fact that the names of the child sitters were often carefully noted by contemporaries indicates that the interest in these pictures hinged on the identity of the child, that the child was recognized whatever the role, and was meant to be so recognized. In a nation which contributed more than any other to the development of Children's Theatre, the difference between the genre-like children's scene for which an individual child sat, and the child portrait with genre features, inevitably disappeared—a principle which *mutatis mutandis* also applies to historical or mythological portraits of adults. No one would maintain that it is essential to know that Hélène Fourment was Rubens's model for the *St. Cecilia* and the *Andromeda*[70] in order to understand these pictures properly. For there the individual woman was merely the perfect, if mortal, representative of a certain type of beauty. But the case is different with a work like Jan de Bray's *Judith and Holofernes*, where the prostrate Holofernes has de Bray's own features and Judith, approaching the bed sword in hand, has those of his wife. If we are to understand the coarse pictorial joke, the likeness must be evident. And it is also probably true that Carel Fabritius in his *Execution of John the Baptist*

[70] Staatliche Museen Preußischer Kulturbesitz, Gemäldegalerie, Berlin-Dahlem.

counted on the spectator recognizing the executioner as the ageing Rembrandt.[71] Reynolds, however, was not satisfied with allusions like these, but carried this tradition to the point where genre painting and portraiture merge, creating an experience familiar to the theatre-goer who sees an actor previously known to him only as a polite conversationalist in fashionable drawing-rooms suddenly transformed into Richard III.

As in portraits of boys, so also in these portraits of girls, the tension created by the displacement in time is only one of several devices used to achieve a heightened effect. Like the miniature Henry VIII, the girl can play a role, which may not so much transpose her in time as raise her to an ideal sphere. For example, the *Heads of Angels* in the Tate Gallery are to be read as portraits of a single child, Frances Gordon. The girl is seen from a number of angles and the spectator is meant to be aware of this identity as his eye passes from one small head to the next. Probably this process is also intended to remind him of those sixteenth- and seventeenth-century portraits in which one person is shown several times, from different angles. This is, in fact, a tradition closely linked with England, for the collection of Charles I contained the first known example of this type of portrait, by Lorenzo Lotto.[72] And Charles I had himself painted in this way by Van Dyck, in order to give Bernini an all-round view of his head.[73] But in both these cases the main objective was a technical one, the painter either helping—or perhaps competing with—a sculptor, whereas Reynolds uses the device as an imaginative conceit; for him the technique is a means of transforming a mortal model into a whole series of supernatural figures.

Yet this heightened representation of girlhood is the exception. Down to the time of John Hoppner, almost the only method of idealizing a young girl was to portray her as an angel. However, with Reynolds there occurs one completely new method of symbolic characterization not found in the portraits of boys—that of bringing out the child's inner qualities in animal terms, by likening her to an animal. A young girl is portrayed with a cat, a mouse, or a bird, and her eyes and bearing are made to resemble those of the animal she is playing with, as in *Felina*, *Muscipula*, and *Robinetta* (Figs. 15–17). Playful physiognomic resemblances are made to achieve a cat-like, mouse-like, or bird-like effect. Since there is

[71] Both paintings are in the Rijksmuseum, Amsterdam. On their interpretation, see F. Schmidt-Degener, *Rijksmuseum. Führer* (1930), pp. 39 and 49. [The tradition in which artists portrayed themselves as the decapitated victim of a *femme fatale* has a long history since the Renaissance. E. Panofsky suggested that Titian represented his own likeness in his head of St. John the Baptist, now in the Doria Pamphili Gallery, Rome (*Problems in Titian mostly iconographic* 1969, pp. 42 ff.), and J. Shearman has analysed Baldinucci's account of Allori's self-portrait as the dejected lover before his mistress as Judith ('Cristofano Allori's "Judith"', *The Burlington Magazine*, CXXI, 1979, pp. 3 ff.). Artists also represented themselves as executioner as well as victim, as is shown by the many images of Caravaggio as David, as well as Giorgione's self-portrait in that role.]

[72] [Following J. Kerr-Lawson's article, 'Portrait of Lorenzo Lotto by Himself', *The Burlington Magazine*, VI (1905), pp. 453 f., the prevailing interpretation for many years of the *Triple Portrait* in the Kunsthistorisches Museum, Vienna, was that the artist had represented himself, seen from three different viewpoints, ironically gambling with himself. After cleaning in the 1950s, it was revealed that the object held by the central figure was not a *giuoco del lotto* but a goldsmith's box with rings. This has led to the plausible identification of the three men as the Carpan brothers, all Trevisan jewellers, who could be counted among Lotto's closest friends, for they are often mentioned in his *Libro di spese*. See J. Grabski, 'Sul rapporto fra ritratto e simbolo nella ritrattistica del Lotto', in *Lorenzo Lotto. Atti del Convegno Internazionale di Studi per il V Centenario della Nascita* (Asolo, 18–21 September 1980), ed. P. Zampetti and V. Sgarbi (1983), pp. 385 ff.; also L. Vertova, 'Lorenzo Lotto: Collaborazione o Rivalità fra Pittura e Scultura?', ibid., pp. 401 ff.]

[73] Her Majesty the Queen, Windsor Castle.

a tradition that in both *Felina* (Theophila Palmer, later Mrs Gwatkin) and *Robinetta* (Magdalena Lewis, later the Hon. Mrs Tollemache) a real child was meant to be idealized in this way, we can probably assume that all these pictures were intended as portraits despite the genre-like quality of the subject.

If Reynolds's theoretical writings require the artist to stir the spectator's imagination by leading him beyond sensuous appearance to a general idea, then it is fair to ask what kind of picture will result. These portraits of girls provide the answer; for the juxtaposition of animal and child automatically stimulates the idea of the species (cat, mouse, or bird). In the case of *Felina* the analogy is particularly emphasized by the full-face pose of the cat's head below the girl's, so that one is confronted with two pairs of slit eyes. In these physiognomical portraits Reynolds was reverting to an old tradition which can be traced back to Leonardo da Vinci and Giovanni Battista della Porta.[74] And the conceit he adapted here clearly fits into the style of portrait painting he himself called 'composite', a style that shaped the individual face according to a general idea, and in which the particular alludes to the general and vice versa.[75]

The peculiarity of these pictures is very evident if we compare, for example, *Felina* with Gainsborough's *Miss Brummell* (at Kenwood), which also shows a girl with a cat (Fig. 18). Here any dramatic link, any physiognomical analogy between the child and the animal, is consciously avoided. The child is seated in her carefully pleated frock, with a white mob-cap on her head, quietly holding the kitten which has settled itself on her lap in such a way that its face cannot be seen at all. She is looking sideways, whereas in Reynolds's pictures the children stare fixedly at the spectator in order to emphasize the mimetic role they are playing.

Portraits of Men of Learning—Hume and Beattie

A difference in the treatment of the sexes in these child portraits has been observed: boys are dignified by the prophetic hint of a future vocation, they are shown as heroes or as gods, whereas girls are likened to angels or kittens, shown as symbols of innocence or motherliness. This contrast is even more marked in the portraits of adults. In the case of a man, the heroic elements were taken from his actual calling: the conditions of his real life in society provided the means for the painter to give him a heightened existence. A woman, when not portrayed as a mother playing with her children—a favourite scene of Reynolds's which he exploited to the limits of its possibilities as genre—was often given an artificial role by the painter, who seized on the device of mythological metaphor, showing her as Juno or Hebe.

It is remarkable that just when eighteenth-century English portraiture revived the

[74] Giovanni Battista della Porta, *De humana physiognomonia* (1586). On the historical tradition of physiognomy, see E. Panofsky, 'Dürers Stellung zur Antike', *Jahrbuch für Kunstgeschichte*, I [XV] (1921), p. 64 note 82; also Wind, 'Leonardo as Physiognomist', *The Listener*, XLVII, 15 May 1952, pp. 787 f. How lively this tradition was in the eighteenth century can be seen from Winckelmann's penetrating critique in *Geschichte der Kunst des Alterthums*, I, iv, 2, pp. 145 f.; also 171 f., on Le Brun's *Méthode pour apprendre à dessiner les Passions* (1667).

[75] On the 'composite style', see Discourse IV, ll. 442 ff.

allegorical and historical portrait to compete with the late Dutch and French tradition, it limited the type to pictures of women. Nattier had painted Madame Adélaïde *en Diane* and Madame Henriette *en Flore*, Largillière had painted a whole range of French ladies of society *en naïade* (Fig. 38); Reynolds likewise painted Lady Blake as Juno (Fig. 39), Mrs Musters and Miss Meyer as Hebe. But for Nattier's *Duc de Chaulnes en Hercule* (Fig. 37) there is no English counterpart.[76] There are, it is true, many degrees of heroic effect in portraits of men. But on the whole a man is glorified in the social position he has attained and not shown outside it; only in the special cases of actors or scholars is allegorical imagination given a freer rein.

Here, too, I shall first take one of these marginal cases, the portrait of a man of learning, then look at portraits of soldiers and professional men, and conclude with some portraits of actors.

Reynolds's portrait of James Beattie is one of the most important documents we have of his own philosophical position (he submitted it with the title *The Triumph of Truth* to the Academy exhibition); and here he uses every heightening effect of learned allegory. When he painted Dr Johnson, he resorted to the infant portrait in order to glorify the man—an idea no doubt partly suggested by the eccentric character of Johnson himself, who in his rumbustious moralizing style would certainly have refused to be made the subject of a serious heroic allegory. But Beattie was quite a different matter. Here was a moralist whose chief method of argument was pedantic fervour, not conversational wit. And the painting, too, is loftily pedantic (Fig. 19). Beattie holds a book which bears the word 'Truth'; this is both an emblem and a specific historical allusion, for Beattie is represented as the champion of Truth because he had written an *Essay on Truth* to refute Hume's scepticism. So this strikingly insignificant and rather timorous-looking man, whose person takes up only half the picture, is shown against a background of metaphysical symbolism which appears somewhat to overwhelm him. Behind him is a winged Genius, obviously the Spirit of Truth, but looking more like Nemesis. Like an angel at the Day of Judgement, he holds a scale which has sunk deep in Beattie's favour. On the other side, clearly weighed and found wanting, three demonic figures are gathered. Truth is thrusting them into the abyss. There are two different theories about the meaning of these figures: one that they embody abstract ideas—wickedness, unbelief, etc.; the other that they are allegorical portraits—of Voltaire, Gibbon, and Hume. A letter from Reynolds to Beattie essentially confirms the second theory, without actually spelling it out. Reynolds openly admitted that he set out to depict Voltaire, and he took gleeful satisfaction in the contemporary view that the figure with 'a tolerable broad back' bending forwards was Hume.[77] But Gibbon is not mentioned in the

[76] Musée du Louvre, Schlichting Collection, Paris, 1747. See P. de Nolhac, *Nattier. Peintre de la Cour de Louis XV* (1925), pp. 191 f.: 'Cet étrange tableau, d'une chaude coloration dorée, sort tout à fait de la manière ordinaire de Nattier.' See also Wind, 'In Defence of Composite Portraits', below, p. 121.

[77] Reynolds to Beattie, 22 February 1774: 'Mr. Hume has heard from somebody, that he is introduced in the picture, not much to his credit; there is only a figure covering his face with his hands, which they may call Hume, or any body else; it is true it has a tolerable broad back. As for Voltaire, I intended he should be of the group.' (*Letters of Sir Joshua Reynolds*, ed. F. W. Hilles, 1929, no. xxv.) See also P. Murray, 'The Source of Reynolds's "Triumph of Truth"', *The Aberdeen University Review*, xxx (1944), pp. 227 ff.

letter, and external evidence proves that this is a relatively late historical construction. After all, Gibbon was a member of the Club and one of Reynolds's closest friends; Reynolds painted an intimate portrait of him; after Goldsmith's death the two men were constant companions, and in his autobiography Gibbon speaks of Reynolds as his friend. Moreover, the interpretation is chronologically untenable. The atheistic genealogy—Voltaire, Hume, and Gibbon—could only be constructed after Gibbon's *Decline and Fall of the Roman Empire* had made him notorious as an 'infidel'; and the first volume appeared only in 1776, some two years after Reynolds had completed the picture.

Oliver Goldsmith, as mediator in the battle of the styles, did not hide his contempt from Reynolds: 'Dr. Beattie and his book together will, in the space of ten years, not be known ever to have been in existence, but your allegorical pictures, and the fame of Voltaire, will live for ever to your disgrace as flatterer.'[78] Goldsmith's pronouncement has a double significance for the controversy over the heroic portrait. He was indignant at the way Beattie had been glorified and the sceptics defamed, and he also, given his view of metaphor, detested pictorial flattery as such.[79]

Hume would never have allowed himself to be painted in this way. The portrait of him in the Scottish National Portrait Gallery (Fig. 20) is by Allan Ramsay, whom Reynolds belittled as an artist, although he liked him as a man: 'There was Ramsay . . . the most sensible among all the painters of his time; but he has left little to show for it.'[80] Reynolds criticized Ramsay's work as being 'weak in execution', finding his hand unequal to his conceptions. Indeed, there is no glorification, no appeal to the faculty of thought or memory in such a portrait. Hume is shown as a man of letters, his left arm resting on two books, in a pose which suggests the easy nonchalance of a man of the world. In his treatment of the head, Ramsay has used shadow to offset what for contemporaries was the most striking feature of Hume's face, the immense mass of fat (one has to imagine the right side of the face, deliberately left in shadow, as matching the left), that visible expression of the indolence he had to contend with all his life, and which appears in his philosophy as the most amiable vice with which Nature has in her benevolence blessed us. Schlegel's theory that laziness is the one divine fragment of a god-like existence left to man from Paradise[81] is clearly anticipated both in Hume's epistemology and in his arguments against theodicy. It is the saving grace which prevents the thinker from putting into practice the abstruse conclusions reached in his study; and although it endangers the belief in divine providence by showing that the deity has bestowed on us, over and above all our other defects, a sluggishness in overcoming them, at least it also prevents fanatics from carrying out their world-shattering plans and strengthens sceptical philosophers in their aspiration to moderation and gentleness. Hume's face seems to symbolize this view of life. For the heavy mass of fat is articulated and touched with meaning, even the broad expanses of the cheeks

[78] Northcote, op. cit., I, p. 300.

[79] See above, pp. 13 ff. and below, p. 123.

[80] William Hazlitt, *Conversations of James Northcote, R.A.*, ed. E. W. Gosse (1894), p. 179.

[81] *Lucinde* (1799), 'Idylle über den Müßiggang' (*Dichtungen*, Kritische Friedrich-Schlegel-Ausgabe, ed. H. Eichner, V, 1962, p. 25).

have no dead spot. Moreover, the mouth, beautifully arched, expresses a mixture of sensuality, melancholy, and wit; the forehead is broad and the eyes calm and clear. In the portrait of Beattie, the pompous treatment contrasted strangely with the slightness of the subject. But here the monumental character of the man strongly emerges from a painting which does not claim to make his achievement visible in symbolic form.

Hume's choice of Ramsay as painter can easily be explained biographically. Ramsay was a Scot and his father, 'Ramsay the poet', was a friend of Hume's. But the choice was also of historical significance, for Ramsay, as Principal Painter to the King, was not only a dangerous social rival of Reynolds, he was also esteemed at Court as an exponent of the tradition in painting which Reynolds broke away from but which Gainsborough perfected. When Ramsay died and a new King's Painter had to be appointed, Reynolds managed to get himself chosen instead of Gainsborough only by means of threats. A pithy sentence in a letter from Johnson to Reynolds sheds light on these manœuvres: 'I am glad a little favour from the court has intercepted your furious purposes. I could not in any case have approved such publick violence of resentment.'[82]

The heroic style, the composite manner that Reynolds stood for, was rejected by the Court because of its extravagance; and if Reynolds was rightly called the 'Painter of the Opposition' for his political leanings, he merited the same title as an artist. The Court was for 'Nature'. The Prince of Wales, later George IV, was the only exception (and in due course it was he who employed Sir Thomas Lawrence).

Characteristically, in the controversy between Hume and Beattie, the Court was for scepticism. There is a touch of comedy when Beattie, in the diary of his London journey, carefully records the complaint of his friends that the sceptical unbeliever Hume drew a fat state pension while he, the defender of religion and morals, had to eke out a frugal life teaching in Aberdeen.[83]

Portraits of Military and Professional Men

It must be admitted that Reynolds's portrait of Beattie gives a very unfortunate idea of his style of heroic portraiture, and his methods should be more fairly illustrated by some examples of greater artistic merit. His portrait of General Heathfield has always been considered one of his most successful achievements (Fig. 22), and we shall look first at this, among a group of military and professional portraits.

Reynolds shows the General in a particular role, as defender of Gibraltar, with the key to the fortress in his hand, against a background of cannon and powder-smoke. A significant moment in the hero's career is chosen to convey his determination and inflexibility. The heroic view of life is expressed through emblems: the key; the cannon; above all, the

[82] Boswell, 2 September 1784 (Hill and Powell, IV, pp. 366 f.).

[83] Cf. M. Forbes, *Beattie and his Friends* (1904), p. 77; and now *James Beattie's London Diary, 1773*, ed. R. S. Walker (1946), pp. 54 f. and 88 f. When the king finally granted Beattie a pension, his friends found it much too small (Forbes, p. 100).

landscape filled with powder-smoke, which sets off the General's red uniform, itself an emblem of his rank. The figure stands alone, in dramatic relationship to the background; for although he clearly belongs to the landscape, he is not integrated into it but contrasted with it—a contrast emphasized by the fact that it is a three-quarter-length portrait and we cannot see how the feet are placed on the ground. In addition, the subject is off-centre; he stands in the right foreground, protagonist of an event whose nature is emphasized by his forceful gesture and the slight turn of his body, which produces a diagonal recession.

These various ways of achieving heroic effect are not, as one might suppose, confined to portraits of soldiers or of men holding public office; there are similar portraits of men in private life, Lord Richard Cavendish, for example (Fig. 21). He, too, is shown in a landscape, in dramatic relief against the background, and the pose and treatment of the landscape are again emblematic. He has his left hand on his hip, his right hand resting on a large rock; his face stands out light against the background of cloud, and the landscape itself gives a sense of unrest and dark adventure (the picture shows a stormy sea; Reynolds allowed the engraver to change this into a desert with a Sphinx).[84] These paraphernalia are not simply used for atmosphere, but are pictorial images which have to be interpreted in detail. Lord Richard was a great traveller, and became particularly famous for his daring adventures in Africa. The rock on which he puts his hand, as it were, proprietorially, is an erratic block and symbolizes the elemental forces which he brought under control. Even the way he looks out at us and yet remains related to the landscape emphasizes his independence and creates a tension between the subject and its background. The relationship between the two is conveyed formally by the slight turn of the body and the resultant diagonal recession. It is dramatized by two details, the hand resting on the rock and the hair which appears ruffled by the same wind that rages in the background.

Gainsborough seems deliberately to avoid such heroic effects. Whereas Reynolds favoured the three-quarter portrait because of the tension it creates between the figure and its background and frame, Gainsborough uses this form only rarely; and when he does so there is no dramatic emphasis, since his figures are mostly seated. Conversely, the full-length portrait which is less usual in Reynolds is favoured by Gainsborough because of the tranquillity it conveys. In the portrait of Dr Schomberg (Fig. 23), the whole figure is set within the landscape, and in the exact centre of the picture. Even though there are features which the portrait shares with those of Reynolds—the slight turn of the body, the clouds in the background which set off the face—it is precisely the way these features are used that points up the difference in purpose. The clouds are not black as in Reynolds's picture, but a soft grey; the coat not a sharply contrasting red, but a subdued shade of brown. The turn of the body has no dramatic foreshortening effect such as is given to the left arm in the Heathfield and Cavendish portraits; it merely serves to show the figure in the round and

[84] 'There is a mezzotint by J. R. Smith, published in 1781, probably taken from the duplicate belonging to the Duke of Devonshire. The background altered to the Egyptian desert with the Sphinx.' (C. Fairfax Murray, *Catalogue of Pictures belonging to His Grace the Duke of Portland, at Welbeck Abbey, and in London*, 1894, p. 37, no. 133.)

suggests, for all the upright bearing of the subject, his freedom from constraint. Schomberg's hands are still, not moving in a dramatic gesture, and instead of a key or other such emblem he holds his hat and walking-stick. Walking is what Dr Schomberg is doing in the landscape, and he is at home there. His expression is not that of a man shown in action at an exceptional, perhaps the most important, moment in his life; rather he conveys a sense of normality, something a man can achieve and maintain in his everyday existence. Significantly, we could not tell from looking at the portrait that he is a physician. But is this perhaps simply due to the character of the profession? And ought we to compare the portrait of a physician with that of a general or an explorer with a view to establishing the difference between the two painters' approaches to human nature? These questions are easily answered. For Reynolds, too, painted physicians, surrounding them with all the trappings of test-tubes, death's heads, and ground-plans of hospitals, with whatever would serve to glorify them in their characteristic activity (as, for example, in the portrait of John Hunter). Conversely, Gainsborough painted generals and admirals—showing them in an ordinary social attitude, standing (Admiral Earl Howe) or out for a casual ride (General Honywood), never at a moment of heroic action.

The contrast is clearest in the case of the celebrated scholar and lawyer Sir William Blackstone, of whom there are three portraits illustrating the three typical methods of portraiture. The portrait in Oxford (Fig. 24) is by Reynolds's imitator Tilly Kettle, who uses all the contrivances for achieving heroic effect—the three-quarter face, the knee-length, the dramatic contrast of the hands, one on the hip, the other pointing to his own books. The dignity of the place is conveyed by a background of pillars and drapes, the dignity of his office by his robes and wig. Yet in faithfully following his master's precepts Kettle avoided the bold foreshortening which gives Reynolds's pictures their life. His portrait is sterile when compared with Reynolds's own portrait of Sir William.[85] Reynolds's portrait (Fig. 25) does not seem at all heroic at first glance and is not in fact in his grand style. And yet even in this relatively simple portrayal the marks of the grand style are distinguishable: the knee-length again, the contrasting movements of the hands, the deliberate choice of typical gestures (the right hand seizing a paper, the left resting on the arm of the chair), and the gaze fixed on the spectator. It has even been plausibly suggested that the left hand grasping the end of the chair-arm is a reminiscence of the classic papal portrait. But all these outward signs of grandeur pale in comparison with the impressive dignity emanating from Gainsborough's portrait of the same man (Fig. 26). And this is only a head-and-shoulders portrait, which deliberately eschews drama. The back of the chair, which in Reynolds's picture contrasts so magnificently with wig and gown, is hardly hinted at here; and the lips, there opened as if to speak, betray no passing emotion. True, Gainsborough also uses wig and gown; indeed, he puts a folded charter in Blackstone's hand, and it even bears the great seal of King John. So all the elements needed for

[85] [National Portrait Gallery, London, no. 388, now merely attributed to Reynolds, cf. *British Historical Portraits* (1957), p. 194.]

emblematic glorification are present. And yet, with Gainsborough, these attributes lose their symbolic meaning. Despite wig and gown, the result is a distinctly informal portrait which depicts the man himself in all the ample solidity of his physical presence; and it is typical that the seal, which in a work in the heroic style would have been meaningfully held and prominently placed (that is, centrally, like the document in Reynolds's picture), is here treated like an object in still life, as are the wig and gown. In fact, the seal is given so little significance that the picture-frame cuts through it. The hand holding the charter has the same unassertive place in the composition of the picture as the hand in Fig. 29 which casually holds the odd book.

Portraits of Actors—Garrick

But the acid test for my thesis—namely, that the fundamental divergence of Reynolds and Gainsborough was over the legitimacy of heightened, metaphorical, and dramatic expression—is the actor portrait. For the actor is, so to speak, a living metaphor; and a painter who tends to emphasize the metaphorical element in portraiture, presenting Beattie as the Defender of Truth, Heathfield as Governor of Gibraltar, and Cavendish as Explorer of Africa, will certainly not miss this opportunity of showing a man playing a part. Indeed, Reynolds's portraits of actors are almost invariably portraits of roles, whereas Gainsborough without exception takes the actor out of his role and depicts him so naturally in his social situation that his theatrical calling goes unnoticed.

Reynolds's portrait of Garrick as 'Kiteley' (Fig. 28) gives a clear idea of his heroic treatment of the actor in a dramatic situation. Garrick is shown as the jealous husband in Ben Jonson's *Every Man in his Humour*; his clothes proclaim him the imposing merchant. Once again Reynolds catches a significant moment in his subject's life: the head is turned as with a sudden movement, the eyes dart a mistrustful side-glance, nose and mouth are nervously tensed, and the whole impression of a man tormented by jealousy is heightened by the hands, one hidden in the cloak, the other visible, tightly clenched and aggressive yet controlled. Admittedly the portrait of Garrick at Knole (Fig. 27) seems not to fit this theory, for Reynolds here portrays him apparently in private life, quite independent of any theatrical role. However, a closer examination proves this view to be mistaken. For although Garrick does not appear in any particular dramatic role, he does appear in a precise and distinctive role appropriate to his career: before him is a sheet of paper on which is written the word 'Prologue', to the right are pen and ink, and in front of them two books piled one on the other. Garrick is here portrayed as the author of 'Prologues'. His hands are clasped, thumbs pressed together, beside the paper on which he is about to write, and he looks straight out of the picture as if apostrophizing the spectator. The author in the act of jotting down some flash of wit presents himself to his admirers, rather provocative and vain.

How different is Gainsborough's *Garrick* in the National Portrait Gallery (Fig. 29)! Hardly a trace of the actor is discernible here, only a more general impression of boundless versatility. The picture is artfully arranged to give a sense of the static—of a man sitting peacefully and gazing out—combined with a vibrant, utterly restless yet elegant vitality. By a remarkable touch, we seem to see him full face when in fact the face has a concealed asymmetry. The right side is foreshortened, the head being turned slightly, as is most noticeable in the curls, which look far smaller on the right than on the left. Asymmetry dominates each individual feature. The right eyebrow is drawn in a single sweep, the left one is nervously puckered. Similarly, the expression of the eyes differs: one looks straight out, the other seems to blink. The smile, too, is more marked on the right side. And finally, the facial asymmetry is continued in the arrangement of the clothes. The casually elegant pose, leaning on one arm with the other effortlessly across it holding a book, angles the whole attire to the right. The careful disorder of the cravat and waistcoat completes the impression of nonchalance. But the easy attitude has its own peculiar tension: the hidden asymmetry, counterpointing the apparently frontal view, both conceals some quality and insistently suggests it. In particular, the way in which the right arm is hidden by the book conveys most forcefully the actor's nimble miming. Where Reynolds tried to capture the actor's power by showing him in a specific dramatic situation, Gainsborough lets us sense it indirectly as an ever-present potentiality.

The versatility which this portrait of Garrick expresses seems from contemporary accounts to have been one of the essentials of his art. Perhaps the best description of his style of acting is to be found in Lichtenberg's letters from London, from which we learn that Garrick used syncopation at the climax of a scene.[86] In an excited speech the rhythm of voice and gesture would be exactly synchronized so that the audience experienced a perfect coincidence of sight and hearing, effecting a direct association between the gesture, the sound, and the meaning of the words, until the very climax, when the gesture was performed clearly and precisely at the expected moment, but the accompanying words were surprisingly missing and spoken, as it were, two seconds late. By this process of indirect intensification Garrick brought his audience to a far greater pitch of excitement than those actors who expressed rising emotion merely by gradually raising their voices. We should also remember Lichtenberg's description of the ghost scene in *Hamlet*, where Garrick playing Hamlet sees his dead father: 'At last he speaks, not at the beginning, but at the end of a breath, with a trembling voice: "Angels and ministers of grace defend us!", words which supply anything this scene may lack and make it one of the greatest and most terrible which will ever be played on any stage.'[87] One aspect of this versatility was Garrick's ability to play both tragic and comic parts without any inner conflict, and with a chameleon-like quality that made him, for all that mastery of both genres was fairly common among actors of his generation, by far the most versatile and universal of them all.

Whereas Gainsborough captures this quality by an almost still-life treatment of the man in his social reality, Reynolds characteristically tries to capture it by means of allegory. His *Garrick between Tragedy and Comedy* (Fig. 32) follows the pattern of 'The Choice of

[86] 'Briefe aus England. An Heinrich Christian Boie', 1 October to 2 December 1775, see now G. C. Lichtenberg, *Schriften und Briefe*, ed. W. Promies, III (1972), pp. 326 ff.

[87] Ibid., 1 October, p. 335, trans. by M. L. Mare and W. H. Quarrell, *Lichtenberg's Visits to England as described in his Letters and Diaries* (1938), pp. 10 f.

Hercules',[88] which was so well known to contemporaries that an indifferent engraver of Reynolds's picture actually entitled it *L'homme entre le Vice et la Vertu.*[89] Shaftesbury had described the theme's possibilities for the artist,[90] and Reynolds's successor as President of the Academy, Benjamin West, treated it again.[91] Any knowledgeable spectator looking at Reynolds's picture is reminded of the Hercules motif; and the wit of Reynolds's composition lay precisely in forcing an interpretation in this allegorical sense: Hercules had to choose between Virtue and Vice, Garrick is trying to decide between the two Muses. Admittedly, the analogy soon breaks down; for the point about the Hercules fable is that he finally chooses Virtue. But how does Garrick choose between the two Muses?

In the literary tradition, two interpretations of the picture can be traced. According to one, the real point of the composition is that Garrick cannot choose between the two Muses who have him in their grasp, and that he is fated to flirt with both:

> Who but has seen the celebrated strife
> Where *Reynolds* calls the Canvas into Life
> And, 'twixt the Tragic and the Comic Muse,
> Courted of both and dubious where to chuse,
> Th' immortal Actor stands.[92]

According to the other view, the words 'courted of both and dubious where to chuse' do not do justice to the picture. For although Garrick is wooed by both he appears to be in no doubt as to his choice. He chooses Comedy: he is quite plainly going off to the left and the movement of his hand implies that he is yielding to the pull of Comedy, while making his excuses to Tragedy who tries to hold him back. And this interpretation is surely strengthened by the fact that Garrick himself adopts a comic pose, and declares his fate by a gesture befitting Comedy.

This view has been taken further still by the suggestion that Garrick in later years increasingly turned away from tragedy, ceased to be a 'mixed' actor, and became a purely comic one. But the facts of theatrical history belie this, and the interpretation as a whole founders on the fact that, however we interpret Garrick's gesture, his dalliance with both Muses cannot be denied. Garrick's twofold talent was meant to be seen humorously as a twofold attachment. Yet it is a fact of observation that Garrick does follow Comedy and is merely held back and admonished by Tragedy. This has to be taken into account in any valid interpretation.

There is, I think, besides the two interpretations already mentioned, a third possibility, which requires a brief digression. A set of portrait drawings by Carmontelle, in the Musée Condé at Chantilly, shows almost all the prominent personalities whom he met as a member

[88] Cf. above, p. 23 note 62.

[89] J. E. Haid, Print Room, British Museum, London.

[90] See E. Panofsky, *Hercules am Scheidewege* (1930), p. 133, and Wind, 'Shaftesbury as a Patron of Art', below, pp. 64 ff. and Figs. 72–3.

[91] Victoria and Albert Museum, London.

[92] From the Epilogue to a comedy by Richard Cumberland, *The Brothers* (first produced in 1769), spoken by Mrs Yates.

of the entourage of the Duc d'Orléans. These drawings were later sorted and given titles by his friend the Chevalier Richard de Lédans. Among them there is, next to the picture of Sterne, one of Garrick.[93] The special feature of these portraits is the unpretentiousness of their invention and execution. They are simple sketches in profile which usually show the subject sitting or standing on a terrace. Only the portrait of Garrick (Fig. 30) is an exception, and a surprising one in that it has an element of fantasy quite at odds with the rest. For Garrick appears twice in the picture: on the left in a declamatory pose, poised to advance, arms outspread and gaze directed upwards; on the right roguishly peering round the door, one hand holding it half open, the other on his knee. To remove all doubt as to the meaning of the picture, Lédans wrote on the back of the sheet: *Le célèbre Garrick, tragique et comique.*[94] In other words, Garrick plays tragedy while Garrick the comedian watches him.

Now with the best will in the world it cannot be claimed that Carmontelle's sketch influenced Reynolds's painting. But when two painters of such different styles who scarcely came into contact with each other both hit on the same idea for a picture of a particular man, it can perhaps be said that the idea originates not so much from the painter as from the subject. This suggests itself the more readily in Carmontelle's case because the picture does not conform to the rest of the set. The interpretation is confirmed by Grimm in his discussion of Carmontelle's portrait in the *Correspondance littéraire*,[95] where he describes how the idea of portraying Garrick simultaneously as tragedian and comedian is founded on Garrick's own theory of acting, namely the doctrine that the tragedian, in order to be able to present the peaks of emotion truthfully, should first and foremost be schooled in the manifold techniques of comedy.[96] It is popularly said that it is only a step from the sublime to the ridiculous. Garrick turned this upside-down, claiming it is only a step from the

[93] F. A. Gruyer, *Chantilly. Les Portraits de Carmontelle* (1902), pp. 320–2. [Louis de Carmontelle was better known as a playwright and producer at the court of the Duc d'Orléans. His *Proverbes dramatiques*, enacted as 'amusements de société', were republished in 1825, followed by *Proverbes et comédies posthumes* in 1828. These works influenced many nineteenth-century authors of *Proverbes dramatiques*, notably Musset; see C. D. Brenner, *Le Développement du proverbe dramatique en France et sa vogue au XVIIIe siècle* (1937). A letter from Garrick to Dr Richard Brocklesby, 20 July 1766 (*The Letters of David Garrick*, ed. D. M. Little, G. M. Kahrl, and P. de K. Wilson, II, 1963, no. 416) shows that Garrick subscribed to an edition of a print made by Carmontelle to raise money for the family of Jean Calais, the Huguenot martyr. T. Crowe has now attributed to Carmontelle a substantial body of anonymous Salon criticism; see 'The *Oath of the Horatii* in 1785. Painting and Pre-Revolutionary Radicalism in France', *Art History*, I (1978), pp. 424 ff. But this attribution may prove controversial according to R. Wrigley, 'Mainly Monographs', ibid., VII (1984), p. 130, who also discusses the general problem of the attribution of such Salon literature in 'Censorship and Anonymity in Eighteenth-Century French Art Criticism', *Oxford Art Journal* VI, ii (1983), pp. 17 ff. See also below, p. 119 note 60.]

[94] The inscription continues: *Carmontelle delineavit ad vivum, 1765. Cette caricature fut faite au Raincy, sous les yeux de M. le duc d'Orléans, et passe pour une des plus parfaitement ressemblantes de l'auteur, dont le mérite connu était la minutieuse fidélité dans la physiognomie.*

[95] July 1765: 'Il prétend qu'on ne saurait être bon acteur tragique sans être excellent acteur comique, et je crois qu'il a raison. . . .' (*Correspondance littéraire, philosophique et critique par Grimm, Diderot, Raynal, Meister, etc.*, ed. M. Tourneux, VI, 1878, pp. 319 f.)

[96] This view that tragic acting rests on a mediated artistic effect in contrast to the immediacy of comedy is also reflected in Lichtenberg's letters on Garrick, where it serves as introduction to his account of Garrick in the role of Archer in *The Beaux' Stratagem* by Farquhar: 'Der Schauspieler sowohl als der Zuschauer sind beide immer mehr im Lustspiel zu Haus, als im Trauerspiel, und was der erstere auch selbst durch die feinste Kunst im Trauerspiel hervorbringt, läßt sich immer, dünkt mich, leichter in Worte fassen, als was die unerschöpfliche Natur im erstern sowohl tut als bemerkt.' (Op. cit., 30 November 1775, p. 351.)

ridiculous to the sublime. For the versatility which the actor has to acquire when playing comedy, in order to make the audience laugh, is just as vital to the creation of those powerful conflicts which inspire pity and fear. The syncopated style Garrick used for heightened expression is one instance of a comic technique adapted to tragic purpose.

A knowledge of Garrick's theory, which is documented in Carmontelle's case, can be assumed in Reynolds. For in the literary Club to which both Garrick and Reynolds belonged, Goldsmith's project for a Dictionary of the Arts and Sciences was discussed—Goldsmith and Johnson were to contribute essays on literature, Reynolds on painting, Garrick on acting. And even though this plan never materialized, those concerned must have often conversed about the theory of their respective arts. Garrick's theory of acting is surely the link between pictures of such differing artistic intentions as the Carmontelle drawing and the Reynolds painting. The comic Garrick looking through the doorway is exactly like a producer looking on from the wings at an actor he has rehearsed. The fact that Garrick the tragedian dominates the scene while Garrick the comedian supports him is a visual statement of the way Garrick saw, and embodied, the relation between tragic and comic acting—and this, I suggest, is also the meaning of Reynolds's painting, where Garrick follows Comedy, and justifies his action to Tragedy.

Of course, the notion of acting as a theoretical programme with the Hercules fable as an allegorical motif does not exhaust this picture's inventiveness. For Reynolds also used particular emblems to convey the themes of subordination and decision. He placed Garrick between Melpomene and Thalia. This idea of portraying an actor between the Tragic and the Comic Muses was a fairly obvious one, for he did, in fact, appear on stage between them. Statues of Melpomene and Thalia stood on either side of the stage in nearly every theatre of the time, and remained symbolically present to the audience while the central scene changed. Hence they are a necessary part of the scene in Hogarth's caricatures of the theatre. The theatre-goer could see actors in this situation every evening, so he could hardly help imagining any actor 'between the two Muses'. This motif became a familiar image in engravings, where actors were shown within a frame as on a stage, a mask suspended from above and two figures at left and right in front of them, one holding Melpomene's attributes of dagger and cup, the other Thalia's mask and flute. Sometimes they appear as *putti*, but more often as the Muses themselves. In 1797 a particularly strange use of this configuration was made by Henry Webber in his monument to Garrick in Westminster Abbey, which was copied in commemorative engravings. Webber played with the idea of making the podium evoke both the stage and a tomb. Garrick stands upon it, parting the curtains as if to speak the epilogue (or take the final bow); above him is the Shakespeare medallion; to his left and right are the Comic and Tragic Muses. But Garrick can also be seen as standing on the podium like a statue, for the way the Muses sit before it recalls the mourning figures on Baroque funerary monuments.

This transformation of the Muses into mourners makes them symbolic figures in a dramatic action, an adaptation for which the way had long been prepared by contemporary pantomime, where statues like Melpomene and Thalia were actually presented as living figures, from which it was only a short step to Reynolds's allegory. A drawing by Gravelot in the British Museum, a sketch for an engraving by Gerard van der Gucht, shows an actor in Fool's costume arguing with the Tragic and Comic Muses on stage (Fig. 31). The engraving served as frontispiece to *The Beauties of the English Stage* (1737).[97] The fact that it was engraved by Van der Gucht shows that it was drawn during Gravelot's stay in England, and it is significant in connection with Reynolds's painting that among Gravelot's English friends Garrick had pride of place.

This completes the range of observations needed to place Reynolds's achievement in its historical context. Only when it is realized how common was the idea of placing any actor between the two Muses can the originality and wit of Reynolds's invention be appreciated. For not only did he transform the simple spatial juxtaposition in the theatre into a dramatic choice by recalling the fable of Hercules, he also used it to symbolize the versatility of this one actor, Garrick. The simultaneous suggestion of these different ideas, in a single image which is at once a dramatic scene and a pictorial puzzle, reveals the artist's eclectic power. This, too, is a sort of heroic portrait: the actor is shown at an imagined moment, so far lifted above his usual situation that the artist can allude to Hercules' choice, and yet so firmly related to it (that is, on stage between the two Muses) that the allusion becomes a witty comparison.[98]

When Garrick was planning the great Shakespeare Jubilee in Stratford in 1769, two paintings were commissioned: one was to be a portrait of Shakespeare himself, the other to symbolize Garrick's relation to him. It is an irony of history that by Garrick's choice the latter commission, initiated and paid for by the Stratford Corporation, went not to Reynolds but to Gainsborough.[99] To think what Reynolds would have done with such a theme! The commission was an invitation to the emblematic imagination; but for Gainsborough this fact was a hindrance. A sparklingly witty letter survives in which he describes what a symbolic Shakespeare-painting should look like: Shakespeare should be standing gazing upwards with a beam of light striking his eye from above to show 'where that inimitable Poet had his Ideas from'.[100] But Gainsborough interrupts this most amiable ridicule of allegorical portraiture to curse himself for even entertaining the thought that he

[97] [See H. A. Hammelmann, 'Gravelot in England', *Book Handbook*, II (1951), p. 189.]

[98] How strongly the emblems in this picture excited the imagination of contemporaries can be seen from the way in which this representation of Garrick deciding between two powers passed into caricature and journalism. When Garrick was attacked for allowing the splendours of décor at Drury Lane to overshadow poetic values (*The Theatres. A Poetical Dissection*, by Sir Nicholas Nipclose, Baronet, 1772; see below, p. 91 and Fig. 74), the artist of the frontispiece immediately thought of Reynolds's allegory and drew Garrick between the Muses and the Tailors and Carpenters. Garrick has to decide between them, is held back by the Muses, but still follows the Tailors:

> *Behold the Muses ROSCIUS sue in Vain,*
> *Tailors & Carpenters usurp their Reign.*

And when another artist wished to symbolize Garrick's death (*Farewell! a long Farewell!* J. Parkinson pinx., Robert Laurie fecit, 1779), he also placed him in a situation of decision: on the one side, the Muses with the bust of Shakespeare; on the other side, Cronos pointing to the blessed land, represented symbolically by the hill of Virtue with its temple—a reminiscence of 'The Choice of Hercules'. Garrick, standing in the middle, hesitatingly decides for Cronos. (Print Room, British Museum.)

[99] See Gainsborough's receipt from the Corporation of Stratford-upon-Avon, first published in Whitley, *Thomas Gainsborough*, p. 67; see also *The Letters of Thomas Gainsborough*, no. 29.

[100] Ibid., no. 28.

might deviate from the 'simple Portrait way'. The picture he painted, for all that it necessarily had—and duly fulfilled—an official function, did in fact become an intimate portrait (Fig. 33). Garrick stands in a garden, leaning in an easy, amiable way against a plinth carrying a bust of Shakespeare; his arm is round the bust, his head turned to look obliquely out of the picture. A peculiarly intimate affinity is suggested in the configuration of heads: not the relationship of the inspired poet to the subordinate actor, but rather of devotion freely bestowed by the interpreter on the model he has made his second nature. Shakespeare's face lacks any noble, sublime expression. The bust is, to quote Gainsborough, 'a silly smiling thing'.[101]

Contemporaries with a taste for the heroical took offence at the intimate character of the picture. In an account of the Stratford Festival by the bookseller Samuel Ireland, whose name has a sad notoriety among Shakespeare scholars, exception is taken to the fact that Garrick is not portrayed in an idealized costume as befitted such an official portrait. The natural air, and the pose suggestive only of a refined gentleman, were found offensive.[102]

How a painter in the heroic style treated such a theme can be seen in a composition by Pine (Fig. 34) which attempted to symbolize Garrick's relationship to Shakespeare when the Jubilee was re-staged at Drury Lane in London. A great many elements in this picture can be traced back to events which actually happened, or were to have happened, in Stratford. It shows the statue of Shakespeare Garrick endowed for the occasion, and it shows Garrick precisely at the moment when he recited the Ode to Shakespeare he had composed for the ceremony. But the histrionic way he apostrophizes the statue must be the painter's own invention, for other engravings and reports of the original event describe Garrick's noble and affecting performance inside the Rotunda specially constructed for the celebrations. The Rotunda is hinted at in the picture, but it is peopled with figures which, Garrick apart, were never in it. For the characters in Shakespeare's plays, as depicted here, formed part of a procession in which the tragic and the comic heroes were to have appeared attending their respective Muses. But the procession was, in fact, rained off. They are seen here in their imagined ranks around Shakespeare's statue—on the right Thalia accompanied by Pan with his flute, and with her Falstaff, Prospero, and Caliban; on the left Melpomene with a Genius blowing a trumpet, and in front of her Hamlet, Macbeth, and Lear with the strangled Cordelia. To heighten the imaginary solemnity of the scene, the air is filled with stage-clouds pierced by the sun's rays, and the picture is partly framed by a stage-curtain—a genuinely 'synthetic' painting which combines all the details of the Jubilee into one imaginary heroic spectacle.[103]

[101] *The Letters of Thomas Gainsborough*, no. 28.

[102] *Picturesque Views of the Upper Avon* (1795), pp. 219 ff.

[103] Even the sun's rays penetrating the clouds could be a reminiscence of Stratford, for the more important buildings of the town were decorated with transparencies, colour-reflecting silk screens, which were illuminated from behind after dark. And just as the theatre showed transparencies of Shakespeare between the Comic and Tragic Muses (according to Ireland, these were designed by Reynolds), 'the humble dwelling, where our Swan of Avon first drew breath, [was not] left undecorated on this occasion: a transparent painting was hung before the windows of the room in which he was born, representing the Sun breaking through a Cloud' (ibid., p. 224). For further details, cf. contemporary newspaper reports and

How popular this kind of theatrical commemoration was is shown by this re-enactment of the Stratford celebrations on the London stage. Garrick wrote a play especially for the purpose, entitled *The Jubilee*. Scenes loosely strung together showed what visitors to Stratford had seen, or not seen: visitors quarrelling about a room, the ceremony in the Rotunda with choirs and recitations, and to round it all off the great festival procession in which, to the Dead March from *Saul*, the Tragic Muse played by the leading tragic actress passed by with her entourage, followed after a short interval and with suitable merry musical accompaniment by the Comic Muse and her motley crew. This pantomime ran concurrently in the two largest theatres in London, Drury Lane and Covent Garden, for more than a hundred performances, and the managers vied with each other to offer the more magnificent production. According to the critic in *The Town and Country Magazine*, Drury Lane won, because they had a triumphal chariot for the Muses.[104]

When the reputation of an actress depends on her being cast as the Muse of her art in a dumbshow, then we can see the extent to which the stage, from being a place where the actor played a part, had become a place where the actor was himself given heroic stature by the part he played. Just how far the cult of the actor was carried in Garrick's own case can be seen in a picture by George Carter, painted immediately after Garrick's death for purposes of engraving.[105] It shows nothing less than an apotheosis of the actor: Garrick is borne up to Parnassus by two winged angels, and awaited by Apollo—with seven Muses. Melpomene and Thalia are missing; they, together with Shakespeare, have already run half-way down the mountain to greet him. At ground level, the surviving actors are seen in their favourite roles and appropriate poses. Carter published a commentary on the picture—an extreme example of the tendency to emblematic glorification which Reynolds refined by his eclecticism and Gainsborough totally opposed.

These productions lead us to ask why Reynolds himself, with his striving for heroic effect, hardly ever succumbed to this overcharged rhetoric. Possibly because he was fully aware of the opposition between the natural and the heroic and strove to resolve it, and to make it visible by introducing elements of contrast into this type of portrait. His various devices—the three-quarter length, marked foreshortening, and strongly contrasting colours which isolate the figures in space and endow them with an ideal setting—are all compositional motifs not found in the stock pictures in theatrical style, where Garrick might be depicted full-length, declaiming before a packed house, or borne on the arms of genii, apparently quite at home up in the air.

Yet it remains remarkable that, when Garrick was asked to award the commission for the commemorative portrait, Reynolds's heroic manner, however cleansed of excess, seemed

A Concise Account of Garrick's Jubilee in 1769 and of the Commemorative Festivals in 1827 and 1830 (Stratford-upon-Avon, 1830). The Jubilee was fully described in various studies published to mark the Shakespeare celebrations of 1964. These make extensive use of the contemporary sources. See, for example, C. Deelman, *The Great Shakespeare Jubilee* (with bibliography); M. W. England, *Garrick's Jubilee*; and J. M. Stochholm, *Garrick's Folly. The Shakespeare Jubilee of 1769 at Stratford and Drury Lane.*

104 *The Town and Country Magazine*, I (1769), p. 550.

105 Collection of the Shakespeare Memorial Theatre, Stratford-upon-Avon.

less worthy to him than Gainsborough's natural style. This preference seems intimately linked with the actor's view of his own art, for his theory that the power of acting heroic parts is rooted in the versatile playfulness of light comedy sounds very much like a reversal of Reynolds's artistic theory. Contemporary reports agree that what set Garrick apart from other actors was the way he replaced declamation with 'naturalness', both in his gestures and in his speech.

Garrick's choice of Gainsborough is particularly significant because a corresponding link can be shown between actor and painter in the heroic style. When, in Mrs Siddons, an actress came to the fore who could give new substance to high tragic expression, declamatory speech, and grand gesture, Reynolds was one of the first to recognize her greatness and to celebrate her as 'the Tragic Muse'.

Portraits of Women

In the case of Gainsborough, portrayal of women was the genre that brought out the natural grace of his style to the full. His women appear in idyllic landscapes, lean with graceful ease against pillars, sit on the grass making flower-garlands, or roam with their lap-dogs through the park. There is no trace here of the different degrees of heightened effect or refinement with which Reynolds by ever new inventions strove to convey the types and moods of women. Idyll, with Reynolds an exception used only in special cases, was for Gainsborough the norm. When Reynolds paints Mrs Sheridan he celebrates her musicianship by making her a St. Cecilia. She sits at the organ accompanying two child angels, while a ray of light breaks through the clouds. Gainsborough's portrait of Mrs Sheridan gives no hint that she is a musician. She is shown in *style champêtre*, sitting in a landscape, her flowing hair ruffled by the breeze. When Reynolds paints Mrs Robinson he represents her in a melancholy pose, gazing out to the open sea, and calls the picture *Contemplation*. Gainsborough paints her in her garden with a little dog by her side. He puts a handkerchief in one hand, and in the other a medallion which is supposed to be a picture of the Prince of Wales, but is so sketchily painted that it is quite impossible to recognize any features.

To give some idea of the range of characterization in Reynolds's portraits of women, we shall distinguish three types of model: the courtesan, the lady of fashion, and the actress.

The portraits of courtesans run the gamut from the heroic to the idyllic. He celebrated Emily Pott as Thaïs. He painted Kitty Fisher now as Danaë receiving the Golden Rain,[106] then as Cleopatra throwing the pearl into the cup (Fig. 81), then shown writing what was recognizably a love letter, and finally in a pensive pose, her hands in her lap, yet with an unmistakable portrait medallion on her breast and in the company of a pair of turtle-doves.

In such cases the subject's way of life and the circumstances of the commission suggested the form of paraphrase. But when the subjects were gentlewomen Reynolds had to cast

[106] [A portrait with this title is listed in the catalogue of paintings then in the Kaiser-Friedrich Museum, Berlin (1912), no. 1637A, a loan from the Kaiser.]

about for it. In the case of Lady Blake, a sitter with no striking characteristics, heroic effect had its problems. Reynolds painted her as Juno (Fig. 39), indeed as Juno standing on Mount Ida and receiving the cestus from Venus, who appears above her in a cloud accompanied by doves. The idea of the 'composite style' is nowhere more clearly demonstrated than here, for although involved in a dramatic incident, the figure keeps a formal portrait stance, while the facial expression retains individual features despite the idealizing arrangement. It is particularly odd that in a painting conceived as a portrait Juno is shown receiving the cestus from heaven, a subordination to the goddess of love befitting Lady Blake, but hardly the goddess she personifies.[107] Again the spectator is meant to recognize and enjoy the tension between sitter and scene, and to see the painting as a scene from mythology and at the same time as a tableau: Lady Blake as Juno—an elevated portrait, Juno as Lady Blake—a restatement of the scene in earthly terms. No doubt the representation of the lady as goddess was intended as a compliment, but to keep it so involves forgetting what followed. For Juno had asked for the cestus so that she might seduce and then deceive her husband. We are meant to concentrate on the gift only, as the symbol of beauty bestowed. This is a limitation characteristic of all Reynolds's allegories. In the allegorical portrait of Garrick, we must likewise not pursue the comparison with Hercules, but merely accept that the two characters are similar only in that they have to choose between two figures, one lively and the other grave. Reynolds's picture is certainly not *Garrick en Hercule* (Fig. 37).[108]

Perhaps this points to the difference between the function of allegory in a society of commoners and in a court nobility like that of France. There the motifs from classical mythology symbolized a real identification—the person portrayed was transformed by his disguise. By contrast there are no mythological portraits of the court of George III, only of ladies belonging to the socially ambitious bourgeoisie or new nobility, who projected themselves into classical roles which had become forced poses.

Yet it is not always in such exaggerated form that Reynolds's tendency to the heroic manifests itself in his portraits of women. Transformation into a goddess is an exception, and even Lady Blake as Juno was, in pictorial terms, subordinate to another goddess. A subordination of this kind—which at the same time implies elevation to a higher sphere—is in fact the norm. Women are shown sacrificing to a goddess or performing some sacred rite—*Lady Bunbury sacrificing to the Graces*, *Lady Talbot sacrificing to Minerva* (Fig. 41). The picture in the National Gallery catalogued as *The Graces decorating Hymen* is of this type (Fig. 8). Sir Walter Armstrong agreed with Leslie and Taylor that the title was unsuitable, invoking the evidence of the Academy Catalogue (1774), where it was entered as No. 216, 'Three Ladies adorning a term of Hymen'. He insisted that the three ladies were

[107] The form of the composition has clearly been determined by Christian iconography: the bestowing of a girdle, in the Christian view, can properly be imagined only as taking place from above. Exactly the same pictorial scheme can be seen, for example, in *St. Hubert receiving the Miraculous Stole* by Theodor van Loon (Musées Royaux des Beaux-Arts, Brussels), where an angel appears in place of Venus.

[108] See above, p. 29 and below, p. 121.

not Graces at all, but the daughters of a Scottish country gentleman, Sir William Montgomery.[109] Now we know the circumstances in which this picture was painted. It is a betrothal picture, for all three daughters, who had been brought up in Ireland, became engaged at the same time, and one of the bridegrooms commissioned Reynolds to paint an 'emblematic' picture of them.[110] The meaning of the allegory is clear—the brides are adorning a term of Hymen; but manifestly the allegory has a further point, for the girls, on account of their number and their beauty, were known as 'the Irish Graces'. Without doubt Reynolds was playing on this allusion and his composition is meant to evoke it. To pose alternatives—either the Montgomery girls or the Graces—is mistaken. It is the Montgomery girls as Graces. The fact that the girls had reason to crown Hymen, whereas the Graces had not, once again shows the internal problems of the allegorical style, but does not alter the fact that an allusion to the Graces was part of the intention.

Another of the portraits of ladies sacrificing is Romney's painting of the actress Mrs Yates (Fig. 40). She, too, is shown making a libation on a burning altar. But in this case there is nothing to indicate which goddess is receiving sacrifice. No Grace, Minerva, Venus, or Hymen stands in the background. However, the subject herself holds a dagger in her hand and is thus identifiable as the Tragic Muse. This results in a contamination of two pictorial ideas: the Muse to whom the sacrifice is made is herself embodied in the sacrificing figure. The actress who plays the Muse is equated with the goddess she represents. Admittedly, in the case of Mrs Yates there was historical cause for this: she was to have played the Tragic Muse in the Jubilee, and it was she who created the role at Drury Lane. An anonymous engraving shows her together with Mrs Abington as the Comic Muse, both in their costumes. So we know what their costumes were like.[111] We cannot, of course, take their gestures as rendering those of the performance itself, since in the production the Muses did not appear walking together but stood each on a triumphal chariot, pulled by sombre or cheerful demons and followed by the main characters from Shakespeare's tragedies or comedies. A screen formerly in the Town Hall at Stratford-upon-Avon shows the procession in a somewhat later production, and the Tragic Muse is easily recognizable by her dagger and cup, black robes and crown, leading the first group of the procession (Figs. 44–5).[112]

These triumphal processions must be borne in mind if we are to understand the historical context in which Reynolds's *Mrs Siddons as the Tragic Muse* was produced (Fig. 46). Significantly, at the time of the painting Mrs Siddons had never yet played the Tragic Muse. Mrs Yates was then the queen of tragedy, with exclusive claims to this symbolic role.

[109] W. Armstrong, *Sir Joshua Reynolds* (1900), p. 91.

[110] Reynolds to the Right Hon. Luke Gardiner, [July 1773], *Letters of Sir Joshua Reynolds*, no. xxii; see also E. H. Gombrich, 'Reynolds's Theory and Practice of Imitation', *Norm and Form* (1966), pp. 129 ff.

[111] British Museum, Burney Collection, *Theatrical Portraits*, x, p. 101.

[112] [For full details of the characters, see *A Descriptive Account of the second Royal Gala Festival at Stratford-upon-Avon . . . 1830: by a member of the Royal Shakespearean Club* (1830). The screen in the Town Hall (cf. F. C. Wellstood, *Inventory of the Civic Insignia and other Antiquities belonging to the Mayor and Corporation of Stratford-upon-Avon*, 1940, p. 24) was destroyed in September 1955 on the authority of the Chairman of the Public Relations Committee. I am grateful to Dr Levi Fox for this information.]

Romney glorified her in this part.[113] His picture was painted a few years after the first production and idealizes both the pose and the costume. In his portrait of Mrs Siddons, however, Reynolds was anticipating theatrical developments and therefore not bound by memories of any actual performance. His Muse, in an attitude clearly reminiscent of Michelangelo, is shown on a throne floating on clouds, flanked by the figures of Pity and Terror, the Aristotelian elements of tragedy. In the pose of the head, Pity repeats exactly the attitude of the figure beside the Prophet Jeremiah on the Sistine Ceiling (Fig. 48). Terror is based on a sketch for a self-portrait; the mouth, open as if arrested in a cry, recalls an ancient tragic mask.[114] The objects the two figures hold, dagger and cup, are the usual attributes of the Tragic Muse, symbols of open and secret murder. When we try to interpret the Muse's pose, we find the same conflict as in Romney's picture of Mrs Yates. The actress is portrayed as if, as an actress, she is being inspired, whereas she should, as a Muse, herself inspire.

Much has been written about the question of whether Mrs Siddons's pose in this picture was prescribed by Reynolds or adopted by her spontaneously.[115] Mrs Siddons in her memoirs claims the latter, but all the internal evidence points to the former. The left arm with the elbow supported and the hand pointing upward is surely imitated from Michelangelo's Isaiah (Fig. 47), and the right arm hanging limply over the side of the chair recalls his Daniel. If further proof is needed, Reynolds used the same pose with a slight modification for another portrait, this time of a lady of fashion (Fig. 50). Is it likely that he would have transferred to someone else a pose Mrs Siddons had invented for herself, a pose that was, so to speak, her copyright? However, any attempt to give a definite answer is not only doomed to failure, but also meaningless, for in the theatre of the time actors as well as painters were consciously influenced by classical examples and once an actress had discovered a formula it would be adopted by others, particularly by women in society. Even so, the controversy is not without interest to the cultural historian. The fact that an actress vied with a painter in claiming credit for a gesture in one of his pictures shows how much importance was attached to their invention.

With his allegorical portrait of Beattie, Reynolds had actively taken sides in the philosophical controversy between sceptical and heroic ethics. With his portrait of Mrs Siddons as Tragic Muse, he actively intervened in the life of the stage, for he had no sooner exhibited the picture (1784) than, under its compelling influence, Mrs Siddons played the Tragic Muse (1785). She thrust out her rival Mrs Yates with such panache that reverberations of the contest can still be sensed in the contemporary pictorial record. There are two engravings of the same scene from *Medea* which show almost too palpably the ousting of the older actress by her younger rival (Figs. 42–3). The first shows Mrs Yates as Medea, in oriental costume, arms rhetorically outstretched, accompanied by her frightened,

[113] Private Collection.

[114] Study for a figure of Terror, Tate Gallery, London, no. 1834.

[115] See Whitley's summary of the various accounts in *Artists and their Friends in England*, II, pp. 4 ff.; and now the excellent discussion of Reynolds's sources for this portrait in R. R. Wark, *Ten British Pictures, 1740–1840* (1971), pp. 43 ff.

crying children. The other picture, Mrs Siddons as Medea, adopts the same composition, attitude, and even details of costume, but simply substitutes Mrs Siddons's head, which is taken straight from Reynolds's portrait. Even the crown and the coiffure are thus somewhat pointlessly those of the Tragic Muse. The fact that a draughtsman who illustrated Medea considered Mrs Yates no longer right for the role and felt compelled to substitute Mrs Siddons is, apart from Mrs Siddons's own achievements, indirectly the effect of Reynolds's portrait. This is confirmed by the fact that the artist, in aiming at tragic expression, could not free himself from the influence of Reynolds's picture. In the history of the stage, Reynolds emerges as the champion of a new, now purely heroic, style of acting. For Mrs Yates was the last tragic actress who, like Garrick, could also play comedy. Significantly, Mrs Siddons consistently failed in comic roles.

When Gainsborough painted this heroine he took great pains to remove every sign of the heroic style from her expression and bearing (Fig. 49). The portrait contains no trace of sublime declamation. It embodies nothing but the ideal of natural social composure—the elegant woman at her ease. Gainsborough had no commission for this picture. A year after it was painted it still hung unsold and evidently unsaleable in his house, and a critic who called on him reflected that 'all the graces of private station' could not popularize this picture, 'so prevalent is the striking form of action and of character over the mild and unimpassioned deportment of private life'.[116]

If we can believe contemporary anecdotes, even the behaviour of the two men to the actresses in their studios was characteristic of the contrast between them. Reynolds greeted Mrs Siddons with the words 'Ascend your undisputed throne, and graciously bestow on me some good idea of the Tragick Muse!', and after completing the picture he put his name on the hem of the garment, declaring that this was how he wished to go down to posterity.[117] There was no high style in Gainsborough's converse with her. When after long struggles he abandoned the picture as finished, there was no bow and elaborate compliment. Instead he threw down his pencil, exclaiming, 'Damn the nose—there's no end to it'.[118]

Archaeology and the Theatre

A number of Reynolds's portraits of women show the heroic style toned down to something nearer to ordinary life. A pose which in a theatrical scene expressed heroic passion would suddenly be modified so as to achieve a lyrical effect. A characteristic example is a portrait of Mrs Stanhope as *Contemplation* (Fig. 51) which adopts almost exactly Mrs Siddons's pose as Tragic Muse, but by a slight turn of the hand and head transforms power into charm. Or again, Mrs Stanhope as *Melancholy* (Fig. 50), where the grand gesture of supported head

[116] Whitley, *Thomas Gainsborough*, p. 237.

[117] Mrs Siddons's 'Memorandum', in Thomas Campbell, *Life of Mrs Siddons*, II (1834), pp. 241 ff.

[118] Whitley, op. cit., p. 370.

and trailing arm is subtly modulated by a moonlit landscape to lapse into psychological impressionism. Pictures like these represent an intermediate stage between Reynolds's heroic portraits of women (*Lady Blake as Juno*) and the informal social portraits which he painted without any mythological pretensions. The subject is still presented 'as' something, but no longer as the embodiment of a supernatural heroine or goddess, purely as an example of an identifiable psychological concept, a mood or state of mind.

Some light is cast on this intermediate stage between heroic and informal styles by the game of *tableaux vivants*, one of the best known and most popular entertainments in eighteenth-century society. The ladies assumed particular roles for masquerades, and from descriptions we can see that the poses they struck on such occasions had often been copied from the stage. Mary Hamilton, a dilettante whose sketch-book is preserved in the British Museum, drew with great care and application the 'attitudes and costumes' which Mrs Siddons displayed during her appearances in Dublin. These watercolour drawings are designed to capture pose and costume at a particular moment in a given role, but do not attempt to catch the facial expression. The face remains blank, from which we can see that the artist was not so much concerned to render the overall impression the actress made as to capture those facets which could be of use to her socially—pose and costume.

One of these drawings shows Mrs Siddons as Hermione in *The Winter's Tale* (Fig. 56). She is standing by a column in a white robe, her arms uplifted, in the act of turning round; it is the moment when she changes from a statue into a living person. An engraving (Fig. 55) shows us her pose before this movement, when she was still supposed to look like a rigid statue. Her attitude is that of an antique sculpture of a Muse (Fig. 53). Thus when a statue had to be represented on stage, a classical example was consciously chosen. And this was not Mrs Siddons's own choice but, as we can see from other pictorial evidence, the traditional formula. Every actress who played Hermione, whether it was Mrs Siddons, Mrs Hartley,[119] or Miss Farren, struck the Muse's pose when it came to portraying the statue; and whenever a painter or engraver had to portray an actress as Hermione, he chose the last scene when the statue comes to life. This moment must have been regarded as the climax of the whole play and the decisive test of the actress's performance. The ability to mime a statue gradually coming to life was perhaps of such paramount interest because it was a stage version of what was commonly played out in the very different setting of the drawing-room charade—the representation of a statue by a living person.

Nor, incidentally, should one assume, from the fact that both the engraving and the drawing are relatively late and show the ample form of Mrs Siddons's ageing figure, that this deliberate imitation of attitudes from antiquity was a phenomenon peculiar to Napoleonic times, with their conscious classicism in costume and pose. Zoffany's portrait of Miss Farren as Hermione (Fig. 57) disproves this, for here, as indeed in the portrait of Mrs

[119] [Cf. *Mrs. Hartley in the Character of Hermione*, an engraving inscribed *Published for Bell's edition of Shakespeare, October 10th 1775*, British Museum, Burney Collection, *Theatrical Portraits*, IV, p. 192. This did not appear in the first or second edns. (1774 and 1785–7), and may have been used for subscription purposes.]

Hartley, the actress strikes the attitude of the Muse even though the classical costume has not yet been introduced. Miss Farren is shown, just as Mrs Siddons was in Mary Hamilton's sketch, at the moment when the statue comes to life. The legs are still crossed as in the Muse's pose, the right arm is still leaning on an altar, but it has moved away from the head, which has been turned to face forwards. In fact, what was originally a pose of quiet contemplation has here become a gesture of unveiling. Classical gesture was thus a much earlier feature of the statuesque pose than classical costume, and in these portraits of Miss Farren and Mrs Hartley we can see it still combined with the flowing dress, the wig, and even the veil of the time.

There is a direct connection between these stage conventions and contemporary portraiture. In the portrait of Lady Louisa Manners, at Kenwood, Reynolds uses this very stage pose of the Muse, legs crossed and elbow supported by a plinth, as a means to heroic effect (Fig. 52). In this case there can be no doubt that Reynolds chose the attitude quite consciously to convey reflection; for this portrait differs strikingly from the countless eighteenth-century English portraits in which men as well as women are shown leaning cross-legged against a pedestal. This latter attitude had come to typify relaxed elegance, and can also be found in portraits by Gainsborough, who was not aiming at heightened expression or even conscious of expressing anything specific such as sentimental contemplation. When Gainsborough shows Mrs Meares leaning against a pedestal (Fig. 54), her crossed legs are merely the obvious consequence of this posture. And the arm is not placed on the pedestal in such a way that her head can rest reflectively upon it; it is lying extended on its edge, completing and emphasizing her totally relaxed stance, and her head is turned freely out with no trace of melancholy. Similarly, Gainsborough's Garrick (Fig. 33) has his legs crossed because he is leaning against the plinth, yet there is no suggestion of the Muse's pensiveness in this characteristically classical pose. It is an expression of social ease and nonchalance.

If we follow the progress of the heroic pose from its origins as a classical motif by way of the stage and the society charade into portraiture, we find that the gestures of antiquity remained a living heritage for the eighteenth century through two agencies, archaeology and the theatre. At root it is an antiquarian interest that influences even the stage and leads actors to mould grand gestures on ancient models; and it is at root a theatrical interest in the classical gestures that leads to such imitations. The combination of theatre and antiquarianism is nowhere clearer than in the case of the famous, or notorious, Lady Hamilton. As mistress, then wife, of a 'collector of antiquities', she impersonated in her villa at Naples the figures of Etruscan vases and paintings newly excavated at Pompeii. Lady Hamilton's 'positions' and 'attitudes' have been preserved by many painters, notably Romney, and in countless drawings.[120] The collection of engravings published by Rehberg, *Drawings Faithfully copied from Nature at Naples* (Fig. I), added to the would-be sculptural

[120] See now K. G. Holmström, *Monodrama, Attitudes, Tableaux Vivants, 1770–1815* (1967); also P. A. Jaffé, *Lady Hamilton in relation to the Art of her Time* (1972), nos. 44 and 45.

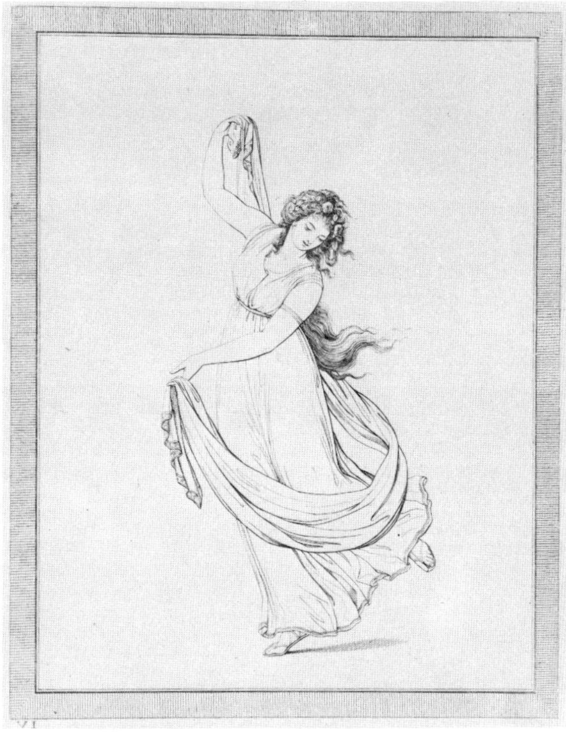

I. Frederick Rehberg and others: Emma Hart, later Lady Hamilton, from *Drawings Faithfully copied from Nature . . . MDCCXCIV* [1797]. Engraving by Thomas Piroli. London, British Library

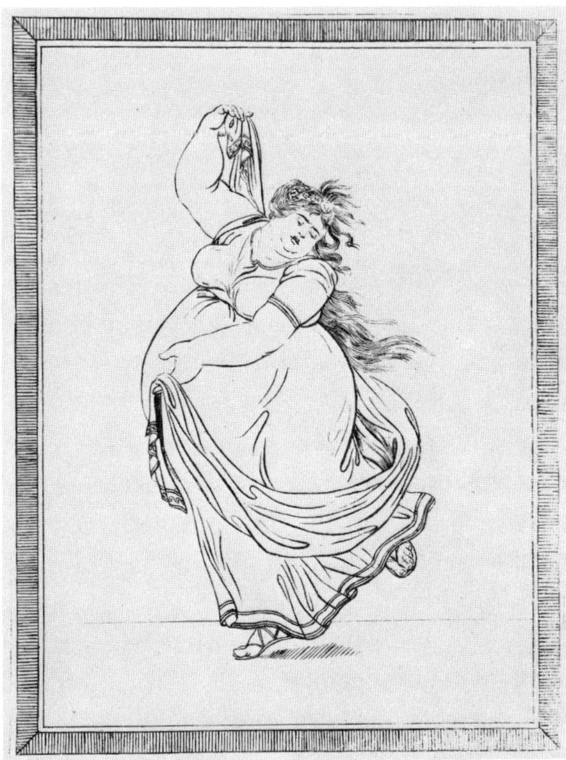

II. Lady Hamilton, from *A New edition Considerably Enlarged, of Attitudes faithfully copied from Nature . . .*, 1807. Engraving. Glasgow, The Mitchell Library

quality of the classical poses a strong *haut goût*, so that even contemporaries felt that the mixture of heroic style with sentimental social appeal went too far. A set of caricatures appeared under the title *A New Edition considerably enlarged, of Attitudes faithfully copied from Nature and humbly dedicated to all Admirers of the Grand and Sublime*, in which the author quotes Homer to support the view that the true mark of divinity, particularly in women, is to be larger than life, as in 'Homer's description of the helmet of Minerva, which was alone capable of containing many regiments of infantry'. Hence he felt it his task to add to Lady Hamilton's attitudes that degree of heroic quality lacking in the original drawings and to show, in line with the subsequent development of the model herself, grace of gesture complemented by corpulence of form (Fig. II).

To interpret these bizarre theatrical excesses we can return to the ideas of David Hume. In his *Dialogues concerning Natural Religion*, he made the point in favour of the dialogue form that it unites the two most agreeable occupations of man, study and society. He chose to combine the two because study by itself drives us into a brooding solitude, for which the only cure is the society of other men; and social life, conversely, becomes a shallow thing if it is not refined by some intellectual pursuit. The dialogue is an example of the balance being struck. Against this (as Hume called it) 'natural' form, the theatre and antiquarianism stand as the excesses of social and intellectual life respectively; for in the theatre the unobtrusive forms of sociability are infringed by display, and in antiquarianism the milder forms of learning are exaggerated into an abstruse specialization. In both cases the balance dear to Hume was lost. Both are always avoided by Gainsborough (the parallel is striking), and consciously cultivated by Reynolds. Gainsborough is innocent of learned allusion or histrionics; with Reynolds theatrical gesture and metaphorical expression are the means *par excellence* of achieving heroic effect.

Self-Portraits

I have tried to show the opposition between Reynolds and Gainsborough in their portraits of children, of men and women, of learned men and actors. The final confirmation is found in the way the two artists saw themselves and captured their own personalities in self-portraits.

From his early youth Reynolds used his own face, often in conscious emulation of Rembrandt, as a model for studies in physiognomy and likeness. And there is hardly one of his many self-portraits that does not show him in a particular role, if not actually at a decisive moment in his life: Reynolds as Doctor of Civil Law, Reynolds as President of the Royal Academy (Fig. 60), Reynolds as learned painter (with horn-rimmed spectacles), even Reynolds hard of hearing (with his hand to his ear). Not one of Gainsborough's self-portraits would even tell us that he is a painter at all. They are for the most part head-and-shoulders portraits, which have a touch of elegance through the careful treatment of contemporary fashion in dress. In a small painting from his Ipswich period, he is shown with his wife and child in a landscape—just like any other English gentleman of the eighteenth century who had himself painted in a conversation piece. Not one picture, as far as I know, shows him in front of his easel; not one emphasizes, as Reynolds would undoubtedly have done, that music was his favourite pastime. A youthful and a mature self-portrait by each artist will point the contrast.

Reynolds's self-portrait as a young man, in the National Portrait Gallery, quite consciously shows him as a painter, at a significant moment (Fig. 58). One hand holds a palette and brush, the other is raised to shade his eyes: the painter looks searchingly at his subject—in other words it is the critical moment of conception. The shadow the hand casts on the face creates a contrast of light and shade intended to remind the spectator of the Caravaggesque self-portraits which Rembrandt painted in his youth. In contrast, Gainsborough's self-portrait as a young man (Fig. 59) shows the artist in a neutral situation observed quite objectively. With some slight pretence to elegance, he has set his tricorn hat at an angle, but the youthful face is rendered very simply and has an almost peasant coarseness of expression. The stories of his taste for crude jokes and amusements become credible when we look at this picture.

In his self-portrait in later years as President of the Royal Academy, Reynolds presents himself wholly emblematically in cap and gown, one hand on a bust of Michelangelo (Fig. 60). And there are also allusions to another spiritual ancestor. His personal bearing, the subtle tones in the treatment of his robe, and the way the figure stands within the frame, are meant to recall Rembrandt. Ultimately, of course, the idea of a portrait containing a bust that has some spiritual significance for the sitter stems from a Venetian tradition which also influenced Rembrandt's *Aristotle contemplating a bust of Homer*.[121] The pride which led Reynolds to depict himself with conscious allusion to his great predecessors, yet without renouncing the credit for his own achievement, fits what Burke tells us about his character. He was, Burke says, a man whose tenacious ambition turned even his weakness into strength, and who bore his title as if born to it, despite his quite uncourtly behaviour. Even the name Joshua Reynolds was made, it seemed, to be prefixed by 'Sir', and as Sir Joshua he has gone down in history.

There is also a self-portrait of Gainsborough in later years which today has an official function, although it was not painted to this end. It hangs in the Royal Academy. This official setting is in strange contrast with the reserved way in which Gainsborough has, here too, kept to the 'simple Portrait way'. The painting restricts itself to the representation of head-and-shoulders in an oval, painted apparently without any personal sympathy. The head in this portrait has sometimes been thought to show a trace of childlike obstinacy; but even this minimum of self-evaluation has gone from our last portrait (Fig. 61). Looking at this gentleman, with his powdered hair, his elegant clothing, and an energetic three-quarter

[121] The Metropolitan Museum of Art, New York. See J. Held, *Rembrandt's 'Aristotle' and other Rembrandt Studies* (1969), pp. 3 ff.

turn to his head, which betrays a courteous rather than a grand manner, one is reminded of Hume's description of the man who needs no supernatural sanctions, refrains from forcing his mental powers, and, in full awareness of his natural limits, leads a life both socially useful and aesthetically agreeable; and one feels inclined to apply the words with which Hume ends his description, 'a model of perfect virtue'.[122]

[122] *An Enquiry concerning the Principles of Morals*, Sect. IX, part i, § 218 (Selby-Bigge and Nidditch, p. 270).

II · Julian the Apostate at Hampton Court

The Painting of the King's Staircase

THE decoration of the King's Staircase at Hampton Court was painted by Antonio Verrio 'as ill as if he had spoiled it out of principle'. The phrase is Horace Walpole's, and it alludes maliciously to a story about Verrio's religious and political sympathies: he was reputed to have been a faithful Catholic and so loyal a servant of the dethroned King that he refused at first to work for William III.[1] It was only in response to the entreaties of his patron Lord Exeter that he condescended to decorate the King's Staircase in that pompous style of allegory which offended the sensibilities of Walpole and was inadvertently immortalized by Pope:

> On painted Cielings you devoutly stare,
> Where sprawl the Saints of Verrio or Laguerre.[2]

These lines were so successful in ridiculing any would-be spectators of Verrio's or Laguerre's paintings that to this day no student in ascending Wren's staircase at Hampton Court (Fig. 62) has ventured or bothered to inquire into the meaning of the forbidding scenes that greet him from the walls and ceiling. The description in the guidebook of 1938 is still couched in the style of Walpole:[3]

> Every kind of subject is represented on this staircase, surrounded by ambrosial but solid clouds, which seem to support the superimposed gigantic figures. Apparently all the gods and goddesses of Roman mythology with their various attributes—such as the peacock of Juno and the rainbow of Iris—are represented at a banquet; also the founder of Rome, Romulus, with his wolf; Aeneas and the twelve great Caesars of the Roman Empire; the soothsayer Spurinna, of Roman fame; Julian the Apostate, attended for some unknown reason by Mercury, the messenger of the Gods; all the signs of the Zodiac; a winged figure of Peace with an olive branch; and a figure supposed to be the genius of Rome hovering over them all.[4]

[1] Horace Walpole, *Anecdotes of Painting in England* (1763), ed. R. N. Wornum, II (1862), pp. 470 f.

[2] 'To Richard Boyle, Earl of Burlington', *Epistles to Several Persons* (*Moral Essays*) (1735), iv, ll. 145–6 (ed. F. W. Bateson, 1951).

[3] '. . . without much invention, and with less taste, his exuberant pencil was ready at pouring out gods, goddesses, kings, emperors and triumphs, over those public surfaces on which the eye never rests long enough to criticise, and where one should be sorry to place the works of a better master: I mean ceilings and staircases. The New Testament or the Roman History cost him nothing but ultramarine.' (*Anecdotes*, p. 467.)

[4] E. M. Keate, *Hampton Court Palace* (1938), pp. 17 f. [emended after the publication of this article in more recent guides by G. H. Chettle, with additions by J. Charlton (6th edn., 1975), pp. 13 f.].

Journal of the Warburg and Courtauld Institutes, III (1939), pp. 127–37 (revised from Wind's notes and correspondence, with the generous help of Professor Raymond Klibansky and Dr R. A. Beddard).

The inclusion in this programme of so sensational a character as Julian the Apostate suggests that Verrio may have regarded his subject as something more than a mere excuse for using ultramarine. There can be no doubt that the name of Julian for one of the figures (Fig. 63) is transmitted correctly, for it appears under an engraved copy in *Apelles Britannicus*.[5] It is moreover evident from the attitude of this figure sitting at a desk and writing under the inspiration of Hermes, and from the position assigned to this scene in relation to the rest of the fresco, that Julian is represented as the author of the dramatic pageant which is unfolded on the main wall (Fig. 66). The simplest way to discover the meaning of the allegory is, therefore, to consult the writings of the Emperor Julian. His satire, *The Caesars*, available at this time in several translations,[6] has actually served as Verrio's text; and the illustration of it is so literal that we can identify from it every figure and correct several modern errors of interpretation. For example, the person generally called Aeneas is really meant to be Alexander the Great; and the old sage dressed in white is not 'the soothsayer Spurinna, of Roman fame', but the Stoic philosopher Zeno. It is now also possible to explain why Julian is attended by Mercury. He himself describes his satire as 'an invention of Hermes—for it was from him I learned what I am going to tell you'.[7]

But more important than these details is the significance of Verrio's choice—or acceptance—of this theme as a suitable subject. It implies that the story of his religious and political scruples must be a myth.[8] Whatever his reason may have been for hesitating to enter the service of William III, it cannot have been loyalty to James II and the Roman Catholic faith; for we shall find that the name of Julian the Apostate was used as a slogan in the Protestant pamphlets against 'popery'. Verrio's own representation of Julian's *Caesars* proves conclusively that this supposedly loyal painter had become a partisan of the Whigs. He glorified the Protestant succession in his allegory and expressed—through the language of antique imagery—the utmost contempt for the Church of Rome.

Julian's Satire of the Caesars

In *The Caesars* the leading and most ludicrous part is played by Romulus. Having been raised to the status of a demi-god, he invites his successors, the Roman emperors, to join the immortals in the banquet of the Saturnalia. While the gods feast in the upper heaven, painted on the ceiling (Fig. 65), a second table like an altar has been provided for the mortals 'below the moon'—Diana with her crescent appears on the borderline between ceiling and wall, hovering over the empty table (Fig. 66), and below her begins the sublunar

[5] Frontispiece to bk. I, 'The Royal Palace at Hampton Court' [1741?], engraved by G. Bickham, jun., from a drawing by Gravelot after Verrio.

[6] The one most commonly quoted is the French version of Ezekiel Spanheim (1683). Gibbon, *The Decline and Fall of the Roman Empire*, II (1781), ch. xxiv, gives a concise and perfect summary of *The Caesars*.

[7] *Caesares*, 307A, trans. by W. C. Wright, *The Works of the Emperor Julian*, Loeb Library, II (1969).

[8] [Dr R. A. Beddard comments: 'Walpole (cf. above, note 1) wrote much later and may not have been reliable on this point. Verrio came to England in the train of Ralph Montagu, Charles II's ambassador to Louis XIV, and later an avid Exclusionist. Equally, Verrio's English patrons were all along of mixed religious persuasions.' On Verrio, see E. Croft-Murray, *Decorative Painting in England, 1537–1837*, I (1962), pp. 50 ff. and 236 ff.]

world. Romulus, standing on a cloud, approaches the table from the right and intercedes for the emperors grouped below him (Fig. 64); but he is opposed by Hercules on the other side, who undertakes to produce a single Greek to outrule all these Romans: Alexander the Great. His claim is supported by Silenus, the jester of the party, riding his ass through the air and apostrophizing Romulus, while Dionysus lays a restraining hand on his head: 'Then Silenus began to rally Quirinus [Romulus] and said, "See now whether all these Romans can match this one Greek".'[9]

It is clear why this particular episode was chosen: William III is the new Alexander with whom the Romans cannot compete. It is particularly fitting that he should be introduced and protected by Hercules; for the lion and lion-slayer were favourite emblems of William III and appear in his medals in the most far-fetched associations.[10] In the planning of Hampton Court Palace itself, the reference to Hercules recurs again and again. The quadrangle of the Fountain Court was designed in such a way that the south façade, facing the main passageway, has twelve bays, thus providing space over the windows for twelve medallions illustrating the twelve labours of Hercules (painted by Laguerre). Each of these medallions, as well as the corresponding circular windows on the two other sides of the quadrangle, is decorated by a wreath of oranges (the emblem of the House of Orange) surrounded by a lion's skin. The pediment of the Garden Front again shows the victorious Hercules trampling on Superstition, Tyranny, and Fury, while Fame leads him to the Arts of Peace (Fig. 67).[11] The entrance on the South Front is flanked by two statues, one representing Hercules and the other Mars; and the same configuration of Mars and Hercules recurs on the fire dogs in William III's First Presence Chamber.

Verrio's fresco conforms to this general plan and extends the allusion to all the minor details. In the picture on the ceiling, the table at which the gods are feasting is supported by three lions; and the zodiac surrounding the heavenly banquet culminates in the sign of Leo.[12] The sublunary table to which the Romans aspire is decorated with festoons of oranges and supported by double eagles.

To stress the contrast between the good and the evil side and the political relevance of the issue, Verrio introduced a number of accessory figures not mentioned in Julian's text. While Alexander (William III) is attended by a figure of Victory pointing upwards, with a laurel wreath in one hand and a palm in the other, the figure of Nemesis[13] descends with flaming sword and bridle upon the Roman emperors, who seem unaware of their impending fate.

[9] *Caesares*, 316C.

[10] Cf. Nicolas Chevalier, *Histoire de Guillaume III . . . par Médailles, Inscriptions, Arcs de Triomphe & autres monumens Publics* (Amsterdam, 1692).

[11] Fury is represented, as usual, by a woman with serpent's hair; Tyranny by a king with crown and sword; and Superstition by a woman holding an incense-burner, while an idol is seen in the background. [Cibber received a round £400 for 'insculpting the Relieve on the Timpon of the great frontisspeece, with Iconologicall ffigures, and for several Journies of himself and men to look after the performance'. Cf. *The History of the King's Works*, ed. H. M. Colvin, v (1976), p. 162.]

[12] The figure of Leo, the emblem of William, is half covered by roses (the emblem of Mary), which are strewn by genii (see Fig. 65). The union of William and Mary is also represented in a separate allegory on the north wall over which Apollo presides, thus forming a counterpart to the scene on the east wall over which we saw Diana presiding.

[13] The name traditionally given to this figure, 'the genius of Rome' (cf. Keate, loc. cit.), is definitely an error.

Caesar, the central figure (directly under Romulus), continues proudly to extol his power; Augustus (on his right) still listens to the counsels of the philosopher Zeno, who, 'by reciting over Octavian a few of his doctrines, in the fashion of those who mutter the incantations of Zamolxis, . . . made him wise and temperate';[14] and at the far right of the group is the figure of Nero, playing the guitar.[15]

The idea of associating Romulus and his successors with the Roman Church had been familiarized by the 'Comparison of Popery and Paganism', a common topic of Protestant declamations. In one of the crudest and most devastating tracts of this sort, the argument culminates in the following passage: '. . . I cannot find any such great difference betwixt *Romulus* and *Ignatius Loyola*; the one having been, in his Time, the Governor of a Den of Thieves, and the other the Captain General of the Modern Banditi: and it is all one to me, whether they worship the Nine Muses, or the Seven Sleepers, for still the Polytheism remains the same. . . .'

The author who used this revolting language was called—*sit venia verbo*—Samuel Johnson. He had been chaplain to Lord Russell, 'the martyr' in the cause of the Protestant succession, and was himself imbued with such a raging spirit that martyrdom came to him as an almost desired reward. The pamphlet from which I have taken the above passage (p. 136) is of considerable interest in our context, for its full title reads: *Julian the Apostate: Being A Short Account of His Life; The Sense of the Primitive Christians about his Succession; And their Behaviour towards him, Together with A Comparison of Popery and Paganism . . .* MDCLXXXII.

Johnson's Pamphlet

Johnson's pamphlet was written to promote the exclusion from the succession of James, Duke of York, who had embraced popery. The principal object of attack was, therefore, the doctrine of passive obedience, 'which when it is taught without any regard to Laws, and is prescribed both without Law and against Law, is not Evangelical but Mahumetane, and the very Turkish doctrine of the Bow-String'.[16]

A whole flood of pamphlets was produced in reply.[17] Some writers answered the Revd Mr Johnson in his own language, describing his production as 'the stinking blast of a Deacon, that had long been grip't and in pain with the Business, till out it came rattling with

[14] *Caesares*, 309C. In Verrio's painting this scene is clearly meant to suggest the figure of a 'priest-ridden' monarch.

[15] In the centre of the picture, directly under the clouds, Verrio represented four impish little spirits with bats' wings and fish-tails (rain-gods rather than wind-gods), one of whom expresses his disdain for the Roman emperors by spitting on them.

[16] *Julian the Apostate*, The Preface to the Reader, p. viii.

[17] *Some Remarques Upon a late Popular Piece of Nonsense call'd Julian the Apostate, &c.* (1682); *Constantius the Apostate: Being a Short Account of His Life and the Sense of the Primitive Christians about his Succession . . . Being a full Answer to a late Pamphlet, intituled Julian the Apostate* (1683); *Jovian. Or, An Answer to Julian the Apostate* (1683); *A Letter of Remarks upon Jovian* (1683); *The Life of Boethius, Recommended to the Author of the Life of Julian* (1683), etc.

the title of *Julian the Apostate*'.[18] But there were others who quietly undertook to expose the historical errors into which the deacon's zeal had led him. They proved that the rules of succession in ancient Rome were different from those of modern Britain, and that, whatever judgement one might pass on Julian the Apostate in other respects, his succession to the throne was strictly legal. They also proved that the early Christians of Antioch who had rebelled against the rule of Julian had also rebelled against Christian emperors, so that their behaviour could hardly be attributed to an over-anxious concern for their sovereigns' religion.

The debate did not cease until after the accession of William III. By that time Johnson had become something of a celebrity, known as 'Julian Johnson'. He had not only replied to his censors with new arguments of increased acerbity;[19] he had also, after the actual accession of James II, tried to foment disaffection in the army. Already imprisoned for libel, because in the second edition of *Julian the Apostate* he had drawn an explicit comparison between Julian and James, he was now deprived of his orders and sentenced to stand three times in the pillory and to be whipped 'from Newgate to Tyburn'. In refusing to remit the sentence of flogging, James declared: 'Mr. Johnson . . . has the spirit of a martyr; and it is fit that he should be one.' Macaulay, in quoting these words, compared them with those spoken a few years later by William III when dealing with a Jacobite rebel: 'He has set his heart on being a martyr; and I have set mine on disappointing him.'[20]

Disappointment also awaited Johnson when William III ascended the throne. The Protestant succession had been accomplished, and Johnson referred to it with satisfaction in a new preface to his final version of *Julian*, now openly dedicated 'to the Ever Glorious Memory of William Lord Russel': 'The present Impression of this Book, was made in the Year 1683, and has ever since lain Buried under the *Ruines* of all those *English* Rights which it endeavoured to Defend; But by the Auspicious and Happy Arrival of the Prince of *Orange*, both They and It have obtained a Resurrection.'[21]

The Commons took up Johnson's case, and his degradation and punishment were declared illegal; but the Commons' request for his ecclesiastical preferment was ignored by the king. To quote Macaulay: 'The King thought it better to give out of his own revenue a liberal compensation for the wrongs which the Commons had brought to his notice than to place an eccentric and irritable man in a position of dignity and public trust.' Johnson was compensated with 'a present of a thousand pounds, and a pension of three hundred a year'.

[18] *Some Remarques* . . . , Preface to the Reader [pp. ii f.]. The profanities of the rejoinder outdid even those of the provocation (which had the excuse of being in biblical style): 'That whereas he sayes . . . that Bishop *Ridley* said the Church of *Rome* was a *Babilonical Beast, and Whore, a devilish Drab, and a stinking Strumpet*: I desire to be satisfied how the Reverend Father came to know, whether a Strumpet stunk or no. . . .' (P. 35.)

[19] *The Account of the Life of Julian the Apostate vindicated* (1682); *Julian's Arts to Undermine and Extirpate Christianity* (1689).

[20] *The History of England* (1848), ch. vi, referring to 1686 (ed. C. H. Firth, II, 1914, p. 762).

[21] *Julian's Arts* . . . , facing title-page. In 1688, the year of William III's expedition, a Dutch translation of Johnson's original pamphlet was printed in Amsterdam: *Julianus den Apostaat . . . Uit het Engelsch vertaald.*

It appears that 'the infirmities of his temper and understanding had increased to such a degree that he was as offensive to Low Churchmen as to High Churchmen'.[22]

Given William's failure to act, it can hardly be assumed that he would have deemed Johnson's pamphlet on Julian the Apostate a suitable text for glorifying his own reign in the principal decoration of his palace. The relation of Johnson's propaganda to the programme of Verrio's fresco is not as close as may at first appear.[23] The most striking difference is in the role assigned to Julian himself. In Johnson's pamphlet Julian is the villain. His apostasy foreshadows the treachery of the modern sovereign who forsakes the true Protestant religion in favour of Romish idolatry. Julian's return to paganism is an allegory for James's submission to Rome. But in Verrio's fresco the attack on popery is based on Julian's own text, whose satire of the Caesars exposes the false pretences of Romulus and his successors. Julian appears here as a pagan witness to Protestant truth. His portrait is introduced as that of the philosophical emperor to whom Hermes reveals the supremacy of Hercules over Romulus, of Virtue over Superstition.

This estimation of Julian belongs to a tradition, already flourishing in England, which can be traced back to Montaigne.

The Montaigne Tradition in England

In his essay 'De la liberté de conscience', Montaigne speaks of 'Julien, surnommé l'Apostat' as 'un très-grand homme et rare', and praises him as the prototype of the enlightened sovereign who opposes the tyranny of the clergy.[24] According to this view, Julian's 'so-called apostasy' foreshadows the emancipation of civil government from the tutelage of the Church.

It can be shown that this opinion of Julian was held and propagated in England by the circle of prominent Whig politicians who supported the Bill of Exclusion. In 1681—one year before the first publication of Johnson's 'Julian' and the year in which the Oxford Parliament met, the prelude to the great Whig conspiracy which ended with Russell's and Sidney's arrest and Shaftesbury's escape to Holland—a pamphlet appeared entitled: *Some Seasonable Remarks upon the Deplorable Fall of the Emperour Julian, With an Epistle of his to the Citizens of Bostra*.[25] In contrast to Johnson's brutal invective, it is written with quiet reasonableness, arguing that the intolerance of the Roman Church was responsible for the apostasy of Julian, who was himself a man of profound learning and charitable disposition. His 'Letter to the Bostrans' is reprinted and translated as a model of toleration. When 'so excellent a person as *Julian*' breaks away from Christianity, it is only fair to seek 'the lamentable causes of it in the debaucht Christianity of those times'.[26] To illustrate the

[22] *The History of England*, ch. xiv, referring to 1689 (Firth, IV, pp. 1653 f.).

[23] See above, note 8.

[24] *Essais*, II, xix (*Œuvres complètes*, Bibl. de la Pléiade, 1962, p. 651).

[25] The author calls himself 'Philaretus Anthropopolita'.

[26] *Some Seasonable Remarks* . . . , pp. 1 f.

debased condition of the Church, the author quotes a long passage from Julian's *Caesars*, which 'though the Learned *Loyolite Petavius* durst not translate to his Catholick Friends, I may to pious Protestants without the least offence, since they derive not their Religion from *Constantine's* Bishops, but from Christ immediately'.[27]

The notoriety of Johnson's pamphlet seems to have obliterated from the minds of contemporaries, as well as of later historians, the memory of its more subtle and gentle predecessor. There can be no doubt that both represent the same interest; but whereas the first 'Julian' was conceived as an indirect defence of Whigs and Protestants, Johnson's 'Julian' is an open attack on Tories and Catholics. This change of tactics—and the corresponding change in the interpretation of Julian's character—may have been due to the temperament of Johnson who, with the leaders of his party either fled or in prison, seems to have been left more or less to his own devices. Courageous and thoroughly unintelligent, yet endowed with a strong instinct for demagogy, he shaped his picture of Julian not after Montaigne and other sceptical authors, but after the Church Fathers, particularly St. Gregory of Nazianzus, who had reviled the Apostate as a messenger of the Devil. This relapse into medieval tradition probably satisfied the political needs of the moment far better than the subtle dialectic of the earlier pamphlet. To defend a traditionally odious character, and draw a parallel between his situation and one's own, is to play into the hands of one's enemy and provide him with a formula for slander. The safer way is to leave the traditional character untouched and impute his odious reputation to the enemy himself.

Thus two 'Julians' emerged from the Protestant camp: one a distinctly revolutionary figure and a philosophical symbol of emancipation; the other the traditional figure of the Apostate and a blatantly demagogical weapon.

There can be no doubt that the men who devised the programme of Verrio's fresco as a glorification of William III revived the first 'Julian' and abandoned the second. Strong reasons can in fact be adduced for the supposition that they were direct disciples of the authors of the first pamphlet.

The Authors of the first 'Julian' and their Disciples

The *Seasonable Remarks upon the Deplorable Fall of the Emperour Julian* was clearly written in defence of the 1st Earl of Shaftesbury and the other Whig leaders who had incurred the anger of Charles II. The date of publication, 1681, leaves little doubt that the pamphlet was intended as a plea for toleration, making the point that despotism and tyranny produce the very apostasy they seek to avoid. There is nothing extraordinary or far-fetched in this

[27] Ibid., pp. 18 f. The passage in question (336A-B) which, according to modern editions (cf. Loeb Library) is supposed to be spoken by Jesus, was attributed in the seventeenth-century editions to Constantius, the son of Constantine, *υἱός* being read for *'Iησοῦς*. It says: '*Ho! whosoever is either Sodomite, Murderer, Rogue or Villain, let him dread nothing but repair hither, with this water I'll make him clean in a trice: And if he shall happen* (as humane Nature is frail) *to repeat the same Crimes, if he will but thump his breast, and box his noddle, I'll warrant him as innocent as the Child unborn.*'

argument. The only surprising point is that so controversial a figure as the Emperor Julian should be chosen to strengthen the plea for toleration. And it is remarkable therefore that the same argument, again illustrated by the apostasy of the Emperor Julian, is repeated a few years later in Locke's letters on toleration.[28] Locke was secretary to the Earl of Shaftesbury, in whose personal and party interest the *Seasonable Remarks* were written; and it is known that, before the Earl's flight to Holland in 1683, Locke was suspected of writing anonymous political pamphlets for him.[29] Circumstantial evidence thus points to Locke as a possible author of this pamphlet. But setting aside the question of authorship, it is clear that in general it accords with his known views on toleration. For the choice of Julian the Apostate as an example—as I have said, a very eccentric choice in a plea for toleration—he may have been indebted to his extensive reading of Montaigne on the one hand, and on the other to his close acquaintance with the Cambridge Platonists, who cherished Julian's writings as an exposition of Neoplatonic mysticism. Cudworth, in particular, assigned to the opinions of Julian a very prominent place in his 'intellectual system of the universe'.[30] It is not impossible that he or another of the Cambridge Platonists had a hand in formulating the *Seasonable Remarks* on Julian.[31]

Locke and Cudworth had a common pupil in the grandson of Locke's patron the 1st Earl of Shaftesbury, namely Anthony Ashley-Cooper, the 3rd Earl. Author of the *Characteristicks*, and more famous as a philosopher than as a politician, the 3rd Earl became, like his teacher Locke, a supporter of and adviser to King William; and, again like Locke, a close associate and friend of Lord Somers, who besides being Lord Chancellor[32] was a classical scholar[33] and a patron of the arts.[34] It was in this circle that the figure of Julian the Apostate was restored to the glory of a philosophical liberator. In his *Letter concerning Enthusiasm* (1708), dedicated to Somers, the 3rd Earl of Shaftesbury expressed his admiration for the Emperor

[28] *A Third Letter for Toleration* (1692), ix, 'Of the Usefulness of Force in Matters of Religion'; and x, 'Of the Necessity of Force in Matters of Religion'. In the relevant passage (*The Works of John Locke*, II, 1714, p. 443), Locke opposes to the legendary 'apostasy after the flood', which his opponent attributes to the absence of coercion, the historical apostasy of Julian, which Locke attributes to the application of coercion: 'This I think as good an Argument against, as yours for Force, and something better; since what you build on is only presum'd by you, not proved from History: whereas the matter of Fact is here well known. . . .' [See J.-P. Larthomas, 'Julien en Angleterre dans le milieu Whig', in *L'Empereur Julien*, II, *De la légende au mythe* (*de Voltaire à nos jours*), ed. J. Richer (1981), pp. 61 ff.]

[29] Cf. Pierre Des Maizeaux, *A Collection of several Pieces of Mr John Locke* (1720). H. O. Christophersen, *A Bibliographical Introduction to the Study of John Locke* (1930), pp. 9 f., questions to what extent the authorship of Locke can be proved.

[30] Ralph Cudworth, *The True Intellectual System of the Universe* (1678), I, iv, pp. 274 f., 409 ff. On p. 500 there is an explicit defence of Julian against St. Cyril.

[31] The pseudonym 'Philaretus Anthropopolita' is also in the taste of the Cambridge Platonists. Henry Moore styled himself 'Philophilus Parresiastes' and 'Alazonomastix Philalethes', while Locke, in the second and third letters on toleration, used the pseudonym 'Philanthropus', of which 'Philaretus Anthropopolita' might be an expansion.

[32] Somers's importance as political counsellor of William III may be judged from Addison's lines in 'A Poem to His Majesty' (1695):

> Britain *Advanc'd, and* Europe's *Peace Restor'd*
> *By* SOMERS' *Counsels, and by* NASSAU'*s Sword.*

This couplet is from the dedication to Somers.

[33] In 1702 there appeared under Somers's direction an anti-French tract attacking Louis XIV, entitled: *Several Orations of Demosthenes, to encourage the Athenians to oppose the exorbitant Power of Philip of Macedon. Englished from the Greek by Several Hands.* Swift compared Somers with Aristides in his *Discourse of the Contests and Dissensions between the Nobles and the Commons in Athens and Rome, with the Consequences they had upon both those States.*

[34] Cf. *A Collection of Prints and Drawings, &c, of the Late Right Honourable Ld Somers, to be sold By Auction . . . in the Little Piazza in Covent Garden* (London, 1717).

Julian in exactly the same terms as those used in the *Seasonable Remarks*. More than that, when he later expanded his argument in the *Miscellaneous Reflections* (1711), he justified his praise of Julian's tolerance (which, he claimed, far exceeded that of the orthodox Christians and extended even to his Christian enemies) by quoting *in extenso* from the 'Letter to the Bostrans', precisely the document which had been appended to the *Seasonable Remarks*. It is highly likely that the memory of the pamphlet of 1681, written in defence of the 1st Earl, was still in the mind of the 3rd Earl of Shaftesbury when he resumed the praise of Julian in his *Letter concerning Enthusiasm* and his *Miscellaneous Reflections*, both of which were incorporated in his *Characteristicks*. The programme of Verrio's fresco, painted in 1701–2,[35] can safely be assigned a place in this tradition. It was designed for William III at a time when Somers, Locke, and Shaftesbury were very much to the fore. However inadequate as a painting, it ought to be cherished as a document. It indicates that as early as 1701, in the infancy of the Age of Reason, William had accepted the image of Julian the Apostate as a symbol of toleration and freedom.

The intellectual fascination of the programme stands in marked contrast to the mediocrity of Verrio's execution. No greater discrepancy can be imagined than that an idea which had lived in the minds of Montaigne, Locke, and Shaftesbury should be entrusted to the hand of Verrio. But this failure was not a mere accident. It reflects Shaftesbury's view, sustained in his writings and expressed in his patronage of art,[36] that the artist is merely a mechanical executant of ideas dictated to him by a philosopher—a view which signifies a fatal breakdown of humanism in an attempt at false domination. A collaboration like that between Rubens and Peiresc, in which the scholar inspired the artist who in turn enlightened the scholar, would have been required to bring out the full splendour of the subject. But Verrio worked exactly as was expected of him: as a mechanical executant, without imagination or convictions of his own. Instead of proclaiming the glory of his theme, he helped to sink it into oblivion.

No painter after Verrio seems to have treated the subject. Apart from a few engravers who provided frontispieces to the editions of Julian's *Caesars*,[37] the theme was forgotten in the visual arts; but it survived in literature and statesmanship. The heritage of Montaigne, Locke, and Shaftesbury fell to Voltaire and Frederick the Great.

From Voltaire to Gibbon

In the correspondence between Voltaire and the young Prince of Prussia, the subject of Julian was introduced by the prince himself, who marvelled at the contradictory character imputed to the emperor:

Rappelez-vous les différents caractères attribués à Julien, surnommé l'*Apostat*. La haine, la fureur,

[35] See now *The History of the King's Works*, ed. cit., v, p. 169.

[36] Cf. Wind, 'Shaftesbury as a Patron of Art', below pp. 64 ff.

[37] The most interesting is that of Bernard Picart in the French edition of 1728.

la rage de vos saints évêques, l'ont défiguré de façon qu'à peine ses traits sont reconnaissables dans les portraits que leur malignité en a faits. Des siècles entiers ont eu ce prince en horreur: tant le témoignage de ces imposteurs a fait impression sur les esprits! Enfin, un sage est venu qui, s'apercevant de l'artifice des moines historiens, rend ses vertus à l'empereur Julien, et confond la calomnie des Pères de votre Église.[38]

The qualification of Julian's epithet as 'surnommé l'*Apostat*' strongly suggests that Frederick had read Montaigne. Voltaire, who was acquainted with the development of Montaigne's ideas in England, was quick to encourage the inclination of his pupil and to impress on him the ideal image of the enlightened sovereign. Julian's apostasy was for Voltaire a great landmark in the struggle against hierarchy, a classical symbol for 'Écrasez l'infâme!'.[39] In his *Examen du discours de l'empereur Julien contre la secte des Galiléens*,[40] Voltaire draws an explicit parallel between Julian and Frederick the Great; and a similar comparison recurs in his *Histoire de l'établissement du Christianisme*:

Nous savons bien qu'il est dans l'Europe un très-grand prince, célèbre par ses victoires, par ses lois, et par ses livres, qui, dans ses États de cinq cents lieues en longueur, a pour ses sujets des papistes, des luthériens, des calvinistes, des moraves, des sociniens, des juifs; qui ne prend parti pour aucune de ces sectes, et qui n'a pas plus de chapelle que de conseil et de maîtresse; mais il est venu dans un temps où la démence des disputes de religion est entièrement amortie dans son pays. Il a affaire à des Allemands, et Julian avait affaire à des Grecs, capables de nier jusqu'à la mort que deux et deux font quatre.[41]

Frederick's pious subjects were alarmed by his heterodox sympathy with Julian the Apostate. Klopstock expressed their concern in a touching ode:

Aber euch sag' ich sie ganz, des vollen Herzens Empfindung,
    Wie das Herz sie empfand.
Ohne des Zweifels versuchenden Ton, so offen ich sage,
    Daß dem Sieger bei Sorr
Julianus zum Muster zu klein, und ein Christ zu werden,
    Würdig Friedrich ist.
Aber das ist ein Gedanke voll Nacht: Er wird es nicht werden.[42]

[38] Letter to Voltaire, 13 November 1737, in *Œuvres complètes de Voltaire*, ed. L. Moland, XXXIV (1880), no. 790, pp. 343 ff. [See now *The Complete Works of Voltaire, Correspondence and Related Documents*, ed. T. Besterman, IV, 1969, Letter D1389. According to Besterman, the 'sage' in question was the Abbé de La Bléterie, whose *Histoire de l'empereur Julien l'apostat* had appeared in 1735; and Voltaire did not share Frederick's enthusiasm for its author, cf. III, Letter D856, 1 April 1735. On La Bléterie's life of Julian, see G. W. Bowersock, *Julian the Apostate* (1978), p. 6.]

[39] In Voltaire's *Dictionnaire philosophique* (1767), two separate articles are devoted to this subject: 'Apostat' and 'Julien'. The second, having exposed the ancient and modern calumnies of Julian, ends with this characteristic apology: 'On demande très-sincèrement pardon aux lecteurs sensés d'avoir parlé d'un ex-jésuite nommé Paulian, et d'un ex-jésuite nommé Nonotte, et d'un ex-jésuite nommé Patouillet; mais, après avoir écrasé des serpents, n'est-il pas permis aussi d'écraser des puces?' (*Œuvres complètes*, ed. Moland, XIX, 1879, p. 547.)

[40] Printed together with his *Portrait de l'empereur Julien* and *Supplément au discours de Julien* as commentaries on the French trans. of Julian's *Contra Christianos* by the Marquis d'Argens in *L'Évangile du jour*, V (1769).

[41] *Histoire de l'etablissement du Christianisme* (1777), ch. XXI, 'Questions sur l'empereur Julien' (Moland, XXXI, 1880, p. 101).

[42] 'Friedrich der Fünfte—An Bernstorff und Moltke', *Oden* (1771).

The fears were unfounded. Although Klopstock was right in suspecting that Frederick would not become a pious Christian, German interest in Julian the Apostate soon became a purely literary affair.[43] In the nineteenth century, even so radical an author as David Friedrich Strauss refused to regard him as a challenging political figure. In his lecture *Der Romantiker auf dem Throne der Cäsaren, oder Julian der Abtrünnige* (1847),[44] he praised Julian for having discountenanced the Christian clergy, but he rejected Julian's own religious excursions as a romantic form of mystification which revealed him as an 'enthusiast'.

In England the same view had been expressed much earlier in Gibbon's *Decline and Fall of the Roman Empire*.[45] Gibbon admired Julian's determination and tolerance, but he had nothing but ridicule for his faith in the allegories of the Neoplatonists, 'the strange allusions, the forced etymologies, the solemn trifling, and the impenetrable obscurity of these sages, who professed to reveal the system of the universe'.[46] He admitted that he found Julian's reasoning 'less absurd than that of some modern theologians, who assert that an extravagant or contradictory doctrine *must* be divine since no man alive could have thought of inventing it'. But he was shocked by the idea that 'the revolution of the sun beyond the tropics, or the separation of the human soul from vice and error' should be explained by the castration of Atys, a tale which 'must inspire a man with pity, an eunuch with despair'. To undermine more thoroughly the respect for Julian, Gibbon reported an incident which was supposed to have occurred during his initiation into the mystic pagan rites: 'When Julian, in a momentary panic, made the sign of the cross, the daemons disappeared (Greg. Naz. *Orat.* iii). Gregory supposes that they were frightened, but the priest declared that they were indignant. The reader, according to the measure of his faith, will determine this profound question.'

A passage like this destroyed for ever Julian's reputation in England as the enlightened sovereign. The fancy evoked by the scepticism of Montaigne was to die from the scepticism of Gibbon. The historical justice and impartiality with which he weighed Julian's follies against his merits dissolved the emperor's ideal features into a myth.

This very British death was perhaps due to the fact that Montaigne's Julian had successfully completed his British mission through the Whig revolution of 1688. Having found his statesmen in William III and Somers, his philosophers in Locke, Cudworth, and Shaftesbury, his slanderer and martyr in 'Julian Johnson', and his renegade painter in Verrio, there was no need for him to haunt the imaginations of minor poets as he continued to do on the Continent, long after the death of Frederick the Great.[47]

[43] See Käte Philip, *Julianus Apostata in der deutschen Literatur* (1929). [On Eichendorff's negative view of paganism, see A. Souyris, 'Julien, épopée en vers d'Eichendorff', *L'Empereur Julien*, II, pp. 213 ff. See also 'The Mænad under the Cross', below, p. 75.]

[44] *Gesammelte Schriften*, ed. E. Zeller, I (1876), pp. 175 ff.

[45] II (1781), chs. xxii–xxiv.

[46] Ibid., ch. xxiii, p. 362; for the following passages, see pp. 361 f. and 366 note 24.

[47] Cf. Philip, op. cit., pp. 61 ff.

III · Shaftesbury as a Patron of Art

With a Letter by Closterman and two Designs by Guidi

THE 3rd Earl of Shaftesbury's philosophy of art has been exposed to extravagant praise and brutal derision. On the whole, his enemies have hurt him less than his friends. No one would be misled by Lamb's mocking remark that 'the peer seems to have written with his coronet on'.[1] But when Herder acclaimed him as a 'virtuoso of virtue' and made him the patron saint of German *Genielehre*, he invented a legend which to this day has obscured the far more sober and conservative lesson expressed in Shaftesbury's writings and exemplified in his life. Continental poets of the eighteenth century looked upon him very much as they did upon Spinoza, whom they hailed as 'Gott-trunken' while refusing to share his predilection for the geometrical method of demonstration. It is time to rescue Shaftesbury from the type of enthusiasm he detested.

The best way to correct the Romantic distortion is to study Shaftesbury's patronage of contemporary art. He did not conceive of the artist as a genius responsible only to his own inspiration, but treated him as a manual executant who carries out in visible material the ideas dictated to him by the philosopher. He himself invented the programme and pattern of all the works of art produced to his order. It was not by accident, therefore, that his most willing and obedient amanuensis was an artist whose own powers of invention were slight: John Closterman.

The best-known product of their collaboration is Closterman's portrait of Shaftesbury (Fig. 71),[2] which was then engraved by Gribelin and prefixed as frontispiece to the second edition of Shaftesbury's *Characteristicks* in 1714. The posture, the setting, and the attributes show the traces of Shaftesbury's invention so clearly that it is possible from reading the emblematic language of the picture to infer that his philosophy was not Platonic in the accepted sense of the word. Shaftesbury's master was Socrates and, of the two disciples who had recorded Socrates' teaching, he placed Xenophon higher than Plato—a

[1] 'The Genteel Style in Writing', in *Last Essays of Elia*.

[2] Private Collection.

Journal of the Warburg Institute, II (1938), pp. 185–8 (with the addition here of a note (3) by Professor Raymond Klibansky on Shaftesbury's annotated copies of Horace and Marcus Aurelius).

view elaborated in his published and unpublished writings,[3] and given touching incidental expression in the translation and preface of Xenophon's *Cyropaedia* written by his brother, Maurice Ashley-Cooper.[4]

It was as part of this devotion to Xenophon that Shaftesbury planned 'Socratic Charts', pictures after the model of the 'Judgement of Hercules' described by Xenophon in the *Memorabilia*.[5] In this spirit he invented the emblems which were drawn in Naples under his supervision by a young artist named Henry Trench[6] and sent with his instructions to Gribelin, who engraved them for the second edition of the *Characteristicks*.[7] In the same

[3] The unpublished writings referred to are the text of a *Pathologia secundum Socraticos* and a fragment of a *Socratic History*, also a set of letters and instructions addressed by Shaftesbury to contemporary artists.

[Evidence of Shaftesbury's philosophical attitude is also to be found in his own copies of Horace and Marcus Aurelius, which he annotated. These two small volumes were acquired by Edgar Wind in 1949 and are now in the Bodleian Library, Oxford: *Quinti Horatii Flacci poemata* (Amsterdam, 1676) (Don. f. 533); and *Marci Antonini Imperatoris de seipso & ad seipsum Libri XII* (title-page and all before fol. A 2 missing) (Don. f. 532).

Professor Raymond Klibansky, who recently examined both volumes, has identified the Marcus Aurelius (wrongly described in *The Bodleian Library Record*, IX, 1974, p. 136) as a copy of the well-known Oxford edition of 1680. He writes:

'The numerous annotations in the Marcus Aurelius are of special interest. In the authoritative biography contained in *A General Dictionary, Historical and Critical* by the Revd Mr John Peter Bernard, the Revd Mr Thomas Birch, Mr John Lockman, *et al.*, vol. IX, London, 1739, we read, Footnote [Q] on page 186: "But his principal study was of the writings of antiquity. Among these writings those which he most admired and always carried with him, were the moral works of Xenophon, Horace, the Commentaries and Enchiridion of Epictetus as published by Arrian, and Marcus Antoninus. These authors are now extant in his library, filled throughout with marginal notes, references, and explanations, all written in his own hand." This article was written by Thomas Birch and based on information given to the author by Shaftesbury's son, the 4th Earl, who revised and approved the article. On the authenticity of the biography as a primary source for Shaftesbury's life, see Andrew Kippis in the 2nd edn. of the *Biographia Britannica*, IV (1789), p. 273.

'The notes in Shaftesbury's copy of Marcus Aurelius are extracts from Xenophon, Plato, and Epictetus; some of the headings in the notes also testify to a special interest in Socrates, and at the same time show how important Xenophon was for him in forming his conception of Socratic philosophy. The marginal notes to the Elzevier edition of Horace show his interest in variant readings of the text. These two books in Bodley thus represent precisely those authors whom Shaftesbury "most admired and always carried with him".

'The provenance of each volume is described on the verso of the first fly-leaf by James Harris, the author of *Hermes*, who received the Marcus Aurelius from his aunt, Lady Jane, Countess of Shaftesbury, Shaftesbury's widow, while the Horace was given to him by his cousin, the 4th Earl of Shaftesbury, over thirty years later in 1769. From James Harris the books passed on to his son, the well-known diplomat who later became the 1st Earl of Malmesbury. It was he who had the volumes handsomely rebound towards the end of the eighteenth century, adding silver clasps, unusual in Britain at that time, to the Marcus Aurelius.']

[4] [In 1981 Closterman's two portraits of Shaftesbury, the full-length figure and the double portrait with his brother Maurice Ashley, were seen to advantage in an exhibition at the National Portrait Gallery, London, *John Closterman: Master of the English Baroque*. Discussing them in the accompanying catalogue (nos. 15 and 17) and in his 'John and John Baptist Closterman: A Catalogue of their Works' (*The Walpole Society*, XLIX, 1983, nos. 86–7), Dr Malcolm Rogers convincingly explains the role of the attendant in the full-length portrait (Fig. 71): he interrupts his master's meditations, arriving with the peer's robes that summon him to public life. While the clearly labelled folio volumes of Xenophon and Plato have long been noticed on the square pedestal beside Shaftesbury, the titles of the other two volumes forming part of a Platonic construction are not shown, nor is their any clue to the identity of the small book in his left hand. It is tempting to suggest that this little pocket edition illustrates what we learned about him from the *General Dictionary*, and is an emblematic representation of those ancient authors 'he most admired and always carried with him'. The double portrait has now been acquired by the National Portrait Gallery.]

[5] II, i, 21 ff.

[6] [On Henry Trench and the misreading of his name, see F. Paknadel, 'Shaftesbury's Illustrations of *Characteristics*', *Journal of the Warburg and Courtauld Institutes*, XXXVII (1974), p. 293 note 18; also E. Croft-Murray, *Decorative Painting in England 1537–1837*, II (1970), p. 287.]

[7] Both the drawings made by Trench in Naples under Shaftesbury's personal supervision and the designs drawn by Gribelin after these models are in the British Museum. Binyon's *Catalogue of Drawings by British Artists and Artists of Foreign Origin working in Great Britain*, II (1900), pp. 248 f., erroneously ascribed the whole set to Gribelin. It is quite easy, however, to distinguish the two groups, especially as Shaftesbury's instructions which accompanied the drawings by Trench are preserved in the Public Record Office, London. See now Paknadel, op. cit., pp. 290 ff.

spirit he also wrote his treatise *A Notion of the Historical Draught or Tablature of the Judgment of Hercules* (1713), which is not (as has been occasionally suggested) a purely aesthetic meditation, but is meant to be understood as a philosopher's 'Advice to a Painter' or, to be exact, as Shaftesbury's advice to the Neapolitan painter Paolo de Matteis, who actually executed his painting of Hercules according to Shaftesbury's directions. This is why the tract was originally written in French, the only language Shaftesbury and Paolo de Matteis had in common. The engraving by Gribelin prefixed to the 1714 edition of the *Characteristicks* reproduced the painting in which Paolo de Matteis had embodied Shaftesbury's ideas (Figs. 72–3).[8]

There is evidence that the picture of Hercules was not the only panel that Shaftesbury intended Paolo de Matteis to paint for him. Among the Shaftesbury Papers in the Public Record Office in London is a remarkable letter, written in French, in which Shaftesbury indicates to Paolo de Matteis how he is to represent him in the character of a dying philosopher.[9]

The letter from Closterman published below is of an earlier date, written when Shaftesbury still lived in England. It provides evidence of Shaftesbury's attempt to secure the collaboration of an Italian sculptor to execute statues representing different Virtues. The intermediary is Closterman, then on his first visit to Rome. He has approached Signor Guidi and is reporting the results somewhat reluctantly to England. He sends two drawings by the Italian master (Figs. 68–9), whom he does not fail to represent in the worst possible light. His criticism was well founded; for Domenico Guidi, at that time the most prominent sculptor in Rome, worked with a shameless routine.[10] The two drawings he made for Shaftesbury—the one representing *Justice*, the other *Prudence*—are clearly quite unrelated to Shaftesbury's particular wishes; for one of them, the figure of Prudence, is a close repetition of a statue that Guidi had already made for a tomb in Sant'Andrea della Valle (Fig. 70). Closterman was not to be fooled. When he saw the drawings he refused to give the order for the statues, declaring that he would have to await further instructions. He shrewdly advises Shaftesbury to employ an English sculptor who would work under Shaftesbury's personal supervision after designs that Closterman would collect in Rome. His reference to what the work would gain from the 'genius' of Shaftesbury is probably more than flattery. His proposition makes clear the part played by Shaftesbury in the

[8] [Acquired by the Ashmolean Museum, Oxford and described in the *National Art-Collections Fund Annual Report 1980*, no. 2870, pp. 86 ff.] There is a modello of the painting at Temple Newsam House, Leeds; cf. W. Wells, 'Shaftesbury and Paolo de Matteis', *Leeds Arts Calendar*, III (1950), pp. 23 ff. E. Panofsky, *Hercules am Scheidewege* (1930), p. 131, accepts the tradition that Paolo de Matteis's picture was *selected* by Shaftesbury as the best of the existing versions. Actually the picture was *ordered* by Shaftesbury and the treatise was written as a prescription for it, not as a commentary on it. For Reynolds's use of the Hercules motif, see above, pp. 35 ff.

[9] [During his last months in Naples, Shaftesbury wrote three letters to Paolo de Matteis. These are contained in the 'Virtuoso Copy-Book' (Public Record Office, London, Shaftesbury Papers, 30/24, Bundle 26, pp. 100–4) and have now been published by J. E. Sweetman, 'Shaftesbury's Last Commission', *Journal of the Warburg and Courtauld Institutes*, XIX (1956), pp. 110 ff.]

[10] See R. Wittkower, 'Domenico Guidi and French Classicism', *Journal of the Warburg Institute*, II (1938), pp. 188 ff.

eclectic process of collaboration by which a work of art was produced: Closterman collected the material and assisted in the designing; Shaftesbury supervised the work and directed its progress so as to make it expressive of his philosophical ideas.

Apart from its relation to Shaftesbury, Closterman's letter is not without value, and even charm, as a document of life in Rome. It gives a vivid picture of an Anglo-German artist first arriving there, enthusiastic about the antiquities and the work of Bernini, but bewildered by the French taste and manners prevalent among the Romans. The letter bears no date, but the mention in it of preparations for the Holy Year (1700) makes it certain that it was written in 1699. This explains why Shaftesbury is not yet addressed on the envelope as Earl (a title to which he succeeded in November 1699), and why Guidi is described as so old that he can no longer work by himself (he died in 1701).

This
For the Right Honnorable
Lord Ashley at his hause
in London[11]

Roma the [no date]

My Lord,

I am at last arrived where I did long for these manny years and it gives me such sattisfaction that I want only judment to recomend things according to ther merits as the[12] deserve. I begin alredy to repent that I must live them behind me, however I hope with industry and deligence and with what monny my small purse can procure to bring some things into England to give your L$^{dp.}$ and others virtuosi a pruffe[13] that I have ben at rome, alredy have optaind fine things here.

Ten days since I recev'd your L$^{dp.}$ favorable letter in answer to mine I only whish that I might give a better acount in this then in my former. I send you here twoo drawings according to your direction of the twoo Virtues done by the best man that is here [Figs. 68–9]. I wil not proside[14] tel fither orders of your L$^{dp.}$ If I can find any body els here that may be for your purpose I'l give you notis. I have seen severall but there manner is so much french that I can have no pleasure to see them. This man Sig$^{r.}$ Guidi has done very fine things that made me to agree with him for them twoo statues, which was a 2000 scudi for them twoo, but after he made me them drawings I told him that I must waith for an answer, besids he is so old that he cant work any more as your L$^{dp.}$ sees by his designs, he employs young men to do his work but it is all very bad,[15] however this may be only an opinion since my eyes see so many noble things of the ancients that I cant like nothing else.

In rome are manny fine monuments of Bernini, admirable fine as part of them (or nere that) woold serve your L$^{dp.}$ but he is gone can do no more. Now my lord if I may give my opinion inparcial (that is what I woold do in this matter) to gett a young man that drawes very well and corect after the ancients and modern things, and will over look them as well as I can and bring them

[11] Public Record Office, London, Shaftesbury Papers, 30/24, Bundle 21, no. 230. [The diplomatic text of Closterman's letter, which reflects his German origin, follows here.]

[12] the] they (and elsewhere).

[13] pruffe] proof.

[14] proside] proceed.

[15] As Wittkower notes (loc. cit.), Guidi's practice of employing craftsmen who would help him to execute works more quickly and cheaply than the *professori* was not, as Closterman thought, a studio method Guidi had recourse to only in his old age, but one which he had already been following for several decades.

in to England with this advantage and the Genius of your L$^{\text{dp.}}$, and with what I can assist in this, I am sure any carver that we have, may and shal do better than any what is in rome at present. I assure your L$^{\text{dp.}}$ that the have no thoughts of the ancients nor good modern masters in there works, and for loosing no time I have alredy agreed with one that is to draw for me the scarge[16] will be but small. I give him 5 pistols a mount.

Pray my lord send me an answer, whether you like my proposition besides it will be a diversion for you to see them done by your direction, which will be better at home than mine here, I durst not pretend to instruct an Italian the think of them selves above all the world. I am extremely glad to here of your thoughts for a family pictor, and that I shall perform it as sune I come home, which I like very well as to promis it (you are only to comand). Your L$^{\text{dp.}}$ knowes very well that I loved all ways to do things for you rather than any body whatsome ever, I promis your L$^{\text{dp.}}$ so for that nobody shall prevent me tell your pictors are done, and to goe in to the contry is better still then we shall have howly[17] our thought together. My humble respects to Mr Ashly and that he improves his thoughts well at home. Pray tell him that I shall not be behind hand with him in the Vatican where I have settlet me to drawing and painting while I am at rome and shall think of nothing els while I am here.

I hope your L$^{\text{dp.}}$ will excuse this long letter and I well waith for an answer of this and then it will time a nof to put them statues in hand, becase the are all very busy at present and non of them can begin tel Easter nex by the reason of the howly[18] yeare. In meantime I remain as I allwayes pretended to be

My Lord

your Most humble
Servant to Command

Jno Closterman

My hum^{ble} Servis to Sr Jno Cropley and
Sr James Ash and My Lady in hopes
to draw a family pictor to her as
soon I come home[19]

[16] scarge] charge.

[17] howly] wholly.

[18] howly] holy.

[19] [The 'family picture' for Lady Ashe does not appear to have been executed unless, as Rogers has suggested (1983, pp. 231 and 240), it can be identified as the recently discovered painting, with the provisional title *Unknown Woman* (ibid., pl. 61). However, Closterman did paint *Lady Ashe as St. Cecilia*, now in the collection of Michael Wells, Esq. Lady Ashe was the half-sister of Shaftesbury's close friend Sir John Cropley, for whom Closterman also worked, but—as Rogers argues (ibid., p. 231)—the allegorical portrait of Lady Ashe was probably painted for Shaftesbury himself.]

IV · 'Borrowed Attitudes' in Reynolds and Hogarth

WHEN Horace Walpole undertook to defend the so-called plagiarisms of Reynolds, he defined them as a particular form of wit. He explained that the quotation of a well-known phrase in a new and unexpected context may be regarded as a spontaneous invention and might even have more wit than the original phrase itself.[1] In applying this argument to Reynolds, Walpole made two tacit assumptions which fit a restricted group of Reynolds's paintings but are not universally true of his adaptations: first, that the spectator was meant to recognize the models; secondly, that the use made of them was so new and unexpected that it could not fail to strike the spectator as witty.

A case in which these two conditions are perfectly fulfilled is the portrait of Master Crewe in the attitude of Henry VIII (Fig. 9), which Walpole shrewdly chose as an illustration of his thesis.[2] Here the model was well known, and nobody would fail to see the humorous intent of having the child mimic the pompous solemnity of the king. The picture is one of Reynolds's freshest attempts at comedy painting. Another example, though of a more recondite sort, is the portrait of *Garrick between Tragedy and Comedy* (Fig. 75). The figure of 'Hercules at the Crossroads', standing between Virtue and Vice, was so familiar to English connoisseurs—especially after the Earl of Shaftesbury had written a treatise on the subject and commissioned the painting by Paolo de Matteis (Fig. 72)[3]—that Reynolds could safely reckon on the spectator's recognizing the pattern and understanding the point of the picture.[4] Like the portrait of Master Crewe, it was a piece of good-humoured parody not without flattery to the person portrayed; although Hercules and Henry VIII are mocked in these pictures, the portrait figures are 'raised' by being represented in attitudes reminiscent of such time-honoured personages.

But Reynolds extended the use of 'quotations' far beyond this droll type of composite portraiture. For he was a master of pictorial burlesque even if he practised it but rarely. On

[1] *Anecdotes of Painting in England*, ed. R. N. Wornum, I (1862), p. xvii note 2, from Horace Walpole's advertisement prefixed to the fourth volume of the original edition (1771).

[2] Loc. cit.: 'Is not there humour and satire in Sir Joshua's reducing Holbein's swaggering and colossal haughtiness of Henry VIII. to the boyish jollity of Master Crewe?' See also above, p. 22.

[3] Cf. above, p. 66.

[4] An inferior engraving after the Garrick picture, by J. E. Haid, was actually printed with the title *L'homme entre le Vice et la Vertu* (Print Room, British Museum). See above, p. 36.

Journal of the Warburg Institute, II (1938), pp. 183–5.

his Italian journey, he had painted a brilliant caricature of English connoisseurs in Italy, mimicking in a setting of Gothic architecture the attitudes of Raphael's *School of Athens* (Fig. 91).[5] Thus, even in designing a satire full of sharp and aggressive wit, he had played with the idea of 'borrowed attitudes'. Later he turned the same device to a serious purpose and used it to 'ennoble' his subject. In painting the portraits of Mrs Hartley, of the Duchess of Marlborough, and of Lady Cockburn, with their children, he gave them attitudes borrowed from Michelangelo.[6]

So far we can still argue, with Walpole, that there is wit and invention in the paradox of introducing a reminiscence of Michelangelo's grandeur into this graceful and elegant portrait style. But when we ascend to Reynolds's more ambitious productions, in particular his mythological portraits, it becomes difficult to sustain this view. Though he dissuaded his pupils from literal copying, there is something curiously literal-minded about the way in which he himself repeats in his *Kitty Fisher as Cleopatra* the attitude of Cleopatra as painted by Trevisani (Figs. 80–1), or in his *Duchess of Manchester as Diana* the pose of Diana as painted by Francesco Albani (Figs. 82–3). Even his greatest attempt at the truly 'grand' style, his *Death of Dido* (Fig. 79),[7] is an almost literal quotation; for the main figure is taken from a fresco of the sleeping Psyche on one of the ceilings of the Palazzo del Te (Fig. 78). True, there is 'invention' in expressing the agony of death by a pose that originally signified the relaxation of sleep, and in veiling the Roman style of the original with a voluptuous display of Correggiesque beauty and Venetian colour. So far, the *Dido* is certainly more inventive, that is, more eclectic, than the mythological portrait patterns directly copied from Albani and Trevisani. But there is one trait which all these more ambitious pictures have in common: recognition of the model from which they are derived does not convey that sensation of witty paraphrase which characterizes the fashionable portrait-groups, where the pattern remains transfigured even after it has been disclosed. Knowledge of the allusion does not increase appreciation of the picture. And here the limitations of Walpole's argument become apparent. Not only are the borrowed figures too little known to be recognized by the spectator (even the figure of Psyche from the Palazzo del Te appears in quite a subordinate picture), but the use made of them has lost that element of paradox or surprise which seems indispensable if a quotation is to be witty.[8]

A malicious critic might argue that the one entails the other and that Reynolds, feeling safe from discovery when he used such recondite models, thought he could afford to copy them more literally. But perhaps the truer, and certainly the more charitable, view is

[5] The National Gallery of Ireland, Dublin. Cf. Wind, 'A Source for Reynolds's Parody of *The School of Athens*', below, pp. 77 f.

[6] The first two portraits (Figs. 6 and 2) are discussed above, p. 20, the third in Wind, 'Charity. The Case-History of a Pattern', *Journal of the Warburg Institute*, I (1938), p. 326.

[7] Her Majesty the Queen, Buckingham Palace. There is a copy in the Philadelphia Museum of Art.

[8] An exception might be claimed for the *Kitty Fisher as Cleopatra*, where the intimacy of portraiture absorbs the formal pattern. Also, the choice of this role, and attitude, for a courtesan was certainly an expression of wit. Even so, it remains true that the discovery of the original pattern does not increase, but rather distracts from, the enjoyment of the picture. It is practically certain that Reynolds did not wish the spectator to know that he was copying Trevisani.

that Reynolds's imagination lost some of its freedom and self-assurance when it ascended from the intermediate to the grander style. As long as he was translating the sublime into the fashionable, he was inventive and sure of his ground; when he reversed the process and attempted to raise the fashionable into the sublime, he became less certain of his powers and therefore more servile in his copying. He himself conceived of the grand style as 'artificial', and as presupposing an 'artificial state of mind' that can be reached only by systematic training, that is, by raising oneself—through imitation—to the level of the Old Masters. It is not impossible that he actually forced himself to perform the experiment that he recommended to his pupils: 'even to feign a relish till [you] find a relish come'—a recipe invented by James Harris for the training of Platonists, and despised by Hume as a formula for mixing enthusiasm with hypocrisy.[9]

It is very unlikely that the conscious use of pictorial quotation, which Reynolds raised to academic dignity by extending it to the grand manner, originated in pictorial burlesque. Apart from the testimony of his own caricatures, there is the evidence to be found in Hogarth. It is known that Hogarth was strongly addicted to puns, and not always of the most subtle or delicate kind. When he asked a learned friend to dinner, he invited him to 'Eta Beta Py' ('eat a bit o' pie').[10] To express his disgust with a stupid and filthy superstition, he drew an appropriate FRONTIS-PISS.[11] These verbal jokes have their pictorial counterpart in the ambiguities and *sous-entendus* of his pictures. One of his most pleasant traits was that he liked to travesty, not only the designs of others, but also his own. In several cases, the use he made of an old plate for a new engraving was dictated, not by a desire to economize, but by an urge to paraphrase the earlier design, which was retained as a conscious allusion—a famous example is the transformation of his *Self-Portrait with Pug* into a lampoon on Charles Churchill.[12] His pictures are full of metaphorical statements which, if read only in their literal sense, produce a distortion of form as well as of meaning. It commonly happens that critics who pride themselves on their purely formal approach to Hogarth fail to see where the chief accents fall in his pictures.[13]

In the last scene of *A Rake's Progress*, the rake, chained in the madhouse at Bedlam (Fig. 86), takes on a statuesque appearance which contrasts strangely with his natural representation in the preceding scenes. This is intentional. For in this figure Hogarth is

[9] See above, p. 21.

[10] [Invitation to Dinner, from Hogarth to Dr Arnold King, etched by C. Cary as a vignette on the title-page of the 2nd and 3rd edns. of John Nichols's *Biographical Anecdotes of William Hogarth* (1782, 1785); see also pp. 63 f. and 415 of the 3rd edn., where Nichols refers to this 'specimen of Hogarth's propensity to merriment'. The original drawing, which Nichols said was in the possession of Dr Wright, remains untraced, cf. A. P. Oppé, *The Drawings of William Hogarth* (1948), no. 107.]

[11] [For Hogarth's contribution to this anti-Hutchinsonian pamphlet, which was never published, see R. Paulson, *Hogarth's Graphic Works*, I (1965), no. 250; II, pl. 294, who bases his account on Samuel Ireland, *Graphic Illustrations of Hogarth*, I (1794), pp. 175 f. The plate was engraved by LaCave and the print is rare.]

[12] Tate Gallery, London. See now Paulson, *Hogarth. His Life, Art, and Times*, II (1971), pp. 3 ff. and 392 ff.

[13] For example, in criticizing the first picture of the Election series, it has become customary to mistake a subordinate figure for the hero and then to complain of Hogarth's lack of clarity in the disposition of space.

alluding to the statue of *Raving Madness*, formerly above the gate to Bedlam hospital, which had become a symbol of its inmates (Fig. 87).[14]

A more important example of such a pictorial allusion is to be found in the first picture of the Election series. It can be shown, I think, that each picture of this series is modelled after a biblical verse. In the first scene, *An Election Entertainment*, the motto appropriate to the feast is: 'He that dippeth his hand with me in the dish, the same shall betray me' (Matthew xxvi. 23). It is a 'Judas' meal. Hogarth elaborates on the theme in a great many telling allusions, of which I can mention only a few. Although the room is crowded with people, the number actually seated for the meal is just thirteen. Only two of them are provided with food, the fainting alderman and the sweating gentleman of the cloth. That one of them will betray the other is made evident by the alderman's gesture: he holds his hand out over the dish while the surgeon (an additional touch of irony) bleeds him. Directly behind him is a group of figures literally modelled after the Betrayal of Judas. At the opposite end of the table the motif is taken up again in three figures (Fig. 85) who enact a parody of the Judas group in Leonardo's *Last Supper* (Fig. 84).[15]

The connection between this burlesque type of quotation and the more refined method adopted by Reynolds is illustrated by a peculiar circumstance. The pattern of 'The Choice of Hercules', which Reynolds parodied in his *Garrick between Tragedy and Comedy* (Fig. 75), had already been burlesqued by Hogarth in two pictures. The central group in his *March to Finchley* (Fig. 76)[16] shows a grenadier torn between two women who seem to have equal claims on him: the old hag admonishing him with a rhetorical gesture tries to hold him back, while the young woman, clearly pregnant, grasps his arm firmly and marches with him.[17] Another variation of the same pattern occurs in the second picture of the Election series, *Canvassing for Votes*. The motto for this scene is: 'Let not thy left hand know what thy right hand doeth' (Matthew vi. 3)—a passage perfectly illustrated in the central group, where a worthy citizen allows himself to be bribed by both parties (Fig. 77). It would be daring to claim without further evidence that Reynolds had been inspired by either of these examples to invent the Hercules-parody in his Garrick picture, especially as other models can be shown to have influenced his conception.[18] There is, however, an

[14] The traditional attribution of the famous statues of the Madnesses to Cibber, based on the evidence of Pope and others, has been called into question by M. I. Batten, who notices that Cibber's name does not appear at all in the Court Books in connection with Bedlam, whereas that of Cartwright is mentioned by Hooke apropos of the statues for the gates; cf. 'The Architecture of Dr. Robert Hooke, F.R.S.', *The Walpole Society*, xxv (1937), p. 92. It seems possible, however, that Hooke refers to the lion and the unicorn on the outer piers.

[15] It would be quite wrong to interpret this group of Hogarth's as an accidental reminiscence of Leonardo. The joke was intended, and it was not by chance that Hogarth selected just the group with Judas. That his imagination was filled with classical patterns which he liked to deride and parody is well shown in his engravings *Enthusiasm deliniated* and *The Battle of the Pictures*.

[16] The exact title of the engraved version is *A Representation of the March of the Guards towards Scotland, in the Year 1745*.

[17] The political implication is made clear by the inscriptions on the engraving. The young woman (who carries, in her basket, a broadside with the text of the ballad *God save our Noble King*) is a Protestant and a government supporter, whereas the hypocritical old woman (who wears a Papist cross on her shoulder and sells newspapers—*The Jacobite Journal*, the *London Evening Post*, and *The Remembrancer*) represents the party of the Pretender. See now Paulson, *Hogarth's Graphic Works*, I, no. 237.

[18] Gravelot, for example, had drawn the figure of an actor on stage between two actresses impersonating Tragedy and Comedy, which was then engraved by Van der Gucht, cf. above, p. 39 and Fig. 31.

external circumstance which ought not to be left out of account. The painted set of Hogarth's Election series, in which this pattern so prominently occurs, was in the possession of Garrick.[19]

His academic dignity permitting, Reynolds was a superb master of character painting, the art that Hogarth had been so anxious to protect against the distortions of caricature. Reynolds delighted in preserving in his intimate portraits those individual defects and deformities that render a person characteristic. His self-portrait as a deaf man holding his hand to his ear is chiefly remembered for Johnson's refusal to have his own failings perpetuated by Reynolds in a similar manner: 'He may paint himself as deaf if he chuses; but I will not be *blinking Sam*.'[20] It has never been suspected that this self-portrait of Reynolds might be another case of a 'witty quotation', a translation from a painting by Greuze.[21] Probably Reynolds himself did not greatly sympathize with the fashionable sentimentalism of Greuze; but that would not prevent him from exploring its possibilities. He would feign a relish till he found a relish come. Those who find this type of adaptation obnoxious should remember that Reynolds was one of the first to study the ambiguities of pictorial expression,[22] and to develop from them a new sophistry of artistic production by which he hoped, like the Sophists of old, to 'make the worse appear the better cause'.[23] He would probably not object if he heard applied to himself the verses by which he characterized to his students the dangerous powers of Correggio and Parmigianino:

> Strange graces still, and stranger flights they had
>
> . . . .
>
> Yet ne'er so sure our passion to create,
> As when they touch'd the brink of all we hate.[24]

[19] See now Paulson, *Hogarth*, II, pp. 344f. Garrick bought the painted series from Hogarth in 1762, and after his death it remained in Mrs Garrick's collection until her death in 1823.

[20] Tate Gallery, London. Cf. Hester Piozzi, *Anecdotes of the late Samuel Johnson* (1786), p. 248; also Boswell's *Life of Johnson*, 12 April 1778 (Hill and Powell, III, p. 273 note 1).

[21] [Wallace Collection, London, no. 403. It is not known in what circumstances Reynolds may have seen the painting by Greuze; nor is there any reproductive engraving. Yet his esteem for Greuze is well documented, for there is the famous passage from the Greuze obituary in the *Journal des Débats* (quoted in A. Brookner's biography, *Greuze. The Rise and Fall of an Eighteenth-Century Phenomenon*, 1972, pp. 88 and 150), referring to 'The celebrated Raynolds [*sic*], whose admiration for Greuze was such that he wished to come to France in order to make his acquaintance . . .', which allows for such a witty quotation. Nevertheless, an alternative explanation (suggested to me by Miss Elizabeth Einberg) is that Reynolds portrayed himself in a characteristic pose, as shown in the George Dance drawing of Reynolds listening to Angelica Kauffmann (Harewood Collection), which might suggest that the resemblance is fortuitous. Even so, in assuming such a pose, Reynolds may well have remembered a Greuze *tête d'expression*.]

[22] Cf. Wind, 'The Mænad under the Cross', below, pp. 74ff.

[23] Oscar Wilde, 'The Decay of Lying, an Observation', *Intentions* (1891), p. 6.

[24] Pope, 'To a Lady. Of the Character of Women', *Epistles to Several Persons* (*Moral Essays*) (1735), ii, ll. 49–52 (ed. F. W. Bateson, 1951), emended by Reynolds in Discourse IV, ll. 491–4 (ed. R. R. Wark, 1975, p. 73):

> Strange graces still, and stranger flights she had,
> Was just not ugly, and was just not mad;
> Yet ne'er so sure our passion to create,
> As when she touch'd the brink of all we hate.

See also Wind, 'Reynolds and Pope on Composite Beauty', *Journal of the Warburg and Courtauld Institutes*, VI (1943), p. 223.

V · The Mænad under the Cross

Comments on an Observation by Reynolds

PERHAPS the shrewdest advice Sir Joshua Reynolds gave his students was that they should take hints from ancient masters and employ them 'in a situation totally different from that in which they were originally employed'. In expounding this rule, Reynolds hit, apparently *en passant*, upon a fundamental law of human expression:

There is a figure of a Bacchante leaning backward, her head thrown quite behind her, which seems to be a favourite invention, as it is so frequently repeated in bas-relievos, camæos and intaglios; it is intended to express an enthusiastick frantick kind of joy. This figure Baccio Bandinelli, in a drawing that I have of that Master, of the Descent from the Cross,[1] has adopted, (and he knew very well what was worth borrowing,) for one of the Marys, to express frantick agony of grief. It is curious to observe, and it is certainly true, that the extremes of contrary passions are with very little variation expressed by the same action.[2]

A drawing in one of Reynolds's sketch-books shows what use he himself intended to make of this lesson (Fig. 121).

Some years ago, the late A. Warburg, without knowing of this passage in Reynolds's *Discourses* or of the drawing in Reynolds's sketch-book, collected material which tended to show that similar gestures can assume opposite meanings.[3] The pagan figure of the dancing mænad (as in Fig. 124, for example) was the central theme of these studies, and their most

[1] [F. Antal, in 'Some Examples of the Role of the Mænad in Florentine Art of the later Fifteenth and early Sixteenth Centuries' (part 2 of 'The Mænad under the Cross', Wind's 'Comments on . . . Reynolds' being part 1; loc. cit., pp. 71 ff.), suggested that in all probability Reynolds's drawing by Bandinelli could be identified with one in the École des Beaux-Arts in Paris (Fig. 122), for it had earlier had an English provenance, having been reproduced in William Young Ottley's *The Italian School of Design, being a series of Fac-similes of original Drawings by the most eminent Painters and Sculptors of Italy* (1823), facing p. 13. In 1981, however, another Bandinelli drawing, with Reynolds's collector's mark, appeared on the London art market (Sotheby Parke Bernet, 9 April, Lot 84) and is now in the collection of Daniel Katz, Esq.; this work may with more certainty be identified as the drawing Reynolds refers to here (Fig. 120).]

[2] *Discourses on Art*, XII, ed. R. R. Wark (1975), ll. 431–42. See also above, p. 19 note 53.

[3] This idea recurs frequently in Warburg's collected writings (*Die Erneuerung der heidnischen Antike. Kulturwissenschaftliche Beiträge zur Geschichte der europäischen Renaissance*, ed. G. Bing in collaboration with F. Rougemont, 2 vols., 1932), and also in his unpublished works (cf. E. H. Gombrich, *Aby Warburg. An Intellectual Biography*, 1970, pp. 247 ff.).

Journal of the Warburg Institute I (1937), pp. 70–1, with the addition of later material on Goethe's interpretation of the Tomb of the Dancer near Cumae, in the environs of Naples.

striking chapter contained the story[4] of how Bertoldo di Giovanni, the early Renaissance sculptor, transformed the mænad into a Mary Magdalene under the Cross (Fig. 123).

Dr Antal later discovered that Florentine reliefs of this early period were actually among the models from which Bandinelli worked;[5] so that Reynolds, inspired as he acknowledges by Bandinelli, continues a tradition that goes back to the fifteenth century.

Even today this formula has not lost its force, as can be seen in a recent book on Grünewald by that experienced physiognomist William Fraenger.[6] It would be difficult to imagine anything further removed from classical antiquity than the figures painted by Grünewald for the Isenheim altar. Yet when Fraenger tries to find adequate words to describe Mary Magdalene under the Cross, the memory of the old symbol makes itself felt and he calls her *eine Schmerzmänade*, quoting a Latin hymn which may have been present in the minds of many Renaissance artists who pictured the saint in the attitude of a Bacchante:

Fac me cruce inebriari
Et cruore filii.

It would be of interest to trace this figure in the history of poetic imagination. In the Romantic period, the mænad under the Cross became charged with the significance of a philosophic symbol. In Eichendorff's novel *Ahnung und Gegenwart*, she makes a most striking appearance in a staged tableau, not in the character of Mary Magdalene, but as an abstract allegorical image. A radiant white figure holding a Cross to Heaven, appeared in the centre of the stage. To one side of her stood

> a beautiful female figure in Greek attire, like an image of one of the ancient goddesses . . . with both arms raised, she carried a cymbal like someone dancing, and held it aloft, displaying the ample splendour of her limbs. Her face turned away, as if overwhelmed by the glory, she was only half visible, yet the most distinct and perfect figure. It seemed as if the joyous earthly beauty, touched by the radiance of the heavenly one, was thus suddenly petrified in her Bacchic posture.

To the Romantic poet, the mænad, transformed into stone at the sight of the Cross, symbolizes the fate of paganism: 'die . . . vor dem Glanze des Christentums zu Stein gewordene Religion der Phantasie.'[7]

[4] Published only in the Italian edition of Warburg's selected writings, *La rinascita del paganesimo antico*, ed. G. Bing and trans. by E. Cantimori (1960), p. 303.

[5] Antal, loc. cit.

[6] *Matthias Grünewald in seinen Werken. Ein physiognomischer Versuch* (1936), p. 92 (reprinted in *Matthias Grünewald*, ed. G. Fraenger and I. Baier, 1983).

[7] Joseph von Eichendorff, *Ahnung und Gegenwart* (1815), II, xii (Neue Gesamtausgabe, ed. G. Baumann, II, 1957, pp. 128 ff.).

[Among Wind's papers there are notes for a revised version of this article with the working title 'The Mænad under the Cross (and the Dance of Death)'. These—here collated—concern Goethe's interpretation of the so-called Tomb of the Dancer near Cumae, Naples (Fig. 119). Drawings of three bas-reliefs from the tomb were first published in 1810 by Andrea de Jorio in *Scheletri Cumani dilucidati* (Figs. 117–18); and later by Fr. K. L. Sickler, in *De monumentis aliquot Graecis, e sepulcro Cumaeo recenter effosso erutis* . . . (1812), and also as a popular exegesis: 'Beschreibung eines sehr merkwürdigen neuentdeckten griechischen Grabmals bei Cumä mit drei Basreliefs über die Bacchische Mysterien-Feier', in *Curiositäten der physisch- litterarisch- artistisch- historischen Vor- und Mitwelt; zur angenehmen Unterhaltung für gebildete Leser*, II (1812), pp. 35–66, where it came to the attention of Goethe. Goethe wrote an open letter to Sickler in which he proposed a 'poetic' interpretation of the reliefs based on Sickler's publication (*Curiositäten*, II, pp. 195–202; see also *Goethes Werke*, Weimarer Ausgabe, XLVIII, 1897, *Schriften zur Kunst 1800–1816*, pp. 143–50).

The first relief, as shown in Sickler's engraved copy, the

dance at the Banquet of the Blessed, is interpreted by Goethe as a reminiscence of an earthly banquet: 'Vom Effecte, den die Künstlerin hervorgebracht und der uns in seinen Abstufungen zuerst mehr angezogen als sie selbst, wenden wir uns nun zu ihr und finden sie in einer von jenen gewaltsamen Stellungen, durch welche wir von lebenden Tänzerinnen so höchlich ergötzt werden. Die schöne Beweglichkeit der Übergänge . . . ist hier für einen Moment fixirt, so daß wir das Vergangene, Gegenwärtige und Zukünftige zugleich erblicken und schon dadurch in einen überirdischen Zustand versetzt werden. . . . Daß die Künstlerin sich als ein Bacchisches Mädchen darstellt . . . daran läßt sich wohl nicht zweifeln.' Goethe saw the second relief of three grotesquely dancing skeletons, one in the mænad's pose, as 'ein geistiges Trauerspiel' (reproduced here in Jorio's version, Fig. 118). The last relief, actually the Arrival in the Beyond (Fig. 117), he interpreted as the dancer exchanging life for death: 'Gewaltsam erscheint sie hier in einer mänadischen Bewegung, welche wohl die letzte sein mochte, womit eine solche Bacchische Darstellung beschlossen wurde, weil drüber hinaus Verzerrung liegt. Die Künstlerin scheint mitten durch den Kunstenthusiasmus, welcher sie auch hier begeistert, den Unterschied zu fühlen des gegenwärtigen Zustandes gegen jenen, den sie soeben verlassen hat. Stellung und Ausdruck sind tragisch, und *sie könnte hier ebenso gut eine Verzweifelnde als eine vom Gott mächtig Begeisterte vorstellen*' (Wind's italics).

The tomb was re-excavated and examined by J. Fr. M. von Olfers ('Über ein merkwürdiges Grab bei Kumae und die in demselben enthaltenen Bildwerke', *Historisch-Philologische Abhandlungen der Königlichen Akademie der Wissenschaften zu Berlin, aus dem Jahre 1830*, 1832, pp. 1–47; see Fig. 119). He claimed—it would appear rightly—that his own new reproductions of the reliefs (made in 1824–6) were more archaeologically correct than those earlier published by Jorio and Sickler, both of whom had imaginatively filled in the missing parts of the compositions to support their theories. Olfers also challenged the generally accepted view that the tomb was that of a dancer, and indeed it had earlier been criticized in a review of Jorio's study (cf. A.-L. Millin, *Magasin encyclopédique*, I, 1813, pp. 200–8). Goethe was irritated by Olfers's criticism of Sickler's illustrations and wrote a brief note in defence of their 'artistic freedom'; this was later published in his *Nachlaß* (*Werke*, XLIX, i, 1898, *Schriften zur Kunst 1816–1832*, pp. 193 f.).

As was noted by Gabriel Peignot, who discussed the tomb on the basis of Jorio's publication (*Recherches historiques et littéraires sur les Danses des Morts*, 1826, pp. xviii–xxii): 'On conviendra que rien n'est plus original que l'idée d'amalgamer deux choses aussi disparates, *danser* et *mourir* . . .' (pp. x f.). The connection is, nevertheless, persuasive, once the train of thought is identified: ecstasy or divine rage is equated with a Bacchic communion with the God, and so with Death. The various interpretations of the imagery of the Tomb of the Dancer are discussed by M. Cluzel, *Mimes et poètes antiques* (1957), pp. 115–38, who reproduces Jorio's illustrations.

Unfortunately, like so many archaeological discoveries in the vicinity of Mount Vesuvius, the Tomb has now been lost. The Soprintendenza alle Antichità in Naples have kindly confirmed that they hold no remains of the reliefs or of the Tomb.]

VI · A Source for Reynolds's Parody of *The School of Athens*

REYNOLDS'S parody of *The School of Athens*, a caricature of British connoisseurs whom he met in Rome, is among the most spirited productions of his youthful genius (Fig. 91). Combining the eccentricity of a Thomas Patch with the solid learning of a future academician, Reynolds appears at the age of twenty-eight as a master of pictorial satire.[1] The insertion of Gothic arches and pinnacles into Raphael's classical hall not only ridicules with one stroke both the neo-Roman and the neo-Gothic tastes, but produces an atmosphere of obscurantism singularly suited for debate between these refined *cognoscenti*. In the foreground to the right, the use of Euclid's compasses to measure the circumference of a pie licked by a dog is in the best Hudibrastic tradition:

> In *Mathematicks* he was greater
> Then *Tycho Brahe* or Erra Pater;
> For he by *Geometrick* scale
> Could take the size of Pots of Ale;
> Resolve by Sines and Tangents straight,
> If *Bread* and *Butter* wanted weight.[2]

In one respect the most surprising feature is the group of musical performers on the left; for in painting these gentlemen as musicians Reynolds showed that he understood the original group in *The School of Athens* better than Vasari. Unable to decipher the tablet in the centre of this group (Fig. 88), Vasari declared it to be filled with astronomical figures, and he assumed (ever unruffled by inconsistency) that the person facing the tablet must be the evangelist Matthew attended by an angel.[3] But these are, in fact, Pythagorean students of music, and the tablet is a diagram of musical harmonies.

Reynolds was an industrious reader of Vasari, but for *The School of Athens* he trusted a better guide. 1695 had seen the publication of Bellori's *Descrizione delle immagini dipinte da*

[1] On Patch's style of caricature, see F. J. B. Watson, 'The Life of Thomas Patch (1725–1782)', *The Walpole Society*, XXVIII (1939–40), p. 24.

[2] Samuel Butler, *Hudibras* (1663), I, i, ll. 119–24.

[3] *Le vite de' più eccellenti architetti, pittori et scultori italiani* (1550), ed. G. Milanesi, IV (1879), p. 332.

Harvard Library Bulletin, III (1949), pp. 294–7, with an additional note (7) from Wind's Raphael papers on the 'Pythagorean' tablet in *The School of Athens*.

Raffaelle d'Urbino nelle Camere del Palazzo Apostolico Vaticano. A new pocket edition of this excellent treatise appeared in 1751, the very year in which Reynolds composed his burlesque.[4] His copy of the book is in the Harvard College Library;[5] his name is inscribed on the title-page, which also bears his stamp, and there are occasional marginal notes in pencil, unmistakably by his hand.[6] This edition has prefixed to it Vasari's Life of Raphael, in which Reynolds especially marked pages xii–xiii, writing 'School of Athens' in the margin. It is the more remarkable that he did not take his clue from Vasari's description, but preferred the account given by Bellori on pages 32–6, where the 'Pythagorean' tablet is described, reproduced, and fully explained as a musical system (Fig. 89), and the group surrounding it as students of music.[7]

Wit and learning were recognized by Horace Walpole as the two chief characteristics of Reynolds as an artist. When Walpole claimed that even Reynolds's eclecticism was a form of wit and that his 'borrowings' should be understood as quotations, he was thinking primarily of the good-humoured portrait of Master Crewe as Henry VIII, in which he had paraphrased Holbein: 'Is not there humour and satire in Sir Joshua's reducing Holbein's swaggering and colossal haughtiness of Henry VIII. to the boyish jollity of Master Crewe?' (Fig. 9).[8] Walpole's argument gains in force from the fact that the earliest known instance of a quotation in Reynolds is his burlesque of *The School of Athens.* The apostle of courtesy and grandeur in the use of classical epithets began as a master of derision.

To borrow 'from the grand', to 'change the purpose of the figures without changing the attitude',[9] these became in later years his favourite prescriptions for giving splendour to a

[4] For the date of the painting, cf. E. K. Waterhouse, *Reynolds* (1941, and later edns.), p. 38.

[5] Harvard College Library, Cambridge (Mass.), Charles Eliot Norton Collection, Nor 3106.

[6] For comparison of handwriting, see Reynolds's note-books in the British Museum and Sir John Soane's Museum, London, and the Metropolitan Museum of Art, New York.

[7] Much has been written on Raphael's 'Pythagorean' tablet: Bellori's account anticipates the observations made in 1879 independently by E. Naumann, 'Erklärung der Musiktafel in Raffaels "Schule von Athen"', *Zeitschrift für bildende Kunst*, XIV, pp. 1–14, and by H. Hettner in his *Italienische Studien*, pp. 190–212; see also A. Springer in *Die graphischen Künste*, V, 1883, pp. 53 ff. But it has not been noticed that Raphael's diagram (Fig. 88) is transcribed from Boethius' *Arithmetica*. This ends (II, 54) with the schema in question, and its explanation as *maxima perfectaque armonia* (Fig. 90). It was readily accessible to Raphael and his circle—for example, in the 1488, 1491, and 1498 edns. of Boethius' *Arithmetica*, in Giorgio Valla's *De expetendis et fugiendis rebus* (1501), and in the *Margarita philosophica* of Gregor Reisch (first published in 1503). It could just as easily have been seen, however, in any of the numerous manuscripts of Boethius' work, which was familiar to all university students of the *quadrivium*. Since any one of these potential sources was accessible also to Zarlino when he wrote his *Istitutioni harmoniche* (1558), it is unnecessary to presume that he hit on the same scheme as Raphael by 'sheer logic' (R. Wittkower, *Architectural Principles in the Age of Humanism*, 1949, p. 109 note 4). Furthermore, since Boethius says in his *Arithmetica* (I, *Praefatio*) that his book is but an abstract of a work by Nicomachus of Gerasa (his 'Pythagorean' source), the three main philosophers in Raphael's group of arithmeticians should be renamed (from right to left) Pythagoras—Nicomachus—Boethius; Pythagoras being characterized as *magister silentii*, and Nicomachus as the teacher from whose book Boethius is seen copying the argument on musical harmony. To honour the copyist by the name of Pythagoras (although common practice) is nonsensical. [This brief mention of the 'Pythagorean' tablet, and the proposed identification of the figures grouped round it, is from a note in Wind's unpublished study of Michelangelo's theological sources. There is a fuller treatment of the connection with Boethius in his unpublished Raphael papers. See also his note on Pythagoras as the brooding seated figure in the left centre of *The School of Athens* (*Pagan Mysteries in the Renaissance*, 2nd edn., 1968, pp. 53 f. note 4).]

[8] *Anecdotes of Painting in England* (1771), ed. R. N. Wornum, I (1862), p. xvii note 2. See also above, pp. 19 f. and 69.

[9] *Discourses on Art*, IV, ed. R. R. Wark (1975), l. 461; XV, ll. 465–6 (hereafter in Wark's edn.).

composition, 'the *gusto grande* of the Italians'.[10] Thus, in his portrait of the Duchess of Marlborough and her daughter (Fig. 2), he managed to combine a posture from one of Michelangelo's lunettes (Fig. 1) with an allusion to Raphael's *Madonna della Sedia*; and even his most effective attempt in the grand style, *The Death of Dido* (Fig. 79), is an elegant adaptation of Giulio Romano's *Psyche* (Fig. 78).[11] In the best of these instances, the full enjoyment of the work presupposes an acquaintance with its models; for only then does the painter's grace or wit become apparent, as he transfers (in Reynolds's own words) 'by a kind of parody, their excellencies to his own performances'.[12]

Malone, who was an executor of Reynolds's will, found among his scattered papers a group of notes for 'a history of his mind, so far as concerned his art, and of his progress . . .'.[13] These included an account of his change of opinion about Raphael's frescoes in the Vatican:

> I remember very well my own disappointment, when I first visited the Vatican; but on confessing my feelings to a brother-student, of whose ingenuousness I had a high opinion, he acknowledged that the works of Raffaelle had the same effect on him, or rather that they did not produce the effect which he expected. This was a great relief to my mind; and on inquiring further of other students, I found that those persons only who from natural imbecility appeared to be incapable of ever relishing those divine performances, made pretensions to instantaneous raptures on first beholding them.—In justice to myself, however, I must add that . . . my not relishing them as I was conscious I ought to have done, was one of the most humiliating circumstances that ever happened to me. . . . Notwithstanding my disappointment, I proceeded to copy some of those excellent works. I viewed them again and again; I even affected to feel their merit, and to admire them, more than I really did. In a short time a new taste and new perceptions began to dawn upon me[14]. . . . The truth is, that if these works had really been what I expected, they would have contained beauties superficial and alluring, but by no means such as would have entitled them to the great reputation which they have so long and so justly obtained.[15]

It would be interesting to know, but there is at present no certain way of deciding, whether Reynolds painted his parody of *The School of Athens* before or after his conversion. Also, it would be instructive to compare this humorous piece with those serious exercises by which he trained himself to overcome his initial resistance. It is not impossible that some of these may yet be discovered. 'Of the few copies which he made while he was at Rome,' writes Malone, 'two are now in the possession of the Earl of Inchiquin, who married his niece, Miss Palmer: St. Michael, the Archangel, slaying the Dragon, after Guido; and the School of Athens, from Raffaelle; both masterly performances.'[16]

[10] Discourse III, ll. 56–7.

[11] See above, pp. 20 and 70.

[12] Discourse VI, ll. 545–6.

[13] *The Works of Sir Joshua Reynolds*, ed. Edmond Malone, I, 2nd edn. (1798), p. xiii.

[14] See the recommendation, quoted from James Harris in Discourse XV, ll. 403–4, 'even to feign a relish, till we find a relish come' (cf. above, p. 21 and note 56).

[15] Malone, op. cit., pp. xiv ff.

[16] Ibid., pp. xix f. note 12. See also A. Graves and W. V. Cronin, *A History of the Works of Sir Joshua Reynolds P.R.A.*, IV (1901), p. 1661: 'A sketch from the celebrated School of Athens of Raphael, 27 inches by 5 feet 6 inches', listed as no. 40 in the second day's sale of the Dowager Marchioness of Thomond's collection, Christie's, 19 May 1821.

When the imagination of a painter is aroused and governed by curiosity, as was habitually the case with Reynolds, the reading of a book may be a decisive aid in removing impediments to perception. In his estimation of Raphael, it is not unreasonable to suspect that his change of mind was hastened by a perusal of Bellori.

VII · Blake and Reynolds

It may seem ungracious to use the bicentenary of Blake as the occasion for reflecting on the marginal notes with which he disfigured his copy of Sir Joshua Reynolds's *Discourses*. Blake's copy is now in the British Library, and the notes are easily studied in the Nonesuch edition of his works where they are included among the marginalia. At first glance they seem like childish exclamations of anger, for the margin is filled with such words as 'Fool', 'Liar', 'Hypocrite', 'Knave', 'A Mock', 'A Sly Dog', 'A Polish'd Villain who Robs & Murders', culminating in the plain remark: 'I certainly do Thank God that I am not like Reynolds.'[1]

As President of the Royal Academy, Sir Joshua regarded it as his official duty to raise the level of art in England. Hence Blake observes: 'This Man was Hired to Depress Art.'[2] And lest the reader should be in any doubt whether his animosity was inspired by a sense of frustration, Blake obligingly assures him that it was: 'Having spent the Vigour of my Youth & Genius under the Opression of Sr Joshua & his Gang of Cunning Hired Knaves Without Employment & as much as could possibly be Without Bread, The Reader must Expect to Read in all my Remarks on these Books Nothing but Indignation & Resentment.'[3] And among the satirical verses he wrote on Reynolds there is the little line: 'These Verses were written by a very Envious Man.'[4]

Blake's splendid frankness in admitting his envy should warn us, however, against assuming that envy was the sole cause of his outbursts. If the worldly success of Reynolds had been obtained by Romney or Fuseli, it is fair to surmise that Blake would not have been quite so furious. What angered and embittered him about Reynolds's career was that it rested on a method diametrically opposed to his own. And the *Discourses* contained the rationale of that method: they recommended a prudence which Blake abhorred, and declined the risks of genius to which Blake was committed.

Reynolds was bent on exposing artistic pretensions of two different kinds. He disdained artists who claimed that they could rely on their inspiration alone, without regular exercise

[1] Blake, *Complete Writings*, with variant readings, ed. G. Keynes (1957), see now the reprint of 1979, pp. 447 ff., 471 f., etc. [Blake's annotations are conveniently arranged in R. R. Wark's second edition of the *Discourses on Art* (1975), pp. 284 ff.; see also D. Bindman, 'The dating of Blake's Marginalia to Reynolds', *The Burlington Magazine*, CVIII (1966), p. 522.]

[2] Blake, op. cit., p. 445 (Wark, p. 284).

[3] Ibid.

[4] 'Florentine Ingratitude', ibid., pp. 541 f.

A talk for the Third Programme of the BBC on the occasion of the Blake bicentenary, later published in *The Listener*, LVIII (28 November 1957), pp. 879–80. The notes now added are from Wind's papers.

to sustain it; but his scorn was equally directed against those who sought distinction by 'some inferior dexterity, some extraordinary mechanical power'.[5] Blake comments: 'The Words, Mechanical Power, should not be thus Prostituted.'[6] And where Reynolds warns that 'mere enthusiasm will carry you but a little way',[7] Blake replies: 'Meer Enthusiasm is the All in All!'[8]

On both these points Reynolds tried to disabuse his pupils of an easy and all too prevalent superstition. Reliance on genius as a mysterious power that comes from heaven like a gift of grace seemed to him as destructive of solid artistic achievement as an undue pride in mechanical facility. The one allowed the artist's gifts to run wild, the other reduced him to a mere artisan. Reynolds believed that the imagination should be disciplined; and if he disdained mechanical proficiency because it was unimaginative, he distrusted enthusiasm because it was anarchic.

Blake, on the contrary, believed that 'Mechanical Excellence' was the 'Chariot of Genius',[9] that disdain of the artisan was destructive of art, and that frenzied visions were not, as Reynolds supposed, a source of vagaries, but secured accuracy in craftsmanship. Thus the two heresies against which Reynolds fought—the heresy of pure inspiration at the one extreme, of mechanical dexterity at the other—were joined by Blake in a single act of faith. He believed that the minutiae of his craft were dictated and sustained by his supernatural visions.

We are faced here, in Reynolds as in Blake, with a conjunction of forces that are commonly thought to be incompatible. If an artist is highly imaginative like Blake, we do not expect him to be a stickler for minutiae; and conversely, if a painter is an academician like Reynolds, who discourses on art from a presidential pulpit, we assume that he must be a pedant. But in the controversy between Reynolds and Blake it is the academician, Reynolds, who opposes pedantry, while Blake, who extols the visionary powers of the imagination, is prompted by his very frenzy to be rigid. How is that to be explained?

Let us first look a little closer at Reynolds's fight against mechanical execution on the one hand, and reliance on genius on the other. In the Sixth Discourse, which Blake disliked particularly because it was written in praise of judicious imitation, Reynolds states that '. . . it is by being conversant with the inventions of others, that we learn to invent: as by reading the thoughts of others, that we learn to think'.[10] Thus an artist can raise his natural stature by entering imaginatively into the spirit and style of Michelangelo or Raphael:

> Our hearts frequently warmed in this manner by the contact of those whom we wish to resemble, will undoubtedly catch something of their way of thinking. . . . The disposition, which is so strong in children, still continues with us, of catching involuntarily the general air and manner of those with whom we are most conversant; with this difference only, that a young mind is naturally

[5] Discourse IV, ll. 377–8.
[6] Blake, op. cit., p. 464 (Wark, p. 303).
[7] Discourse II, l. 192.
[8] Blake, op. cit., p. 456 (Wark, p. 295).
[9] Ibid., pp. 453 f. (Wark, p. 292).
[10] Discourse VI, ll. 169–71.

pliable and imitative; but in a more advanced state it grows rigid, and must be warmed and softened, before it will receive a deep impression.[11]

Reynolds drew the logical conclusion that rigid imitation would defeat his purpose. In order to render the imagination of his pupils more supple and awaken their dormant powers of invention, he produced those fanciful exercises in imitation which are described and recommended in the *Discourses*: compete with the model, follow it partially, parody it, change the function of a figure while retaining the attitude, reverse the meaning of a given gesture. All these precepts have one negative rule in common: literal copying is worthless, or, as Reynolds put it, 'a delusive kind of industry', since 'those powers of invention and composition which ought particularly to be called out, and put in action, lie torpid, and lose their energy for want of exercise'. Nothing significant, Reynolds repeats, is attained 'by servilely copying the model before you'.[12] But Blake, in his marginal comment, brands this reflection as 'Contemptible! Servile Copying is the Great Merit of Copying'.[13] And again: '"Copy for Ever" is My Rule.'[14]

The compulsive nature of Blake's visions explains his belief in servile copying: his visions demanded minute execution. 'The Man who asserts', says Blake, '. . . that everything in Art is Definite & Determinate, has not been told this by Practise, but by Inspiration & Vision, because Vision is Determinate & Perfect, & he Copies That without Fatigue, Every thing being Definite & determinate.'[15] The visions Blake saw in hallucinations, or which he projected on a piece of paper, were of an obsessive nature, elaborate, articulate, precise. Hence Blake resented Reynolds's suggestion that in order to acquire the grand style in painting it is essential to eliminate particularities—his claim that fresco painting 'excludes attention to minute elegancies'.[16] Blake notes in the margin: 'Fresco Painting is the most Minute. Fresco Painting is like Miniature Painting: a Wall is a Large Ivory.'[17]

Blake's actual fresco painting, as seen in the examples in the Tate Gallery, is indeed close to miniature painting, because it is carried out on a small scale. Nevertheless, the convergence of the minute and the monumental in Blake's thought is again symptomatic of his obsessive vision. Unable and unwilling to perform an act of detachment, he denies the distinction made by Reynolds between the heroic and the familiar style, between grandeur and neatness of execution.

Difficult though it may seem, we must accept it as a fact that pedantic insistence on exact copying can be inspired by imaginative frenzy; whereas a liberal admission of artistic licence, which allows the artist the fullest freedom to transform his model according to his fancy, can be sanctioned by academic training. Blake did not find Reynolds too didactic: he found him too soft. The transformation of a mechanical into a liberal art he deplored as deterioration. Thus it was precisely because Blake was a compulsive artist, possessed by his

[11] Ibid., ll. 176–85.
[12] Discourse II, ll. 135–6, 139–41, 155.
[13] Blake, op. cit., p. 456 (Wark, p. 294).
[14] Ibid., p. 446 (Wark, p. 285).
[15] Ibid., p. 457 (Wark, p. 296).
[16] Discourse V, ll. 126–7.
[17] Blake, op. cit., p. 466 (Wark, p. 305).

visions and positive about every detail he saw, that he believed in neat execution. While Reynolds relied for the refinement of his art on 'that friendly intercourse which ought to exist among Artists of receiving from the dead and giving to the living',[18] Blake was far too certain of his personal vision to concede that creative powers may be increased by conversing, by selecting, by a slow process of growth, or by what Reynolds called 'digesting, methodising'.[19] Truth must strike directly, or it is not truth at all. 'He who does not Know Truth at Sight', says Blake, 'is unworthy of Her Notice.'[20]

Having heard what both Blake and Reynolds have to say about copying, nothing is more revealing than to see in what manner they actually do copy. In every instance Reynolds transfers his model to such a new setting that it is either unrecognizable, or is recognized with a shock: a Michelangelo posture in the Duchess of Marlborough (Figs. 1–2); Giulio Romano's *Psyche* transformed into Dido (Figs. 78–9); Kitty Fisher mimicking Trevisani's *Cleopatra* (Figs. 80–1); Holbein's *Henry VIII* enacted by a child (Fig. 9); an ancient mænad as a mourning Magdalene (Figs. 120–2). 'Such imitation', to quote Reynolds himself, 'is so far from having anything in it of the servility of plagiarism, that it is a perpetual exercise of the mind, a continual invention. Borrowing or stealing with such art and caution, will have a right to the same lenity as was used by the Lacedemonians, who did not punish theft, but the want of artifice to conceal it.'[21]

Blake was naturally revolted by Reynolds's cunning. He himself copied straight, emphatically straight; for in copying he was completely caught up in his subject, absorbed by the details he saw before him. Hence, when he copied Michelangelo, as in his engraving *Joseph of Arimathea among The Rocks of Albion*, a figure derived from the frescoes of the Cappella Paolina (Figs. 92–3) which Blake knew 'from an old Italian drawing', he did not change the model in the least; in fact he stressed the borrowing in the inscription.[22] But while 'servilely copying' Michelangelo's figure, he endowed it verbally with 'Gothic' mysticism.[23] The legend that Joseph of Arimathea had landed in Albion hovers over the design as a poetic aura, yet the visual scheme to which the aura is attached remains a literal piece of copying. The mechanical performance supplies the core around which the fantasy clusters.

Blake proceeded in the same way with the Laocoön.[24] His copy of the ancient group is a

[18] Discourse XII, ll. 318–20.

[19] Discourse III, l. 93.

[20] Blake, op. cit., p. 474 (Wark, p. 314).

[21] Discourse VI, ll. 471–5.

[22] On a unique impression of the first state (Keynes Collection) Blake later added in ink: *Engraved when I was a beginner at Basires / from a drawing by Salviati after Michel Angelo*. Such a drawing by Salviati is otherwise unknown, and Keynes, *Engravings by William Blake. The Separate Plates* (1956), no. I, pp. 3 ff., refers to an unfinished print by Beautrizet as Blake's possible source. Blake kept the plate (his first independent engraving at the age of sixteen), and retained the original date when he reworked it some thirty years later: *Engraved by W Blake 1773 from an old Italian Drawing*.

[23] The full inscriptions on the second state (*c.* 1810) reads: *This is One of the Gothic Artists who Built the Cathedrals in what we call the Dark Ages / Wandering about in sheep skins & goat skins of whom the World was not worthy / such were the Christians / in all Ages. / Michael Angelo Pinxit.*

[24] Blake first drew the Laocoön *c.* 1815 from a cast in the Royal Academy in preparation for a simple line engraving, to illustrate Flaxman's article on sculpture in Rees's *Cyclopaedia*, Plates, IV (1820); see Keynes, *Blake's Pencil Drawings*, Second Series (1956), no. 30; and now M. Butlin, *The Paintings and Drawings of William Blake* (1981), no. 679, where the original drawing is illustrated, pl. 898A.

literal one, and so insistently literal that the image acquires a ghostly quality which is increased by the mystical words that surround the group. They suggest that the ancient sculptors merely echoed a lost biblical revelation and, following Virgil, gave secular features to a visionary triad representing Jehovah with his two sons, Satan and Adam.[25] There is no need to discuss here the sources of this apocryphal speculation; but it is important to notice that, like a crystal-gazer, Blake needs a crystal-ball to concentrate his vision. His fantasy is tightly fastened to a design of the utmost accuracy.

There is a curious fascination in observing how often Blake and Reynolds admired the same men. For example, they both admired Plato, and declared themselves to be his followers. But when Reynolds interprets Plato's theory of inspiration he refuses 'to understand literally these metaphors or ideas expressed in poetical language':[26] for if Plato says that inspiration comes from heaven, the student of Plato must beware of thinking that he has direct access to heaven; his place is on earth, and he must patiently work in order to aspire to that higher region. Blake writes in the margin: 'Plato was in Earnest.' Plato, according to Blake, 'believe'd that God did Visit Man Really & Truly & not as Reynolds pretends'.[27]

The words 'Plato was in Earnest' are likely to startle any reader of Plato today, since we do not find it quite so easy to make out when Plato was in earnest and when not. Perhaps Reynolds was nearer the truth than Blake, since he understood Plato's use of irony. Blake's Plato is a zealous prophet.

A similar contrariety occurs in the admiration of both artists for Michelangelo. The very grandeur of Michelangelo made Reynolds feel that he should not copy him exactly. He chose him as a guide and recommended him as a model—which the student might approach but could not hope to attain. Blake tried to take complete possession of Michelangelo's art by bringing it close to his own and appropriating his figures with all their idiosyncrasies. Historic distance, so congenial to Reynolds's tact, produced in Blake a *soif de l'impossible*.

It is a pity that Stendhal did not know Blake, for his remarks on Reynolds's relation to Michelangelo state the disparity in Reynolds's worship so sharply, and so sarcastically, that one wonders whether Blake would have fared any better. It seems certain, however, that Stendhal's conclusion, cast in the form of a reflection on Dante and Virgil, would have offended Blake far more than Reynolds, since it inadvertently argues Reynolds's case: 'No one loved Virgil so much as Dante, and nothing is less like the *Aeneid* than the *Inferno*.'[28]

[25] For this later version of the engraving, with elaborate inscriptions printed in all directions, see Keynes, *Engravings by William Blake*, no. xx, pl. 37, and now D. Bindman, *The Complete Graphic Works of William Blake* (1978), no. 623, who dates the *Laocoön c.* 1822, seeing a connection in thought with no. 622, *On Homers Poetry* and *On Virgil*. For the inscriptions, see Blake, *Complete Writings*, pp. 775 ff.

[26] Discourse VII, ll. 98–9.

[27] Blake, op. cit., p. 473 (Wark, p. 313).

[28] *Histoire de la peinture en Italie* (1817), clxxii (ed. P. Arbelet, II, 1924, p. 283).

VIII · Johnson and Demonax

DR JOHNSON, in one of his outbursts against David Hume and the sceptics, employs a daring image for the deviation from truth: 'Truth, Sir, is a cow that will yield such people no more milk, and so they are gone to milk the bull.'[1] Johnson's spontaneous remark was inspired by a literary reminiscence that was current in the eighteenth century. Kant, for example, though clearly unlike Johnson in temperament and style, used the same image in the *Critique of Pure Reason*; but in place of the bull he introduced a he-goat and expanded the joke (on the authority of the ancients) by making the milkers use a sieve:

What is truth? . . . To know what questions may reasonably be asked is already a great and necessary proof of sagacity and insight. For if a question is absurd in itself and calls for an answer where none is required, it not only brings shame on the propounder of the question, but may betray an incautious listener into absurd answers, thus presenting, as the ancients said, the ludicrous spectacle of one man milking a he-goat and the other holding a sieve underneath.[2]

Although Kant refers to his source vaguely as 'the ancients', he had a particular passage in mind—a jest in Lucian's *Demonax*: 'On seeing two philosophers very ignorantly debating a given subject, one asking silly questions and the other giving answers that were not at all to the point, he [Demonax] said: Doesn't it seem to you, friends, that one of these fellows is milking a he-goat (*τράγον ἀμέλγειν*) and the other is holding a sieve for him!'[3]

The joke, codified in Erasmus' *Adagia* under the heading *Mulgere hircum*,[4] must have been imported into England at the time of Erasmus and More, who both translated Lucian's dialogues into Latin and sometimes emulated his wit in their own writings.[5] A century later, in 1635, the jest was recalled by Francis Quarles, who succeeded in depriving it of its humour. His book of *Emblemes* (I, xii) shows 'a child sucking out of paps in a globe,

[1] Boswell's *Life of Johnson*, 21 July 1763 (Hill and Powell, I, p. 444).

[2] *Kritik der reinen Vernunft*, 2nd edn. (1787), I, 2. Theil, *Transszendentale Elementarlehre*, Einleitung, iii (*Kant's gesammelte Schriften*, Preußische Akademie der Wissenschaften, III, 1911, p. 79); trans. by N. Kemp Smith (1956), p. 97, 'The Division of General Logic into Analytic and Dialectic'.

[3] *Demonax*, 28, trans. by A. M. Harmon, *The Works of Lucian*, Loeb Library, I (1913), p. 159.

[4] *Adagiorum chiliades tres* (1508), s.v. *mulgere hircum*.

[5] In his 'Life of Lucian', Dryden wrote: 'I know not whom Lucian imitated, unless it might be *Aristophanes*. . . . But he who best imitated him in Latin is *Erasmus*. . . .' (*The Works of Lucian. Translated from the Greek by several Eminent Hands*, I, 1711, p. 43.)

A revised version of 'Milking the Bull and the He-Goat', first published in the *Journal of the Warburg and Courtauld Institutes*, VI (1943), p. 225.

another holding a sieve below'. From the long argument explaining the hideous image (inscribed *Inopem me copia fecit* and engraved by William Marshall) it suffices to quote two lines:

> Alas, how poorly are thy labours crown'd!
> Thy liquor's never sweet, nor yet thy vessel sound.

Fortunately, Johnson drew from a purer source. It is evident from his conversation that he read Lucian,[6] a fact that did not escape his contemporaries. In 1780, according to Boswell,[7] 'the Reverend Dr. Francklin having published a translation of "Lucian", inscribed to him the *Demonax* thus: "To DR. SAMUEL JOHNSON, the Demonax of the present age, this piece is inscribed by a sincere admirer of his respectable talents, THE TRANSLATOR".'[8]

Boswell complained that 'upon a particular comparison of Demonax and Johnson, there does not seem to be a great deal of similarity between them'; and it must be admitted that Johnson's wit is terser than that of Demonax. The double absurdity of the he-goat and sieve was too involved and artificial for his taste. With his robust instinct for simplification he changed the he-goat into a bull and omitted the sieve altogether, and the joke gained in vigour what it lost in intricacy.

As early as 1711 Steele had already thought of adapting Lucian's he-goat and sieve to British taste, but he erred on the side of genteelness. In reflecting on certain 'ingenious Gentlemen' who 'can turn what little Knowledge they have into a ready Capacity of raising Doubts', his essay in *The Spectator* displays, if not a weak memory of Lucian's text, a marked reticence in transcribing it: 'It was of two Disputants of this impertinent and laborious kind that the Cynick said, *One of these Fellows is Milking a Ram, and the other holds the Pail*.'[9] It is not a memorable version, but it deserves a place between Quarles and Johnson in the history of the British Demonax.

[6] See Boswell, 3 April 1776 (Hill and Powell, III, p. 10), where Johnson alludes to Lucian's *Jupiter tragoedus*. See now also *The Sale Catalogue of Samuel Johnson's Library*, with an introduction and notes by J. D. Fleeman (1975): no. 73, Francklin's *Lucian*; no. 158, Lucian's works in 5 vols.

[7] Boswell, 1780 (Hill and Powell, IV, p. 34).

[8] *The Works of Lucian, from the Greek*, by Thomas Francklin, II (1780), p. 63. The piece is inscribed to Johnson by 'a sincere admirer of his truly respectable character' (Boswell misquotes here). On the use of 'respectable', see Hill and Powell's comments in Boswell, III, p. 241 note 2.

[9] *The Spectator*, no. 138, 8 August 1711 (ed. D. F. Bond, II, 1965, p. 46). In *The Works of Lucian ... by several Hands*, IV, p. 135, the translator of *Demonax*, Captain Ayloffe, succeeded in purifying the passage almost out of recognition: 'Who would not believe now (he said) that they are both thick of hearing; or, at least, that one talks in a Language the other knows nothing of?'

IX · The Revolution of History Painting

THE commemoration of contemporary events in a monumental style of painting, combining heroic grandeur of design with historical accuracy in costume and portraiture, was censured by the Academies of the eighteenth century as an offence against good taste. There can be no doubt that the much abused Academicians had Reason and Nature on their side when they declared that the grand style and the faithful portrait manner were incompatible with one another, and that artists ambitious to paint heroes should take care not to make them look too much like themselves or their neighbours, or the soldiers they saw walking about the streets. If it is true that familiarity breeds contempt, then the advice that the hero should not be represented as a familiar figure was sound.

Yet the advice ran counter to the trend of eighteenth-century history writing and to the technique of Enlightenment criticism, which made a point of approaching traditional heroes with an air of familiarity. Voltaire, Hume, and Gibbon were bent on destroying the exaggerated reputations of heroes, saints, and other pretenders to supernatural glory.[1] By exposing what Swift called 'the mechanical operations of the spirit', they advocated a view of life which would seek its glory in humane simplicity, and reject as ridiculous the pretence that men can play the role of demi-gods. In substance, this view would seem to coincide with the academic warning that the grand style and the intimate manner should be kept apart. But the conclusion drawn from this distinction was the reverse. While the Academicians refused to depict ordinary men as heroes, the new historians refused to believe in heroes who could not be depicted as ordinary men.

When in 1771 Benjamin West provided the controversy with a test case by painting *The Death of General Wolfe* in contemporary costume and setting (Fig. 95), he explicitly appealed to the law of the historian:

> . . . the event intended to be commemorated took place on the 13th September, 1758 [*sic*], in a region of the world unknown to the Greeks and Romans, and at a period of time when no such

[1] Although Voltaire's early phase of history writing was not exactly anti-heroical (on his *Histoire de Charles XII, roi de Suède*, 1731, see A. Blunt, 'The Criminal-King in a Nineteenth-Century Novel', *Journal of the Warburg Institute*, I, 1938, pp. 248 f.), he was with some justice regarded by his younger contemporaries as the predecessor of Hume and Gibbon. So firmly was this 'genealogy' fixed in the English imagination that the faces of these 'unholy three' were recognized in the group of infidels in Reynolds's *The Triumph of Truth* (Fig. 19)—a premature glorification of Dr Beattie, for which Reynolds was rebuked by Oliver Goldsmith (see above, pp. 29 f.).

Journal of the Warburg Institute, II (1938), pp. 116–27.

nations, nor heroes in their costume, any longer existed. The subject I have to represent is the conquest of a great province of America by the British troops. It is a topic that history will proudly record, and the same truth that guides the pen of the historian should govern the pencil of the artist.[2]

It is recorded that Reynolds tried to dissuade West from committing so flagrant a breach of academic rules, and that to give due weight to his argument he enlisted the support of the Archbishop of York. But on seeing West's painting he admitted that West had won his point and had managed to treat a contemporary subject in the grand style without offending a sense of decorum. It is curious, however, that neither Reynolds nor even West himself (to say nothing of modern historians of art) seems to have commented on the peculiar device by which this *tour de force* was accomplished: the scene was laid in a very distant region, the Canada of the Colonial Wars. The Academicians had insisted with some justice that heroism is not an everyday occurrence, and they feared that the presentation of heroic death in everyday dress would destroy respect for the hero and deprive the spectator of that feeling of reverence and astonishment, that sense of the extraordinary and marvellous, with which he ought to approach heroic deeds. But in West's picture this sense of the marvellous is preserved. The figure of an American Indian is prominently placed opposite the dying hero. He fulfils the function of a *repoussoir*, and by leading the imagination into a distant land effectively offsets the shock of seeing the hero die in a modern uniform. Instinctively West had followed a rule which Racine had laid down a century earlier in the preface to *Bajazet* (1672):

Some readers may be astonished that I have dared to put on the stage so recent a story; but I have found nothing in the rules of dramatic poetry of a kind to deter me from my enterprise. It is true that I would not advise an author to take for the subject of a tragedy so modern an action as this, if it had taken place in the country where he wishes the play to be performed. . . . Tragic personages should be regarded in a different light from that in which we regard those whom we have seen close at hand. One may say that the respect that one has for the hero grows in proportion as he recedes from us: *maior e longinquo reverentia*. Distance of country compensates in some sort for nearness of time: for people do not distinguish between that which is, if I may venture to say so, a thousand years, and that which is a thousand miles away from them.[3]

The notion of distant glory which Racine had associated with the 'marvels of the east' was evoked for an eighteenth-century English audience by scenes from the Colonial Wars. They had the advantage of being at once actual and remote. They offered to history painters and theatrical producers the opportunity to develop a style of *mitigated realism*. Scenes represented in this manner were neither as plainly realistic as Diderot's *tragédies intimes* nor

[2] John Galt, *The Life and Works of Benjamin West, Esq., President of the Royal Academy of London, Subsequent to his Arrival in this Country; composed from materials furnished by himself*, Part II (1820), pp. 48 f. See also below, p. 124.

[3] Trans. by Roger Fry in *Characteristics of French Art* (1932), pp. 24 f.

as frankly unreal as declamatory tragedies of the grand style. Their intermediate position rendered them more palatable, but also more dangerous, to the academic hierarchy than a truly radical attempt at reform; for they veiled the distinction between the old and the new, and bridged the gulf between the conservative and the revolutionary parties.

The growth of this intermediate style was largely due to the newly revived passion for patriotism. Students of theatrical history have observed,[4] and historians of painting can confirm,[5] that the desire to be historically accurate in the designing of costumes was first inspired by a patriotic interest in the past, and originally confined to subjects of national history. In the London season of 1762–3, *Henry the Fourth*, *Richard the Third*, and *Jane Shore* were produced 'With Characters new Dress'd in the Habits of the Times', but *Tamerlane*, *Hamlet*, and *Barbarossa* were not.[6] The same can be observed in other countries. In Germany, for instance, the first plays performed 'im genauesten Kostüm' were Elias Schlegel's *Hermann* (1766) and Ayrenhoff's *Hermann und Thusnelda* (1768).[7] On the Paris stage, Dormont de Belloy's 'tragédie patriotique', *Le Siège de Calais*, had a sensational success in 1765; and in the same year, Lépicié exhibited his provocative painting *La descente de Guillaume le Conquérant sur les côtes d'Angleterre*.[8] With their topical allusions, both are expressive of what has been called 'le culte du moyen âge national'. National period costume served as a disguised expression of contemporary feeling.

About 1770–1 stage designers, dramatists, and history painters seem to have rediscovered for themselves Racine's law that remoteness of setting is as conducive to grandeur as distance in time, and that it allows what the classical rules of the grand style had forbidden: the representation of contemporary events without disguise. Historical accuracy was extended from patriotic scenes of the remote past to contemporary patriotism in distant countries. The colonial officer and the explorer assume the role of the tragic hero. Accordingly, West's *Death of General Wolfe* has its counterpart in contemporary drama; for the first tragedy in verse in which the hero appears as a modern character without relinquishing any of his traditional grandeur was *La Veuve du Malabar* by Lemierre (first produced in 1770, revived and printed in 1780), a piece acclaimed as a daring innovation[9] but remarkable for no other reason than that its hero, as in West's picture, is a colonial

[4] See Wilhelm Kopp, *Das Tragödienkostüm von Betterton bis Kemble in seiner Entwicklung zur historischen Treue 1660–1817* (1929), pp. 68 ff.; also D. T. Mackintosh, 'New Dress'd in the Habits of the Times', *The Times Literary Supplement*, 25 August 1927, p. 575; and for further material, Lily B. Campbell, 'A History of Costuming on the English Stage between 1660 and 1823', *University of Wisconsin Studies in Language and Literature*, II (1918), pp. 187 ff.

[5] See J. Locquin, *La Peinture d'histoire en France de 1747 à 1785* (1912), pp. 161 and 280; F. Antal, 'Reflections on Classicism and Romanticism', *The Burlington Magazine*, LXVI (1935), pp. 159 ff.; also A. Neumeyer, 'The Early Historical Paintings of Benjamin West', ibid., LXXIII (1938), pp. 162 ff.

[6] *The Public Advertiser*, 25 October 1762, quoted by Kopp, op. cit., pp. 68 f.

[7] Kopp, op. cit., p. 70 note 1. See also W. Klara, *Schauspielkostüm und Schauspieldarstellung. Entwicklungsfragen des deutschen Theaters im 18. Jahrhundert* (1931), pp. 3, 37, and 19 note 91. At exactly the same time Johann Heinrich Tischbein painted his *Hermannsschlacht*, cf. Neumeyer, loc. cit.

[8] Locquin, op. cit., p. 280; F. Gaiffe, *Le Drame en France au XVIIIe siècle* (1910), pp. 410 ff.

[9] Klara, op. cit., p. 119.

officer.[10] In 1771, the very year in which West exhibited his *Death of General Wolfe*, Garrick called Loutherbourg, the Alsatian painter and scene designer, from Paris to London, where he introduced at Drury Lane a revolution in stagecraft by which stage illusion and scenic effect became a predominant part of every ambitious production.[11] Romantic evocation was here combined with a naturalistic 'attack' on the audience. The climax of Loutherbourg's theatrical career was his production (at Covent Garden in 1785) of the exotic pantomime *Omai; or, a trip round the world*, 'the costumes being designed from studies made by John Webber, R.A., the painter who was with Captain Cook on his last voyage, and who painted *The Death of Captain Cook*'.[12]

This theatrical fashion of combining military valour with exoticism found a suitable subject in Shakespeare's *Othello*. In 1788, Ludwig Tieck saw with approval how the German actor Fleck adopted the 'latest English manner'[13] and played the part of Othello in the modern uniform of a colonial general:

> His rendering of Othello was powerful—simple, true, human and noble at first, and then terrifying in the development of passion. It was remarkable how he played the role in a kind of modern costume, which was similar to a general's uniform, and complete with a star and order-ribbon. In a sense, therefore, he modernised the part, but did so without in any way lessening the effect. Othello can, in fact, be regarded as a middle-class play, and therefore such an outward approximation to our own times was appropriate.[14]

For all its absurdity, the suggestion that *Othello* can be regarded as 'ein bürgerliches Stück' expresses in a paradoxical way the same sentiment as animated the sober and plausible argument of Benjamin West when he defended the use of modern costume in the grand style of painting. There is in both the same desire to divest the tragic hero of his illusory glamour and reduce him from an imaginary to a familiar figure. Exotic naturalism is the twilight of the grand style of history painting. It represents marvels as matters of fact.

Once the rules had been breached in this surreptitious manner, the breach was quickly

[10] The exact reverse of the colonial hero is the exotic critic, the Chinese, Nubian, Turk, etc., who is introduced as *ingénu* into European society and finds its behaviour ridiculously artificial. The 'noble savage' generally combines both roles: the messenger of distant marvels and the critical apostle of naturalness.

[11] A satire on this 'reformation' appeared in 1772 under the title: *The Theatres. A Poetical Dissection*, by Sir Nicholas Nipclose, Baronet. The caricature on the title-page (modelled after Reynolds's *Garrick between Tragedy and Comedy*, Fig. 75) shows the Muses united on one side of Garrick, trying in vain to hold him back from the Tailors and Carpenters on the other (Fig. 74). Cf. above, p. 39 note 98.

[12] Campbell, op. cit., p. 210. See now the catalogue of the Greater London Council Exhibition at Kenwood, *Philippe Jacques de Loutherbourg, RA, 1740–1812* (1973), no. 84.

[13] Klara, op. cit., p. 121. Cf. J. Boaden's description of Kemble's costume as Othello in 1785 (*Memoirs of the Life of John Philip Kemble, Esq., including a History of the Stage, from the Time of Garrick to the Present Period*, 1, 1825, p. 145).

[14] R. Köpke, *Ludwig Tieck. Erinnerungen aus dem Leben des Dichters nach dessen mündlichen und schriftlichen Mittheilungen* (1855), 2. Theil, p. 230 (quoted by Klara, op. cit., p. 122):

'Mächtig wirkte . . . sein Othello, den er einfach, wahr, menschlich edel im Anfang spielte, und furchtbar in der Entwicklung der Leidenschaft. Eigentümlich war es, daß er ihn in einer Art von modernem Costüm darstellte, welches mit einer Generalsuniform Aehnlichkeit hatte; er trug Stern und Ordensband. Er modernisierte dadurch die Rolle im gewissen Sinne; aber die Wirkung litt dadurch nicht. Man kann Othello vielmehr ein bürgerliches Stück nennen, und da war eine solche äußere Annäherung an unsere Zeit an ihrer Stelle.'

widened to abolish the academic precept that distance is essential to dignity. When West exhibited *The Death of General Wolfe* in 1771, he showed an event which had taken place twelve years before in a distant colony. When Copley painted *The Death of the Earl of Chatham* during the years 1779–81, he was representing an event that had taken place a year earlier in the House of Lords in London (Fig. 99).[15] Copley himself was, even more than West, an expert in the painting of West Indian *mirabilia*. His *Brook Watson and the Shark*, with the Negro leading the rescue party in Havana harbour (Fig. 100), is a far more impressive example of American marvels than West's idyllic commemoration of innocent Indians selling their valuable land to William Penn for cloth (Fig. 94).[16] What West achieved through sentimentalism Copley reinforced by dramatic action. But in *The Death of the Earl of Chatham* Copley threw off the disguise and renounced the restrictions under which West had introduced modern costume into history painting. What had begun in the setting of *mirabilia* took on the character of plain *reportage*. To secure the greatest possible authenticity for his picture, Copley made all the Members of Parliament who had attended the memorable meeting at which Chatham died sit for their portraits (with the notable exception of Lord Mansfield, who refused). He thus transferred to the grand style of history painting a convention that had been common in the intimate genre of the conversation piece: the group portrait in a commemorative setting. Moreover, when exhibited in Spring Garden, the picture made its impression as a touching contemporary document. Visitors who went to see it behaved as if in the presence of death itself: 'the Room, whatever Number of People it contains, is silent, or the company whisper as at the Bed of a Sick Person.'[17]

Compared with the popular appeal of this picture, the interest aroused by West's *Death of General Wolfe* seems a restricted academic affair. West had provided a test case for Academicians, but never meant to deprive history painting of the dignity associated with the 'grand style'.[18] Copley succeeded in transforming it into the broadly popular form of art which it has remained to the present day.

After *The Death of the Earl of Chatham*, Copley painted *The Death of Major Peirson* and received an order for a great 'portrait painting' of the battle of Gibraltar. In Reynolds's

[15] See now J. D. Prown, *John Singleton Copley in England, 1774–1815* (1966), pp. 275 ff. and pls. 392–417.

[16] S. Isham, *History of American Painting* (1936), p. 26, gives details about Watson and his adventures with the shark; see now also Prown, op. cit., pp. 268 ff. and pls. 368–81. On *William Penn's Treaty with the Indians*, see Neumeyer, loc. cit., and the exhibition catalogue, *Benjamin West, 1738–1820*, Philadelphia Museum of Art (1938), with a preface by Fiske Kimball.

[17] *The St. James's Chronicle*, 9–12 June 1781; see W. T. Whitley, *Artists and their Friends in England 1700–1799*, I (1938), pp. 356 ff., and now Prown, op. cit., pp. 284 f., for details of the exhibition and critical comments on Copley's painting.

[18] In fact, the greater part of West's *œuvre* is devoted to the grand style in its classical or fantastic form. His *Apotheosis of Nelson* (Fig. 102) even bears a deceptive resemblance to the anti-historical school of 'spiritual portraiture' (see below, p. 99). Although its sober frenzy is more akin to the 'enthusiasm' of Carter's *Apotheosis of Garrick* (see above, p. 41) than to the obsessions of Barry and Blake, the oscillation in West's performance between the historical and the fantastic style is sufficiently rich to deserve closer study. [On West's religious paintings see Grose Evans, *Benjamin West and the Taste of his Times* (1959).]

portrait of Lord Heathfield, the defender of Gibraltar (Fig. 22),[19] the hero was placed against a background filled with cannon-smoke, holding the key to the fortress in his hand; but he was not represented as actually engaged in directing the battle. This literal form was chosen by Copley. As a piece of history painting his *Siege and Relief of Gibraltar*[20] is far more ambitious than Reynolds's portrait. Yet as a piece of portraiture it is less heroic. There is an element of genre-like intimacy in Copley's manner of composing the battle scene as a group portrait in which the general is the most prominent of a number of figures who all 'posed' for their portraits.

The ingredients of this type of painting are more complex than would appear on the surface. Two of its historical antecedents have been mentioned: the documentary painting of distant marvels (*mirabilia*), which enters into history painting as well as into stage design; and the conversation piece, which, whenever heightened by solemn pathos or an exotic setting, passes almost imperceptibly from portraiture to history painting.[21] The group portraits painted by Zoffany in India illustrate and confirm both points.[22] As in West's picture *William Penn's Treaty with the Indians*, the exotic setting tends to increase the documentary interest, and surrounds the portrait figures with an aura of distant marvels. When the event commemorated is a formal act of state, such as *William Watts negotiating the Treaty of Bengal*,[23] the representation assumes quite definitely the form of history painting—even though it was meant to serve as a token of private remembrance.

Zoffany was in sympathy with West's and Copley's attempts to paint contemporary history in a grand style. He himself tried his hand at it in his *Death of Captain Cook*,[24] which again combines heroic pathos with an exotic setting, in a manner closely related to Loutherbourg's stage designs.[25] But Zoffany seems to have been by nature too straightforward to attempt the sentimentalism of West, and too amiable to achieve the pointed effects of Copley. Adaptable though he was, he could not acquire the robustness of approach that transforms a portrait figure into a dramatic slogan. A faithful chronicler of Garrick's stage before the advent of Loutherbourg, he never quite adjusted his conventional style to the aggressive fervour of the reformers. For that very reason his East Indian pictures, like West's *Death of General Wolfe*, are also valuable as a test case. They confirm the view that one of the sources of modern history painting lies in the painting of *mirabilia*. They also show how much the conversation piece prepared the way for modern history

[19] See above, pp. 31 f.

[20] Tate Gallery, London.

[21] The gestures in *Penn's Treaty* by West (Fig. 94) are typical of the English conversation piece. It is imprudent, therefore, to quote them (as Neumeyer does, loc. cit.) in support of the thesis that the composition of the picture derives from Masaccio.

[22] Zoffany was in India from 1783 to 1789. This phase of his work has not received the attention it deserves, although much material has been collected by Lady Victoria Manners and Dr G. C. Williamson, *John Zoffany, R.A. His Life and Works* (1920), pp. 80–113. [The accuracy of the research in this volume has been challenged recently by M. Webster in her catalogue of an exhibition held at the National Portrait Gallery, *Johan Zoffany, 1733–1810* (1977). On Zoffany in India, see pp. 74 ff.; also Webster's article in *Country Life*, CLIII (1973), pp. 588 f.]

[23] Reproduced in Manners and Williamson, op. cit., following p. 112, and in S. C. Hill, *Bengal in 1756–1757*, II (1905), following p. 399.

[24] National Maritime Museum, Greenwich.

[25] See above, p. 91.

painting, and how closely this new branch of art was related to the revolution in stagecraft. At the same time, they prove decisively that the new species did not come into existence until these elements were successfully merged in a formula which had the direct appeal of a heroic slogan. The ingredient lacking in Zoffany's pictures is one that Copley's paintings inherited from a species generally regarded as an inferior branch of art: the 'news'-picture.

'Pictorial news' is not an invention of the twentieth century. Executions of state criminals and ordinary convicts have always aroused the classic emotions of pity and terror[26] and created a demand for pictorial commemoration in addition, and often in preference, to that created by less sensational affairs of state. When the Jacobite conspirator Simon Lord Lovat was transferred to London before his execution in 1747, Hogarth visited him *en route* to draw his portrait, from which he made an etching and an oil sketch.[27] He did the same honour to Sarah Malcolm, a very ordinary murderess. Nor was Hogarth the first to draw pictures of common criminals. Sir James Thornhill, his teacher and father-in-law, had excelled in this type of picture long before Hogarth ever attempted it. In 1724 (nine years before Hogarth's *Sarah Malcolm*) he had drawn for an engraving a portrait of the criminal Jack Shepherd, whom he visited in prison. It is also on record that, while Hogarth sketched the face of Sarah Malcolm, he conversed with Thornhill who was sitting beside him in the prison cell and, presumably, also drawing.[28]

This is the same James Thornhill who painted the grandiloquent frescoes at Greenwich and Hampton Court, and decorated the cupola of St. Paul's with scenes from the life of the saint. His sketch-book in the British Museum shows allegorical designs in the grand classical style on the same page as intimate sketches in the 'truly Dutch manner'. Both coexisted in his mind without interfering with each other; just as his greater contemporary, John Vanbrugh, could plan and build heroic castles and at the same time write licentious comedies. If we think of Thornhill only as the painter of the large frescoes at Greenwich and Hampton Court, the fact that Hogarth was his pupil seems puzzling. But the relation between the two becomes very plain if we observe Thornhill's predilection for that type of intimate character study which Hogarth defined as 'comedy painting'. This province, for Thornhill a subject of unofficial interest, became the official field of Hogarth's occupation. His intense exploration of it brought to light treasures that Thornhill's more dispersed vision could not discern. However, the difference in quality should not blind us to the affinity of style.

In this common medium of intimate portraiture, Thornhill and Hogarth produced—jointly and separately—a number of topical group pictures which ought to be regarded as *incunabula* of history painting. Some of them commemorate the sittings of parliamentary committees, the most famous being Hogarth's *A Committee of the House of Commons* or *The*

[26] In his dissertation *Of the Sublime and Beautiful*, 2nd edn. (1759), I, XV, Edmund Burke reflects on the weakness of even the best tragic performances on stage as compared with a public execution.

[27] R. Paulson, *Hogarth's Graphic Works*, I (1970), no. 166.

[28] See now Paulson, *Hogarth. His Life, Art, and Times*, I (1971), pp. 307 ff.

Bambridge Committee, of which there is a dramatic sketch in the Fitzwilliam Museum, and a formalized portrait drawing in the National Portrait Gallery.[29] The hero is a criminal, a malicious gaoler who had ill-treated the prisoners in his charge. The painting is a memorial to the Committee that enforced prison reform. Another picture, painted jointly by Thornhill and Hogarth, represents *The House of Commons*, with the Prime Minister, Sir Robert Walpole, and Speaker Onslow as the chief figures.[30]

Whether or not these pictures can be called conversation pieces depends largely on what is understood by that term. Inasmuch as they represent portraits in a commemorative setting on a small scale, one might so describe them. On the other hand, they lack one trait which ought perhaps to be included in a complete definition of the conversation piece: the privacy of the setting. They represent the transition from the private group portrait to a form of pictorial *reportage* that needs only a more dramatic (preferably tragic) subject and a larger scale to become history painting in contemporary costume. It is obvious that this dramatic type of painting is foreshadowed more clearly in the representation of criminals than of respectable, well-behaved citizens. It is therefore neither exaggerated nor cynical to say that Copley's historical paintings of contemporary events conferred on the national hero a type of honour which had hitherto been conceded chiefly to the criminal.[31]

There is evidence that Thornhill, who made the commemoration of criminals one of his hobbies, refused to chronicle a great contemporary event of state in a strictly historical style. When commissioned to paint *The Landing of George I* for the Painted Hall at Greenwich, he was forced by circumstances to consider those problems which West and Copley were to set themselves deliberately. The event to be glorified lay within the memory of living men. Was it permissible to assert historical truth at the expense of heroic grandeur? On the other hand, could heroic grandeur be presented with an air of probability if it deliberately ignored those features which were commonly known to be historical truth?

The course of Thornhill's deliberations is preserved in a sketch in the British Museum (Fig. 98).[32] It is inscribed with a list of *Objections wh. will arise from ye plain representation of ye K[ing]. landing Sepr. 18th*, 1714, *as it was in fact and in ye modern way and dress*. They show the mind of a conscientious chronicler who has taken pains to ascertain the exact historical truth before deciding to deviate from it for reasons of aesthetics and policy:

I *First of all it was Night, wch to represent would be hard and ungracefull in Picture. No ships appearing and boats make a small figure.*

[29] Ibid., pp. 197 ff.

[30] The National Trust, Onslow Collection, Clandon Park, Surrey, see ibid., pp. 200 f. A monograph on Thornhill and a *catalogue raisonné* of his work are urgently needed. See *DNB*, s.v. Thornhill [and now W. R. Osmun, 'A Study of the Work of Sir James Thornhill' (Ph.D. Thesis, University of London, 1950); E. Croft-Murray, *Decorative Painting in England, 1537–1837* (1962), and E. de N. Mayhew, *Sketches by Thornhill in the Victoria and Albert Museum* (1967)].

[31] It is worth remembering in this connection that the growing taste for patriotic self-advertisement was watched by enlightened men of the period with dismay. 'Der Patriotismus verdirbt die Geschichte' was Goethe's verdict. Dr Johnson was even more outspoken: 'Patriotism is the last refuge of a scoundrel.' On the relation of hero-worship to criminal-worship, see Wind, 'The Criminal-God', *Journal of the Warburg Institute*, I (1938), pp. 243 ff.

[32] L. Binyon, *Catalogue of Drawings by British Artists and Artists of Foreign Origin working in Great Britain preserved . . . in the British Museum*, IV (1907), p. 182, no. 23.

2 *Then who shall be there to accompany him, if the Real Nobles that were there, then, some of them are in disgrace now & so will be so much Party in Picture.*

3 *To have their faces and dresses as they really were, difficult.*

4 *The King's own dress then not gracefull, nor enough worthy of him to be transmitted to Posterity.*

5 *There was a vast Crowd, wch to represent would be ugly and not to represent would be false.*

He decides, therefore, to *Take Liberty of an evening sky, & torches, . . . to bring in ye yatch* [sic] *that brought over ye King and ye Barg, & Guns firing, etc.*, to *Make only 5 or 6 of ye chief Nobles, ye rest in ye crowd*, to *Inquire their dresses ye best you can & get their faces from ye life*, to make the King's dress *as it now is & should have been rather than what it was*, and to *Take ye Liberty to lessen ye crowd as they ought to have been then.*

Thus Thornhill, the adept of the grand style, argues with Thornhill, the chronicler of contemporary events. It is not surprising that he rejected the alternative which was to become fashionable fifty years later.[33] The remarkable point is that he contemplated it at all. His normal way of representing such a subject is seen in his sketch for the *Landing of the Prince of Orange at Torbay* (Fig. 97).[34] The formal pattern is almost literally the same, but the semi-historical setting is replaced by a purely allegorical one: Neptune and the other gods of the sea deliver up their charge and entrust the future king to Cybele, goddess of the Earth, and Britannia, who together receive him.

It appears (from another sketch in the British Museum) that a similar allegorical plan was ultimately adopted also for *The Landing of George I.*[35] Even the guarded realism of Thornhill's first compromise seems not to have answered all the 'objections'. No doubt, his final version was calculated to satisfy even the most ambitious and conscientious tastes. According to his own inscription on the sketch, the allegory consists of Father Thames and the Nymphs or Graces welcoming the king; Fame flying before him, carried by Zephyr; Apollo and the Muses singing the king's praise; Tyrannic Power and Despair fleeing before Liberty; Liberty, Religion, Commerce, Courage and Justice introducing the king on a 'triumphant Chariot'; Princely Prudence and Britannic Power protecting him. Thornhill, who was so anxious to avoid 'Party in Picture', seems to have regarded the eagle on the chariot as a particularly subtle stroke of diplomacy: *Shews ye Power of Jove, as well as ye German Ensign*—so it would please those favouring the Hanoverian succession without offending others who might not wish to think of it.

The fact that Thornhill, about 1720, considered and rejected the form of history painting which West and Copley were to introduce with signal success between 1770 and 1780 makes it evident that the problem had occupied men's minds for some time before the right artists appeared at the right moment to bring the new type into existence. The question of why West and Copley found the task more congenial than Thornhill, and why the period of 1770

[33] Thornhill's designs must have been made after 1714, the date of George I's landing, and before 1727, when Thornhill ceased working in Greenwich.

[34] Binyon, op. cit., p. 182, no. 22.

[35] Ibid., no. 24b.

to 1780 was more propitious than that of 1720, is answered by two remarkable facts which hold the key to the historical situation. First, the artists who produced and propagated the new type were without exception Americans; and second, the period was that of the American fight for independence, a struggle which divided public opinion in England itself with regard to colonial policy. As a claim to democratic self-government, the American cause had many English sympathizers who vigorously criticized their own government and frankly favoured the 'rebels'. The prominence given to West and Copley at the very time when the struggle was at its height shows that the position of an American in London, which one would have expected to be somewhat precarious, was socially advantageous if not positively glamorous.[36] Neither West nor Copley made any secret of their democratic convictions and American loyalties. There is a rather pleasant anecdote about Copley. After he had heard the king's speech acknowledging American Independence, he returned home and, having on his easel the portrait of an American lady with a ship in the background, he at once decorated the mast with the new American flag.[37] His great picture *The Death of the Earl of Chatham* (Fig. 99) is also related to the American cause, for it represents the dramatic moment when the Duke of Richmond requested the British Government to take all measures necessary to end the American war and concede the demand for independence—a motion the Earl of Chatham was prevented from opposing by the stroke that ended his life.[38]

West was even more sensational in his democratic utterances, especially after he had succeeded Reynolds as President of the Royal Academy. When, after the Peace of Amiens, he went to Paris to visit the Musée Napoléon, he sought an interview with Napoleon himself and advised him to follow the example of George Washington who had also 'liberated a continent', a remark which West's English colleagues found so out of keeping with his position as President of the Royal Academy that some thought fit to oppose his re-election. He resigned, but was returned to office the following year.[39]

In addition to being a loyal democrat, West was descended from Pennsylvania Quakers. He therefore had his own quaint form of solemnity and found it quite in keeping with his dignity as a man to become the favourite court painter of the very king from whose tyranny he had been liberated. By accepting the Academy's highest office, he showed himself immune from the academic scruples he had already ignored in his paintings. His ways were always more plain and mellow than those of Copley. He had begun his career as a prodigy—a child who came from the distant woods of Pennsylvania and could paint. The type of interest he aroused, first in Italy and then in England, is well illustrated by his interview

[36] This is not disproved by the exceptional case of Trumbull, who was arrested on a suspicion of treason but soon released. Trumbull fought actively in the American War of Independence, and to judge by his connections in Paris it is not impossible that the British authorities were justified in their suspicions. On the question as a whole, see L. Einstein, *Divided Loyalties. Americans in England during the War of Independence* (1933).

[37] Cf. Isham, op. cit., p. 26; also *Dictionary of American Biography*, s.v. Copley.

[38] In this picture Copley retained an incident which caused much comment at the time. When Chatham collapsed, the Earl of Mansfield, his inveterate opponent, remained seated while the other Members rose. Copley, whose own attitude was defined by his American loyalties, glorified the Earl in a separate portrait of great vigour (National Portrait Gallery, London). See now Prown's discussion of this puzzling matter, op. cit., p. 285.

[39] W. Sandby, *The History of the Royal Academy of Arts*, I (1862), pp. 267 f.

with the blind Cardinal Albani, who asked whether the boy's skin was really as fair as his own—a question that caused great amusement because, while West's complexion was 'of more than the usual degree of English fairness', Albani's was 'of the darkest Italian olive'.[40]

The exotic charm of West's personality explains the otherwise quite unaccountable fact that a painter of his very slight powers should have won immediate and lasting success, not only with the public but also with his far more gifted colleagues and competitors.[41] West's very person appealed to them as a living token of those distant marvels they so admired in some of his pictures. It was natural for Reynolds to permit him to deviate from the academic standards which he considered binding for himself. West had a shrewd sense of his own privileged position. He reinforced it by inducing Copley to come over from America; and Copley in his turn was followed by Trumbull.[42] Thus the final breach with the Academic rules of history painting was produced by the impact of democratic ideas proclaimed by a group of American artists.

The relation of the American to the French Revolution has been the subject of much discussion. A similar study ought to be made of the relation between American and French revolutionary painting.[43] Taken in the abstract, the affinity between them is obvious. David's *Death of Marat* is perhaps the boldest attempt ever made to combine the grandeur of tragedy with the intimacy of portraiture and fuse both in a striking piece of *reportage*. Yet the medium of amalgamation was the exact opposite of that employed by West: it was not the Romantic naturalism of distant marvels nor the solemn pageantry of crowded acts of state, but the ancient Roman ideal of noble simplicity, that West had rejected as insufficiently topical. It was not until the Napoleonic and post-Napoleonic period that the two types of revolutionary painting coalesced. When West visited Paris in 1802, he was hailed (like Trumbull who had preceded him) as a representative of the *avant-garde* of history painting. The most impressive testimony to the influence of their style is found in the Romantic revolutionary paintings of Delacroix. His *Massacre at Scio* expresses in a contemporary exotic setting the same demand for liberation as his *Liberty guiding the People* proclaims in the direct language of naturalism. For sheer power and beauty these paintings are as much superior to those of the American School as Baudelaire's translations from Longfellow are to the original *Hiawatha*. Of course, other sources would have to be taken

[40] Galt, *The Life and Studies of Benjamin West, Esq. President of the Royal Academy of London, Prior to his Arrival in England; compiled from materials furnished by himself* (1816), p. 103.

[41] His case is similar to that of Angelica Kauffmann, who also began her career as a prodigy. She made her name in rural Schwarzenberg when at the age of sixteen she painted frescoes, after designs by Piazzetta, for the parish church. Her vogue in Rome coincided with West's. A portrait drawing inscribed *Mr. West / Drawn at Rome by / Angelica Kaufmann / 1763* is a document of their encounter (National Portrait Gallery, London, Inv. No. 1649).

[42] [For a bibliography on Trumbull, see the 1950 edn. of Theodore Sizer's *The Works of Colonel John Trumbull*, and for all important publications since 1950, see I. B. Jaffé, *John Trumbull, Patriot-Artist of the American Revolution* (1976). See also Sizer's edn. of *The Autobiography of Colonel John Trumbull. Patriot-Artist* (1953).]

[43] The presuppositions for it are given in the remarkable studies by Locquin, cited above (see also 'Le retour à l'antique dans l'école française avant David', *La Renaissance de l'art français et des industries de luxe*, V, 1922, pp. 473 ff.), and by Fiske Kimball, 'Les influences anglaises dans la formation du style Louis XVI', *Gazette des beaux-arts*, L (1931), pp. 29 ff. and 231 ff.

into account for a full interpretation of Delacroix's paintings. But it is remarkable how many of these sources are English. The theme of the *Massacre at Scio*, the landscape of which was repainted after Delacroix had seen Constable's pictures in the Salon of 1824,[44] was certainly inspired by Byron's poetry, which Delacroix greatly admired.

But the Romantic period, while bringing history painting into full bloom, also produced its most virulent antidote. A picture such as Blake's *Spiritual Form of Nelson guiding Leviathan*[45] is a powerful indictment of any attempt to honour a hero by dramatizing his earthly adventures. In his conception of 'spiritual portraiture' Blake followed the precedent of James Barry, whom he placed above all other painters of the eighteenth century. After West exhibited his *Death of General Wolfe*, Barry painted in protest a canvas in which the same theme was treated in a 'spiritual form', all figures being represented as nudes.[46] A relentless worshipper of the sublime, Barry had no patience with West's mitigated realism and its popular appeal. It is interesting, therefore, that his sponsor and patron was Edmund Burke, the author of the dissertation *Of the Sublime and Beautiful*, who, though originally an eloquent supporter of the American cause, ended his literary and political career with the violently reactionary *Reflections on the French Revolution*.[47]

With the spread of liberalism in nineteenth-century Europe and the wide acceptance of constitutional monarchy as the most effective compromise between patriotic intransigence and *laissez-faire*, the style of history painting which had been introduced as revolutionary assumed the vulgar stateliness of an official art. Today academies of painting have made a tradition of what their predecessors distrusted as democratic infringement. 'I should never have believed', wrote Baudelaire, 'that our country could march with such speed on the road of *progress*.'[48]

[44] 19 and 25 June 1824, *Journal de Eugène Delacroix*, ed. A. Joubin, I (1932), pp. 115 f. [For a reconstruction of Delacroix's debt to Constable and the complicated history of the *retouches*, see M. Florisoone, 'Constable and the "Massacres de Scio" by Delacroix', *Journal of the Warburg and Courtauld Institutes*, XX (1957), pp. 180 ff.]

[45] *c.* 1805–9, Tate Gallery, London. [Cf. M. Butlin, *The Paintings and Drawings of William Blake*, 2 vols. (1981), no. 649, pl. 876.]

[46] [With regard to the finished work, Wind (following Locquin, op. cit., pp. 154 f.) was here in error; see his later remarks on the subject (below, p. 103 note 20). The painting was long lost, and it was believed that all the figures were nudes—Barry having been attacked by contemporary criticis on this account. When Dr J. Clarence Webster discovered Barry's *Death of General Wolfe* ('A Study of the Portraiture of James Wolfe', *The Transactions of the Royal Society of Canada*, XIX, 1925, pp. 47 ff.), it was found that the painting (New Brunswick Museum, Saint John) contained only one nude figure, a dying Indian in the foreground; while the General was represented merely with his chest partly bared. This could be construed as a compromise on Barry's part, as Wind remarks in the above-mentioned note. More recently it has been argued that nineteenth-century critics did not even see the original painting (see D. Irwin, 'James Barry, and the Death of Wolfe in 1759', *The Art Bulletin*, XLI, 1959, pp. 330 ff.). A preliminary pencil study by Barry for the figures on the left-hand side of the painting reveals that he began his compositions in the academic tradition, with unclothed figures; but the drawing also has a few delicate indications of clothing (see D. B. Brown, *Ashmolean Museum Oxford. Catalogue of the Collection of Drawings*, IV, *The Earlier British Drawings*, 1982, no. 336). Perhaps critics were shocked by the General's bare chest which they equated with nudity, or perhaps they saw the picture in an unfinished state, when it was closer to the Ashmolean drawing. A scientific examination of the painting might well shed light on the problem.]

[47] I would not claim to have penetrated the very puzzling relation which connects Burke with Barry and thereby indirectly with Blake. Their political allegiances are fraught with contradictions, and their aesthetic sympathies are equally complicated.

[48] 'Projets de Préface pour *Les Fleurs du mal*', I (1859/60): 'Malgré les secours que quelques cuistres célèbres ont apportés à la sottise naturelle de l'homme, je n'aurais jamais cru que notre patrie pût marcher avec une telle vélocité dans la voie du *progrès*.' (*Œuvres de Baudelaire*, Bibl. de la Pléiade, ed. Y.-G. Le Dantec, I (1938), p. 581.)

X · Penny, West, and *The Death of General Wolfe*

THE legend that Benjamin West's *Death of General Wolfe* (Fig. 95) was the first battle piece ever painted in contemporary costume has been four times refuted within recent memory.[1] Unanimously, the blame for originating it has been put on John Galt, the first biographer of West, who gave a fanciful account of how Reynolds first questioned, and then conceded, the propriety of West's indecorous attempt 'to exhibit heroes in coats, breeches, and cock'd hats'.[2]

Poor Galt was a rhapsodic writer, a little careless in his use of the word 'innovation', and not quite trustworthy in minor details. West's *Death of General Wolfe* is remarkable for the relative absence of 'cock'd hats'. But if we allow for a tinge of poetic licence, it would seem that Galt understood the issue of the debate between West and Reynolds far better than his critics; for he was well aware that this was an argument between two Academicians, discussing whether the 'greatness' or 'dignity' of a 'heroic' subject was 'impaired' by the 'modern garb of war'—a question which, phrased in this doctrinal form, presupposes the academic hierarchy of genres. To point to Velázquez or Vasari as having preceded West as painters of battle-pieces in contemporary costume is to disregard the context of the argument. Indeed, if early instances were to be quoted, why was no reference made to Leonardo's *Battle of Anghiari*, Uccello's *Battle of San Romano*, or the 'Battle-pieces of Bourgognone' which Reynolds himself mentions as an inferior genre in his Third Discourse to the Academy?[3]

In 1769, two years before West exhibited *The Death of General Wolfe*, the Royal Academy had been opened in London with a speech by its first President, Joshua Reynolds, who discussed the rules of academic instruction. These included a commitment to the grand

[1] J. Clarence Webster, *Wolfe and the Artists. A Study of his Portraiture* (1930) [a revised version of his earlier article, see above, p. 99 note 46]; Oskar Hagen, *The Birth of the American Tradition in Art* (1940), pp. 113 ff.; Charles Mitchell, 'Benjamin West's *Death of Wolfe* and the Popular History Piece', *Journal of the Warburg and Courtauld Institutes*, VII (1944), pp. 20 ff.; Agnes Addison, 'The Legend of West's *Death of Wolfe*', *College Art Journal*, V (1945), pp. 23 ff.

[2] John Galt, *The Life and Works of Benjamin West, Esq., President of the Royal Academy of London, Subsequent to his Arrival in this Country; composed from materials supplied by himself*, Part II (1820), pp. 46 ff.

[3] *Discourses on Art*, ed. R. R. Wark (1975), III, l. 327 (hereafter in Wark's edn.).

Journal of the Warburg and Courtauld Institutes, X (1947), pp. 162–5.

style—'that grand style of painting, which improves partial representation by the general and invariable ideas of nature'.[4] The argument was resumed and expanded in the Third Discourse (1770), in which Reynolds warned his listeners against 'the neglect of separating modern fashions from the habits of nature':[5]

> . . . the whole beauty and the grandeur of art consists, in my opinion, in being able to get above all singular forms, local customs, peculiarities, and details of every kind.
>
> . . . the painter . . . must divest himself of all prejudices in favour of his age or country; he must disregard all local and temporary ornaments, and look only on those general habits which are every where and always the same. He addresses his works to the people of every country and every age; he calls upon posterity to be his spectators, and says with Zeuxis, *in æternitatem pingo.*
>
> The prejudices in favour of the fashions and customs that we have been used to, and which are justly called a second nature, make it too often difficult to distinguish that which is natural, from that which is the result of education; . . . almost every one is apt to be guided by those local prejudices, who has not chastised his mind, and regulated the instability of his affections by the eternal invariable idea of nature.
>
> Here then, as before, we must have recourse to the Ancients as instructors.[6]

In the same year in which this pronouncement was made, West, who was a member of the Academy, began to design a painting in the grand style, the hero of which was shown wearing a contemporary British uniform, thus challenging the academic rules advocated by Reynolds.[7] That the picture created a stir when it was exhibited at the Academy in 1771 is attested by Galt, and implicitly confirmed by Reynolds's own Discourse at the end of that year, when he reiterated his principles with a somewhat pointed insistence, declaring that the grand style would be degraded by the vulgarism of ordinary life in any country:

> As it is required that the subject selected should be a general one, it is no less necessary that it should be kept unembarrassed with whatever may in any way serve to divide the attention of the spectator. . . .
>
> An History-painter paints man in general; a Portrait-painter a particular man, and consequently a defective model.
>
> On the whole, it seems to me that there is but one presiding principle which regulates, and gives stability to every art. The works . . . which are built upon general nature, live for ever; while those which depend for their existence on particular customs and habits, a partial view of nature, or the fluctuation of fashion, can only be coeval with that which first raised them from obscurity. Present

[4] Discourse I, ll. 66–7.

[5] Discourse III, l. 242. See Wind, 'In Defence of Composite Portraits', below, pp. 120 ff.

[6] Ibid., ll. 103–5, 234–41, and 248–57.

[7] That West aimed at the academic grand style in this picture is proved by its general pathos, also by Mitchell's excellent observation that he adopted 'expressions' from Le Brun (1944, p. 31). [West was unrepentant in this debate, and in 1806 he was to vary and elaborate these ideas when he conceived the 'Epic Composition' of *The Death of Nelson*, an intended, though much later, companion-piece to *The Death of General Wolfe*. Cf. Mitchell, 'Benjamin West's *Death of Nelson*', *Essays in the History of Art presented to Rudolf Wittkower* (1967), pp. 265 ff.]

time and future may be considered as rivals, and he who solicits the one must expect to be discountenanced by the other.[8]

From these words one forms the impression that Galt may have exaggerated when he described Reynolds's rather rapid conversion, and how he said: 'Mr. West has conquered.'[9] At the same time, it is true that Mr West did conquer. For the type of history-piece that he introduced into the Academy against Reynolds's opposition became an authorized style of academic painting; witness the dismal succession of academic battle pieces, with all the generals painted in their correct uniforms.

It is difficult to see why the discovery of two or three paintings of *The Death of General Wolfe*, executed in 1763 and representing him in his uniform, should deprive West's *Death of General Wolfe* of the doubtful honour of having started an academic revolution. These pictures may be of interest for a study of West's pictorial sources, but as evidence against the assertion that West introduced an 'innovation' into the Academy[10] they are irrelevant, since they were painted six years before the Academy came into being.[11] Nor is the argument improved by the observation that Edward Penny, who painted two of these pictures,[12] later became Professor of Painting in the Academy. If this constitutes him a history painter, what may we not infer from the fact that Wilson was the Academy's Librarian, Turner and Constable supervisors of the Life Class? A safer guide to Penny's character as an artist is provided by the titles of his exhibits in the Academy. The following examples may suffice: *The benevolent physician*, *The rapacious quack*, *Widow Costard's cow and goods distrained for taxes, are redeemed by the generosity of Johnny Pearmain*, *A boy taken out of the water, drowned*, *The Boy by proper means recovered.* Vide *Dr Franklin's sermons to the Humane Society . . .*, *The Profligate punished by neglect and contempt*, *The Virtuous comforted by Sympathy and Attention.*[13] It is true that occasionally he grew more poetical and took a subject from Thomson's *Seasons* ('Lavinia gleaning') or even quoted Shakespeare's *King John*, though the quotation would hardly establish his title as a grand history painter:

> I saw a smith stand with his hammer, thus,
> The whilst his iron did on the anvil cool,
> With open mouth swallow a tailor's news.

To a 'little master' of such moral scenes, contemporary history offered the opportunity to

[8] Discourse IV, ll. 34–8, 414–16, and 505–13.

[9] Galt, op. cit., p. 49.

[10] Mitchell (1944), p. 21.

[11] The date of these paintings (1763) coincides with the year in which Robert Monckton, second-in-command under Wolfe during the siege of Quebec, returned from Canada to England. West himself became acquainted with Monckton and exhibited a portrait of him in 1764 (see *DNB*, s.v. West) with the Society of Artists, that is, at the very same exhibition in which Penny exhibited his *Death of General Wolfe*; see Mitchell (1944), p. 30. It is of considerable interest that upon his first contact with the subject West produced a portrait rather than a history piece.

[12] There are two versions of *The Death of General Wolfe* by Penny, one in the Ashmolean Museum, Oxford (Fig. 96), the other at Petworth. A third picture of the same subject, now lost, was painted by Romney; cf. Mitchell (1944), p. 30, also John Romney, *Memories of George Romney* (1830), pp. 45 ff.

[13] For these and the following titles, cf. Algernon Graves, *The Royal Academy of Arts*, VI (1906), pp. 101 f.

impart humane lessons. In 1772, after West had exhibited his *Death of General Wolfe*, with the picturesque Indian in the foreground (a motif he carried further in *Penn's Treaty with the Indians*, Fig. 94), Penny followed the fashion[14] and exhibited a picturesque colonial history piece, into which he managed to introduce his personal sentimental note: *Lord Clive explaining to the Nabob the situation of the invalids in India; at the same time showing him a deed whereby he relinquishes Meer Jaffier's legacy, five lacke of rupees, to the Honorable East India Company, for the support of a military fund.*[15]

One cannot help suspecting that the full title of Penny's *Death of General Wolfe*, if known, might turn out to be a similar combination of patriotic *reportage* with moral sentiment, a didactic rather than a heroic piece. The painting (Fig. 96) certainly looks like a typical example of Penny's sentimental genre: the wounded general raises his eyes to heaven as he expires, knowing that the day is won.[16] The fact that one of the two versions shows only the sentimental group, without the battle scene in the background,[17] suggests that Penny did not conceive the design primarily as a battle piece, but rather as an appealing *document humain*.

It seems, from an essay of James Barry's, that Penny was usually compared with Hogarth[18]—an aberration unflattering to Hogarth, but interesting as an estimate of Penny, since it proves that he was classed as a painter of moralistic subjects. The comparison did not, however, please James Barry, himself a successor to Penny as Professor of Painting in the Royal Academy, and he rose to his colleague's defence: 'The line of art pursued by my very ingenious predecessor and brother academician, Mr. Penny, is quite distinct from that of Hogarth, and is of a much more delicate and superior relish; he attempts the heart, and reaches it, whilst Hogarth's general aim is only to shake the sides.'[19] But even from this singular defence it would appear that Penny emerges as a sentimental painter of small subjects, an English counterpart of Greuze, and that Barry, who had himself attempted to improve on West's *Death of Wolfe* by painting the dying general as a classic nude,[20] did not admit the 'very ingenious' Mr Penny to the same class as West or himself.[21] Barry's remarks gain in interest, however, from the fact that they were read by Charles Lamb and apparently 'shook his sides':

The impression caused in me by reading this passage, was an eager desire to know who this Mr.

[14] For the fashion of the colonial history-piece, see above, pp. 89 ff.

[15] Graves, loc. cit.

[16] Is it permissible to suppose that Romney's lost picture of *The Death of General Wolfe* (see above, note 12) may have been composed in a similar vein?

[17] Mitchell (1944), p. 30 note 3.

[18] *An Account of a Series of Pictures, in the Great Room of the Society of Arts, Manufactures, and Commerce, at the Adelphi* (1783), see *The Works of James Barry, Esq., Historical Painter*, II (1809), p. 386.

[19] Ibid.

[20] While it is certainly untrue that Barry's picture, discovered by Webster (op. cit., pl. XXV), shows all the figures nude, it is equally erroneous to describe Wolfe as shown 'in modern dress'; for the coat has been partly removed from the dying hero so as to give him a semi-classical appearance. The compromise is similar to that chosen by Joseph Wilton for his monument to Wolfe in Westminster Abbey (unveiled 1773), in which Wolfe is represented nude while the attending officers wear uniforms (Fig. 101). [See above, p. 99 note 46.]

[21] Even so, this class of Penny's paintings—sentimental histories—should be added to the list of *incunabula* in 'The Revolution of History Painting' (above, pp. 94 f.).

Penny was. This great surpasser of Hogarth in the 'delicacy of his relish', and the 'line which he pursued', where is he, what are his works, and what has he to shew? In vain I tried to recollect, till by happily putting the question to a friend who is more conversant in the works of the illustrious obscure than myself, I learnt that he was the painter of a *Death of Wolfe* which missed the prize the year that the celebrated picture of West on the same subject obtained it; that he also made a picture of the *Marquis of Granby relieving a Sick Soldier*; moreover, that he was the inventor of two pictures of *Suspended and Restored Animation*, which I now remember to have seen in the Exhibition some years since, and the prints from which are still extant in good men's houses. This then, I suppose, is the line of subjects in which Mr. Penny was so much superior to Hogarth. I confess I am not of that opinion. The relieving of poverty by the purse, and the restoring a young man to his parents by using the methods prescribed by the Humane Society, are doubtless very amiable subjects, pretty things to teach the first rudiments of humanity; they amount to about as much instruction as the stories of good boys that give away their custards to poor beggar-boys in children's books. But, good God! is this *milk for babes* to be set up in opposition to Hogarth's moral scenes, his *strong meat for men*?[22]

When I see Penny's *Death of General Wolfe* cited as evidence against West's 'innovation', I may be forgiven for asking: Is this 'milk for babes' seriously proposed as a sample of eighteenth-century grand style, as an anticipation of West's case against Reynolds? The debate between Reynolds and West had a real point on which both men had reasons to feel strongly. The question was whether Reynolds was right in claiming 'that the great style is always more or less contaminated by any meaner mixture', and whether, in order to ennoble a subject, it is necessary to change 'the dress from a temporary fashion to one more permanent, which has annexed to it no ideas of meanness from its being familiar to us'.[23] Diderot, though himself in revolt against the grand manner, recognized its force as a dramatic convention and, in a dialogue written between 1770 and 1773, still hesitated to predict its demise: 'C'est une formule donnée par le vieil Eschyle; c'est un protocole de trois mille ans.'—'Et ce protocole a-t-il encore longtemps à durer?'—'Je l'ignore. Tout ce que je sais, c'est qu'on s'en écarte à mesure qu'on s'approche de son siècle et de son pays.'[24] The greatness of the issue rather than of the work seems to have secured the fame of West's *Death of General Wolfe*, a renown due less to its pictorial merit (which seems slight) than to the breach produced by its intrusion into the precincts of the grand style. But can this conflict have had any meaning for the painter of *Suspended and Restored Animation*?

Penny seems to be one of those minor artists who have the misfortune to be periodically disinterred, apparently for no other purpose than to diminish the lustre of a brighter name. That it was possible to associate him on the one hand with Hogarth, on the other with West, speaks against his having had a very marked physiognomy. And perhaps Lamb may be forgiven for not knowing that he ever painted a *Death of Wolfe*.

[22] 'On the Genius and Character of Hogarth; with some remarks on a passage in the writings of the late Mr. Barry', *The Works of Charles and Mary Lamb*, ed. E. V. Lucas, I (1903), pp. 81 f.

[23] Discourse IV, ll. 459 and 464–6.

[24] *Paradoxe sur le comédien* (1773), *Œuvres complètes de Diderot*, ed. J. Assézat, VIII (1875), p. 372.

XI · The Sources of David's *Horaces*

The First Version

A THEATRICAL experience is said to have inspired David to paint *The Oath of the Horatii*. He had seen Corneille's *Horace* produced at the Comédie-Française, with Brizard in the role of the father. His first plan, we are told, was to represent the scene in the last act where the elder Horatius defends his son, the victor over the Curiatii and the saviour of Rome, who in an excess of patriotic fervour has killed his own sister Camilla. A sketch by David in the Louvre is supposed to depict this scene (Fig. 104).[1] It shows the elder Horatius pleading with the crowds while his son, standing erect and proud at his side, is about to be apprehended by the approaching lictors. Sabina, his wife, sits weeping on the steps by the body of the slain Camilla. In the background, two judges attended by lictors watch the spectacle from their seats.

In view of the persistent assertion that this drawing is based on Corneille, it is somewhat disconcerting to find that the last act of Corneille's *Horace* is played, like the entire drama, within the quiet precincts of a private house, 'dans une salle de la maison d'Horace'; that there is no populace present to which the elder Horatius could address his discourse, no lictor who has come to seize his son, not even a tribune to convict him. The matter is decided by the king, Tullus Hostilius, who unexpectedly visits the Horatii in their house and by his august presence prompts the father to comment on the modesty of the dwelling:

> Ah! Sire, un tel honneur a trop d'excès pour moi;
> Ce n'est point en ce lieu que je dois voir mon roi.[2]

[1] 'Sa première idée', says Cantinelli in discussing this drawing (*Jacques-Louis David, 1748–1825*, 1930, pp. 15 f.), '. . . fut de porter sur la toile le cinquième acte des *Horaces* de Corneille.' The same opinion pervades the entire literature on David. See J. Locquin, *La Peinture d'histoire en France de 1747 à 1785* (1912), pp. 221 f.; J.-L. Jules David, *Le Peintre Louis David, 1748–1825. Souvenirs & documents inédits*, I (1880), p. 25; Charles Saunier, *Louis David* (n.d.), p. 27; Walter Friedlaender, *Hauptströmungen der französischen Malerei von David bis Cézanne*, I (1930), pp. 18 f. [emended in R. Goldwater's trans., 1952, pp. 14 ff.]. Most of these authors cite as their authority Alexandre Péron ('Examen du tableau du *Serment des Horaces* peint par David; suivi d'une notice historique du tableau', *Annales de la Société libre des Beaux-Arts*, IX, 1839, pp. 239 ff.), who said that his source was 'M. Debret, peintre d'histoire, parent et élève de David, qui n'avait alors que quatorze ans' (pp. 285 f.). [A. Calvert has since discovered a new series of preparatory drawings ('Unpublished Studies of *The Oath of the Horatii* by Jacques-Louis David', *Master Drawings*, VI, 1968, pp. 37 ff.), one of which predates the Louvre sketch by a year, and reveals that David's first idea for the *Horaces* was not an oath subject.]

[2] Act V, scene ii.

Journal of the Warburg and Courtauld Institutes, IV (1941), pp. 124–38.

One might argue that David, in order to give the scene more pictorial grandeur, had taken the liberty of transferring it to a public forum, and that this poetic licence in the pictorial adaptation of the text does not disprove that the picture is actually based on Corneille's play. But this convenient argument is contradicted by two observations:

1. The changes concern not only the general setting, but also the dramatic details. The work was officially listed among the projected pieces for 1783 under the title *Horace condamné*,[3] and David himself, in his description of this version, speaks explicitly of the young Horatius as 'condamné à mort',[4] whereas there is no death sentence pronounced in the play, just as there are no lictors to apprehend the culprit. His only accuser is Valerius, who loved Camilla and therefore demands that her death be avenged. But as Horatius has not been convicted, there is no need to appeal to the people for his release. From the first, the father addresses his plea to the king, who quietly pardons the son.

2. The points on which David deviates from Corneille are to be found in Charles Rollin's *Histoire romaine* (1738) from which every detail in David's design can be explained.[5] The two judges in the background, attended by lictors and watching the spectacle from their seats, are the *duumviri* to whom the king consigned the case. They have pronounced the death sentence against which the elder Horatius appeals to the people:

> Le Roi, qui ne voulait pas prendre sur lui les suites d'une affaire si odieuse, en laissa la connaissance aux *Duumvirs* qu'il nomma pour cet effet. Ils ne purent s'abstenir de condamner le coupable à mort, le crime étant manifeste. Déjà le Licteur se mettoit en devoir d'exécuter la sentence, & le supplice auroit suivi de près son triomphe, si le père d'Horace, s'avançant dans l'assemblée, n'eût pris la défense de son fils. . . . Il conclut, en déclarant qu'il en appelloit au Peuple. . . . Le Peuple ne put tenir ni contre les larmes du père, ni contre la constance du fils. . . . Il le renvoia donc absous.[6]

It still remains possible that a theatrical performance of Corneille's play first stirred David's imagination. In that case the play served as a catalytic element, but it certainly did not supply the argument for the sketch, nor did it suggest the pictorial setting. To say that, on his return from the theatre ('rentré chez lui'), David spontaneously sketched ('jeta sur le papier')[7] a scene like the one he had witnessed on the stage is to distort Corneille and to ignore the real source from which David drew his material. At best one may suppose that, excited by Corneille, he came home and re-read his Rollin. Ultimately it was Rollin, not Corneille, after whose words he shaped his image.

The difference between the two sources is considerable. In Corneille, the final decision over the life of the hero rests with the king. In Rollin, it is placed in the hands of the people. By following Rollin rather than Corneille, David asserted his democratic convictions. It is

[3] Cf. Locquin, loc. cit., and pp. 46 f. note 6.

[4] His full title reads: *Horace, vainqueur des trois Curiaces, condamné à mort pour le meurtre de Camille, sa sœur, défendu par son père au moment où les licteurs l'entraînent au supplice, et absous par le peuple touché de ce spectacle et du grand service qu'il vient de rendre à sa patrie.*

[5] Rollin's *Histoire romaine* was the most popular text-book of Roman history in David's youth. In the account of the story of the Horatii it follows Livy (I, 23 ff.) more closely than Corneille had done. [On Livy rather than Rollin as David's source, see now L. D. Ettlinger, 'Jacques-Louis David and Roman Virtue', *Journal of the Royal Society of Arts*, CXV (January 1967), pp. 105 ff.]

[6] Rollin, *Histoire romaine*, I (1738), I, ii. 3, pp. 166 ff.

[7] Locquin, op. cit., p. 221.

difficult to imagine that he ever read with approval the lines of Corneille's *Horace* which deny that the *vox populi* is the voice of the gods:

> Et la voix du public n'est pas toujours leur voix;
> Ils descendent bien moins dans de si bas étages
> Que dans l'âme des rois, leurs vivantes images,
> De qui l'indépendante et sainte autorité
> Est un rayon secret de leur divinité.[8]

Even less can David have been in sympathy with Corneille when, in the fifth act of *Horace*, which is supposed to have inspired David's drawing, he makes the elder Horatius conclude his great speech by addressing these words of advice to his son:

> Horace, ne crois pas que le peuple stupide
> Soit le maître absolu d'un renom bien solide:
> Sa voix tumultueuse assez souvent fait bruit;
> Mais un moment l'élève, un moment le détruit; . . .
> C'est aux rois, c'est aux grands, c'est aux esprits bien faits,
> A voir la vertu pleine en ses moindres effets:
> C'est d'eux seuls qu'on reçoit la véritable gloire;
> Eux seuls des vrais héros assurent la mémoire.[9]

The vigour of David's first sketch, far from reflecting Corneille's convictions, seems inspired by a protest against them. It shows David, seven years before the Revolution, as a champion of popular justice. The moral of the scene in which the hero's father appeals to the people against the *duumviri* could be summarized by the title that Rollin gave to his account of this episode: *Le peuple sauve Horace.*

When, after the final defeat of Napoleon, the aged David was exiled by the Bourbons and took refuge in Brussels, his disciples in Paris used to assemble regularly for a dinner in honour of their absent master.[10] For one of these occasions his pupil Debret made a lithograph which he distributed to the guests. It was a reproduction of David's sketch for the first version of the *Horaces*, the *Horace condamné*. Apparently the condemnation of the painter was seen by his disciples in the same light as that of his hero: 'Horace fut condamné par les Duumvirs, mais il fut absous par le peuple.'[11]

The Oath

If it was a mistake to claim that the first version of David's *Horaces*—the *Horace condamné*—derives directly from Corneille, it is even more confusing to repeat this claim for

[8] Act III, scene iii.

[9] Act V, scene iii.

[10] Cf. Saunier, loc. cit.

[11] Voltaire, in his preface to Corneille's *Horace* (*Théâtre de Pierre Corneille, avec des commentaires* [par Voltaire], *&c. &c. &c.*, II, 1764), attributes these words strangely enough to Corneille himself: 'On sait que quand on le menaça d'une seconde critique sur la tragédie des *Horaces*, semblable à celle du *Cid*, il répondit: "Horace fut condamné par les Duumvirs, mais il fut absous par le peuple."' But is it likely that Corneille, even for the sake of a *bon mot*, would have defended his play by quoting a version that contradicts it?

the final version, *Le Serment des Horaces* (Fig. 105). To say that, in choosing this theme, David 'decided to represent another and more precise scene from Corneille's drama'[12] is a strange error of fact; for in the entire *Horace* of Corneille there is no scene of an oath. Nor does Corneille ever suggest such a scene by implication. When the young Horatius, the hero of the play, takes leave of his father, he is not accompanied, as in David's painting, by his two brothers (they never appear on stage at all), but by Curiatius, the opponent whom he must fight, although he is his friend and betrothed to his sister Camilla. The father dismisses them both very briefly:

> Allez, vos frères vous attendent;
> Ne pensez qu'aux devoirs que vos pays demandent.

And when Curiatius lingers, he ominously commends him to his fate:

> Faites votre devoir, et laissez faire aux dieux.[13]

Throughout this scene of departure, the presence of Curiatius, the most amiable character in the play, tempers the heroic abandon of the Horatii and provides a tragic undertone. The situation on the stage is psychologically far more precarious and subtle than in David's painting, where the tender and conflicting emotions are represented only by the sobbing women. With the opponent present in the role of a friend, the elder Horatius has to express his patriotic hopes within the bounds of propriety. The dedication of the sword, the ritual of the oath, even if they were demanded by the traditional story, are hardly permissible in the presence of the victim.

In fact, no ancient author seems to mention the oath. It does not appear in Livy, Valerius Maximus, or even in Dionysius of Halicarnassus, although Dionysius goes out of his way to give a very elaborate account of how the sons ask their father's permission to fight, whereupon he places the decision once more in their hands and they themselves decide in favour of combat:

> Their father greatly rejoiced to find them in this disposition; and lifting up his hands to heaven, said he thought himself under a great obligation to the gods for having given him children of such worth, and bravery. Then, taking each of them in his arms, and embracing them in the most tender manner, 'You have my consent, also', says he, 'my brave children, go to Tullus, and return him both a pious and a generous answer.' The youths went away pleased with the exhortation of their father; and, going to the king, they accepted the combat.[14]

Here is every imaginable detail except the oath and swords of David's painting. And as the oath is not mentioned by the ancient authors, it is not recorded either in Rollin's *Histoire romaine*. The text from which the first version of the *Horaces* was derived is certainly not the source of the second.

[12] Friedlaender, loc. cit. [but see above, p. 105 note 1]; see also J.-L. Jules David, op. cit., pp. 25 ff.

[13] Act II, scene viii.

[14] *Antiquitates romanae*, III, 17; see *The Roman Antiquities of Dionysius of Halicarnassus*, translated into English, with Notes and Dissertations, by Edward Spelman, II (1758), p. 36.

Did David then invent the scene? This is improbable in view of the fact that, when he announced his picture as *Le Serment des Horaces*, everyone seems to have understood what he meant.

Noverre's Horaces—*Garrick's* Roman Father

A partial clue to the problem is to be found in the configuration of the three swords held in David's painting by the elder Horatius. This configuration occurs in Corneille's play, but in an entirely different scene and in connection with a different character. When Horatius, returning victorious, has the tragic encounter with Camilla, in which she provokes and insults him so bitterly that he draws his sword against her, he is accompanied by Proculus, who carries the three swords of the Curiatii (Livy's *trigemina spolia*[15]) as a symbol of victory. Corneille's own stage direction says: 'Procule porte en sa main les trois épées des Curiaces.' The unmotivated introduction of this character was criticized by Voltaire: 'D'où vient ce *Procule*? à quoi sert ce *Procule*, ce personnage subalterne qui n'a pas dit un mot jusqu'ici?'[16] The answer is that he serves as a herald of victory, raising up the three swords of the vanquished Curiatii at the very moment when the victor dishonours himself by drawing his sword against his sister. The tragic irony of this scene made it the most telling episode of the play to Gravelot. He chose it for his frontispiece to the edition of 1764 (Fig. 103) which clearly shows the three swords in the hand of Proculus, the figure in the centre.

Voltaire was offended by the introduction in the fourth act of a character who had not been seen or heard before. Yet Proculus was to perform the most impressive pantomimic gesture of the whole play. There can be no doubt that there is a connection between the three swords in the hand of Proculus and the three swords which David placed in the hands of Horatius. To be sure, Voltaire's criticism would hardly be met by doubling the scene to which he objected and producing on stage two parallel tableaux, one preceding the battle and showing Horatius raising the three swords of his sons, the other following the battle and showing Proculus raising the three swords of their vanquished opponents. Yet this type of symmetrical balance might have appealed to a producer revising the whole drama not from a literary, but a purely pantomimic, point of view; and this was the case with Jean-Georges Noverre, who, on 21 January 1777 in Paris, produced *Les Horaces* for the first time as a pantomime, a *ballet tragique*.[17]

Noverre is remembered in the history of music as the ballet-master who staged the

[15] Livy, I, 26.

[16] Note (h) to Act IV, scene vi of Corneille's *Horace*, an additional note which appears in the 2nd edition of the *Théâtre*, II (1774), with Voltaire's commentary.

[17] Émile Picot, *Bibliographie Cornélienne (1876), no. 1122.* The scenario is described by Grimm in *Correspondance littéraire, philosophique et critique par Grimm, Diderot, Raynal, Meister, etc.*, XI (1879), pp. 411 ff. [An earlier pantomime than Noverre's, based on Act IV of Corneille's *Horace*, was written by the Abbé Jean-Joseph Mouret for the fourteenth of the *Grandes Nuits de Sceaux*. For a contemporary description of these spectacles, see Nicolas de Malézieu, and the Abbés Charles-Claude Genest and Guillaume Amfrye de Chaulieu, etc., *Les Divertissemens de Seaux* (1712). Miss Anne Marie Meyer drew my attention to this early pantomimic version of the *Horaces*.]

compositions of Gluck.[18] Paralleling Gluck's reform of the opera and vying with him in his austere admiration for the antique, Noverre revolutionized the traditional style of dancing by introducing the *ballet d'action*. Originally a master of '*coloratura*' ballet and celebrated for his dazzling productions in that medium, he discovered new and purely dramatic potentialities for the dance when, on his visits to England at Garrick's invitation, he observed Garrick's mastery of 'dumb show'.[19] It could be inferred from his letters on Garrick addressed to Voltaire[20]—although he may have been exaggerating—that Garrick was the first to arouse in him an interest in dramatic pantomime. In 1751, four years before Noverre's first visit to England, Diderot had published his *Lettre sur les Sourds et Muets à l'usage de ceux qui entendent et qui parlent*, in which he recommended studying an actor's art by shutting one's ears to his declamations and watching only his gestures. This was the first glimpse of Diderot's theory of pantomime as a visual language related to painting and more immediate in its emotional appeal than the spoken word.[21] Noverre can hardly have been unaware of this doctrine. When he saw Garrick act, he probably felt what Diderot himself discovered later: that here was the perfect embodiment of his ideal of an actor who has mastered the language of gesture. For the benefit of his foreign friends whose imperfect knowledge of English would prevent them from enjoying the subtlety of his diction, Garrick frequently condescended to give a purely mimed performance of a character or of a sequence of passions. There are records of the enthusiasm he aroused in the Paris *salons* by exhibiting this virtuosity.[22] But Noverre, whose knowledge of English was presumably equally rudimentary, had the benefit of his ignorance, so that when he saw Garrick at

[18] He shared this function with his rival Angiolini. An excellent account of Noverre is given in Martin Cooper, *Gluck* (1935), 'The Reform of the Ballet', pp. 79 ff. For fuller details, see C. E. Noverre, *The Life and Works of the Chevalier Noverre* (1882); H. Abert, *Noverre und sein Einfluß auf die dramatische Ballettkomposition* (1909); and H. Niedecken, *Jean Georges Noverre, 1727–1810. Sein Leben und seine Beziehungen zur Musik* (Halle, 1919), which contains (pp. 45 ff.) a musical analysis of the score of *Les Horaces*. Noverre chose as his collaborator a minor composer, Josef Starzer (see Lisebeth Braun, 'Die Ballettkomposition von Josef Starzer', in *Studien zur Musikwissenschaft*, XIII, 1926, pp. 38 ff.). On Noverre's reform of the ballet, see his own *Lettres sur la Danse, et sur les Ballets* (1760; trans. by Lessing in 1769), and his *Lettres sur les Arts imitateurs en général, et sur la Danse en particulier*, 2 vols. (1807). [Anita Brookner's recent monograph, *Jacques-Louis David* (1980), contains a perceptive evaluation of the various art-historical theories about *The Oath of the Horatii*. She sees Wind's article as marking the turning-point in the study of David's painting (cf. pp. 71 ff.), but questions whether Noverre's ballet (discussed here in the following pages) could have influenced David, as it was an elaborate jewelled production in contrast to the Spartan nature of the painter's imagery. Furthermore, David's contemporaries failed to make the connection. Wind himself remarked this fact; see below, note 45.]

[19] See F. A. Hedgcock, *A Cosmopolitan Actor. David Garrick and his French Friends* [1912], which includes a chapter on Noverre's relations with Garrick (pp. 127 ff.).

[20] See letters XIV ('À Voltaire, sur Garrick') and XV ('Sur le même sujet') in *Lettres sur les Arts imitateurs*, op. cit., II, pp. 180–213. These letters are an expansion of a section on Garrick in letter IX, in *Lettres sur la Danse, et sur les Ballets*, pp. 195 ff., cf. especially pp. 209 ff.

[21] Cf. Diderot's *Discours sur la poésie dramatique* (1758), XXI, 'De la pantomime'. Starting from the priority of the visual impression, even in the spoken drama ('Je vois le personnage; soit qu'il parle, soit qu'il se taise, je le vois; et son action m'affecte plus que ses paroles'), he leads up to his postulate: 'Appliquez les lois de la composition pittoresque à la pantomime, et vous verrez que ce sont les mêmes. . . . De quel secours le peintre ne serait-il pas à l'acteur, et l'acteur au peintre?' (*Œuvres complètes de Diderot*, ed. J. Assézat, VII, 1875, pp. 380 and 385.) For Diderot's relation to Garrick, see R. L. Cru, *Diderot as a Disciple of English Thought* (1913). Noverre's admiration for Diderot is evidenced by his *Lettres sur la Danse, et sur les Ballets*, op. cit., where the last letter (XV) ends with a glorification of Diderot (pp. 462 ff.). See also in Noverre's *Lettres sur les Arts imitateurs*, II, pp. 47 f., the clear restatement of Diderot's doctrine: 'La scène muette est épineuse, c'est la pierre de touche de l'Acteur' (cf. *Lettres sur la Danse*, p. 474).

[22] Cf. Hedgcock, loc. cit.

Drury Lane he did not require any artificial abstractions to see 'pure pantomime' in a spoken performance.[23]

Given Noverre's admiration of Garrick[24] and the importance he attributes to their collaboration for the development of his own pantomimic style, there can be no doubt that he was acquainted with the English adaptation of Corneille's *Horace* which had been made for Garrick by William Whitehead, and which was at the height of its success when Noverre visited London.[25] The play was called *The Roman Father*, and Garrick himself acted the leading part. In his preface to the printed version, Whitehead acknowledges his debt to Corneille and at the same time justifies the freedom with which he has remodelled the play.[26] Though far inferior to the original in poetic diction, the adaptation has dramatic construction and shows a surprising subtlety in subduing or motivating those features which had seemed offensively harsh in Corneille's plot. The murder of Camilla by her brother was always felt to be so outrageous a deed that even the most eloquent defence by the father could not reconcile the audience to it and atone for the hero's crime. In Garrick's version,[27] Camilla herself desires death.[28] Having suffered the horror of losing her betrothed at the hands of her own brother, she contemplates suicide; but being deprived of all implements, she decides to make her brother, who killed Curiatius, the unwilling instrument of her own death. She provokes him with frenzied insults into a fury and he loses self-control. Having suffered the blow, she absolves him from guilt, revealing that it was her intention to die by the hand that had killed her lover.

Camilla's death in this version is essentially a suicide. The tragic ruse by which she satisfies her grief increases her stature to that of a heroine who actively induces the catastrophe, whereas in Corneille's play she merely suffers it as a victim. But in proportion as her stature rises, her brother's decreases in dramatic complexity. A rash but profoundly sympathetic youth, he becomes the instrument rather than the agent of disaster. His vigorous simplicity is his outstanding trait and proves in its fateful implications as touching as the devotion of his shrewder sister. In this psychological interplay, which injects into the austerity of the Roman setting the more intimate interest of a domestic tragedy, the central position is held by the Roman Father who, concerned for his daughter and proud of his son,

[23] See his *Lettres sur Garrick*, op. cit.

[24] The admiration was mutual. Garrick is said to have referred to Noverre as the 'Shakespeare of the ballet'.

[25] It was first produced in 1750 and was still performed in the early nineteenth century. On William Whitehead, see A. Bitter, *William Whitehead—Poeta Laureatus. Eine Studie zu den literarischen Strömungen um die Mitte des 18. Jahrhunderts* (1933), which also gives the theatrical history of the play; D. F. Fisher, *Corneille and Racine in England* (1904).

[26] By some mistake the play is not included among the adaptations of Corneille's *Horace* in Picot's *Bibliographie Cornélienne*, though he does list (no. 1421) W. Freeman's *A Comparison between the Horace of Corneille and the Roman Father of Mr. Whitehead* (1750).

[27] I call it Garrick's version because it is obvious that Whitehead wrote his adaptation of Corneille not only at Garrick's request, but also under his supervision. The feebleness of the style, which is certainly Whitehead's own, stands in strange contrast to the shrewdness of the dramatic construction. That Whitehead executed Garrick's dramaturgical orders with precision is shown by a letter of 1 November 1767 in which he undertakes to introduce for a revival of the play some quite fundamental alterations which Garrick had prescribed (*The Private Correspondence of David Garrick with the most Celebrated Persons of his Time*, ed. J. Boaden, I, 1831, pp. 271 f.).

[28] She is called Horatia in the English version and has acquired some of the features of Corneille's Sabine.

combines patriotic abandon with protective tenderness.[29] He is not as pure and consistent a Stoic as Corneille's Horatius. Such a reckless patriot could have little appeal for an author who in one of his poems rejected the lessons of Roman imperturbability as 'Stoic fustian'.[30] Nor would Garrick have been interested in playing a character devoid of overtones. His Roman Father is assailed by forebodings arising from the tenderness he feels for Camilla. Even in his most heroic moods there is a suggestion of gloom; and when he prepares his son for battle, the concluding words of his exhortation—which Garrick must have phrased with great beauty—sound a note of doubt:

'Tis Glory calls,
And humbler Passions beat Alarms in vain.[31]

To choose the path of glory with regret is an idea that would hardly have occurred to Corneille as befitting a Roman hero. Nor could it have appealed to David, whose taste for Roman austerity was even more excessive. What makes *The Roman Father* important as a play—its achievement in overcoming the severities of French classicism by a judicious injection of English sentiment—would decidedly not have endeared it to David or even to Noverre. Yet the play anticipates David's composition in two essential features. In contrast with Corneille's tragedy, the interest shifts from the young Horatius to his father, who dominates the play exactly as he dominates the painting. David's three sons form a quasi-anonymous group against which the father stands out as a distinct personality. The picture, like the play, could be called *The Roman Father*, and David himself felt the need to stress the role of the father in the original title, the full text of which reads: *Le Serment des Horaces entre les mains de leur Père*.

The second resemblance is even more striking. In deviating from Corneille, the English version introduces a motive which David was to elaborate to the full. In the first scene of the second act, the father prepares the son for battle by bringing him a sword:

Come, let me arm thee for the glorious toil.
I have a sword whose lightning oft has blaz'd
Dreadfully fatal on my country's foes;
Whose temper'd edge has cleft their haughty crests,
And stain'd with life-blood many a reeking plain.
This shalt thou bear; myself will gird it on,
And lead thee forth to death or victory.

Noverre, who was no judge of English poetry, was probably impressed with the gesture. At the climax of this act, the father enters sword in hand. He finds his son vigorously rejecting

[29] He could almost be described as a cross between the relentless Roman father in Corneille and Diderot's *Père de famille*.

[30] 'On the Death of a Relation' (*Plays and Poems*, II, 1774, p. 173):

Let dull unfeeling pedants talk by rote
Of CATO's soul which could itself subdue;
Or idle scraps of Stoic fustian quote,
And bravely bear the pangs they never knew.

[31] Act II, scene i.

the tender reproofs of Camilla; placing himself between his two children, he has on one side the determined young fighter, on the other the despairing woman—a triple configuration strongly resembling the situation in David's painting.

Noverre's ballet brought the scene even closer to David's picture by having Camilla attended by her women, who catch her when she faints ('elle tombe dans les bras de ses femmes'),[32] and by introducing her three brothers ('ses trois frères'), instead of only one. Naturally, they required three swords, and it seems that these were brought on—*mirabile dictu*—by Proculus, that 'personnage subalterne' who had annoyed Voltaire by appearing unexpectedly in Corneille's fourth act. Noverre, anxious to please Voltaire and to improve on Corneille,[33] introduced him into the very first act as a friend of the elder Horatius, whom he follows like a shadow and whose powers he partly usurps. It is he, for instance, who invites the young Horatii to fight ('Procule les invite à combattre').[34] His prominence in this scene confirms the suspicion that the pantomime with the three swords that motivated his original appearance near the end of the play was balanced by a similar pantomime at the beginning.

Painting and Pantomime

Although Noverre repeatedly visited England, he does not seem to have acquired Hogarth's conviction that 'serious dancing' is 'even a contradiction in terms'.[35] His productions were of an extravagant pretence, and he himself proclaimed their virtues 'avec le sérieux imperturbable qui caractérise les danseurs'.[36] Yet even Grimm, who severely criticized *Les Horaces*, had to admit that the spectacle was impressive: 'Tout ce que le poëte n'a pu mettre qu'en récit, se trouve ici en action et forme un spectacle magnifique.' The scene of departure, in which the three sons presumably received their swords—an expansion of the sword scene in Garrick's *The Roman Father*, was followed by the great scene of the oath, a theatrical innovation which must have recommended itself by its pantomimic possibilities within the medium of the *ballet d'action*.[37] But we must not imagine this scene as too literal an anticipation of David's painting. In the inflated version of the ballet-master, the departure (in the house, with the women crying) and the oath (on the battle-field) were produced as two separate pantomimic scenes. David himself seems to have thought of them

[32] I follow the description in Grimm, *Correspondance littéraire*, see above, note 17. The scenarios reproduced in C. E. Noverre (*The Life and Works of the Chevalier Noverre*, p. 39) are unreliable. He says himself that he has preserved the original translation 'in most instances'.

[33] Cf. Grimm, loc. cit.; also for the following quotation.

[34] He does so with the most foolish motivation, namely that he has promised his daughter to the victor. Though this clumsy device may have had the technical advantage of bringing another female dancer on to the stage, Noverre could hardly flatter himself that he had satisfied Voltaire's demand and produced a sufficient dramaturgical reason for introducing Proculus so soon.

[35] *The Analysis of Beauty* (1753), xvii, 'Of Action' (ed. J. Burke, 1955, p. 159).

[36] Noverre, in his preface to *Les Horaces*, cf. Picot, op. cit., no. 1122. See also Noverre's *Lettres sur la Danse, et sur les Ballets*, p. 47: 'La Danse sérieuse & héroïque est sans contredit la seule qui puisse convenir à un Théâtre où tout respire la décence & la grandeur.'

[37] Picot, op. cit., no. 1115, lists an Italian opera, *Orazio e Curazio* by Bertoni, which was produced in Venice as early as 1746. It is possible that this already contained the scene of the oath, which remained a favourite operatic motif down to Verdi's *Aïda*.

originally as two separate events; for the drawing reproduced here (Fig. 106) shows the scene in which the three sons receive their swords, but the oath is not yet part of the action. The concentration of the two scenes into one episode seems to have been an afterthought, and must have occurred to him after he had already contracted the double roles of Proculus and the elder Horatius into one character.[38] The gathering of literary friends that met at Monsieur Trudaine's could advise him on such matters of artistic economy. It was they who had suggested that he should replace the *Horace condamné* by the *Serment des Horaces*.[39]

Trudaine's circle included two men whose literary orientation must have strongly confirmed David in his artistic impulse. One was Sedaine, who had practically adopted David as a son.[40] He is known today as the author of *Le Philosophe sans le savoir*, generally acknowledged to be the best of the *drames bourgeois*. As well as plays he wrote operatic libretti, and arranged English plays for the French stage. The other was Ducis.[41] He had become famous by adapting Hamlet, Othello, Lear, and other Shakespearian roles to the requirements of Brizard, the very actor whom David is supposed to have admired in the part of the elder Horatius.[42] It was a circle in which the influence of the English stage was as vital as the interest in the *théâtre lyrique*, and where the enthusiasm for pantomimic art was associated with Diderot's revolution of feelings.[43] In advising David to abandon the original version of the *Horaces*, they are said to have argued that painting ought not to represent a scene which depends primarily on the spoken word.[44] Although the source for these *entretiens* is slightly suspect, being also the source of the false opinion that the *Horace condamné* was a direct transcript from Corneille, there can be no doubt that this particular argument corresponds to the actual findings. The difference between the *Horace condamné* and the *Serment des Horaces* is precisely that between a design derived from literature, following a given text (Rollin) and suggesting the spoken word, and a composition conceived from the start as a pantomime.[45]

[38] This type of reduction is characteristic of David's early style. In modelling his *Andromache mourning Hector* after Gavin Hamilton, he reduced the number of figures from thirteen to three.

[39] The earliest sketch of the *Serment* is dated 1782 (cf. Locquin, op. cit., p. 221). In 1780, Fuseli had painted a patriotic Swiss subject, *The Oath on the Rütli* (Fig. 107), which was possibly known in David's circle. If so, David's painting, seen from the point of view of English art, holds a middle position between Fuseli on the one hand and Flaxman's *Seven Chiefs against Thebes*, 1795 (Fig. 108) on the other. [In 1961, R. Rosenblum ('Gavin Hamilton's *Brutus* and its Aftermath', *The Burlington Magazine*, CIII, pp. 8 ff.) suggested that the first painting which inspired this succession of oath pictures, both in England and on the Continent, was Hamilton's *Oath of Brutus* (1763–4). Later, Rosenblum rediscovered Beaufort's *Brutus* ('A Source for David's *Oath of the Horatii*', *The Burlington Magazine*, CXIII, 1970, pp. 269 ff.) a painting inspired by Hamilton and also known to David. As additional confirmation of his argument, Rosenblum noted two drawings in David's Versailles Sketchbook (citing V. Lee, 'Jacques-Louis David: The Versailles Sketchbook', *The Burlington Magazine*, CXI, 1969, pp. 197 ff. and 360 ff.) one after Beaufort's *Brutus*, the other after Hamilton, which David had considered, but rejected, as compositional ideas.]

[40] See *Vie de Sedaine*, in J.-F. Ducis, *Œuvres*, ed. L. S. Augur, III (1826), pp. 455 ff.

[41] On Ducis's relations with Garrick, see Hedgcock, op. cit., pp. 291 ff.

[42] See above, p. 105.

[43] It was Trudaine who commissioned David to paint *The Death of Socrates* (The Metropolitan Museum of Art, New York), the very scene which Diderot had written as a pantomime.

[44] Cf. Locquin, loc. cit., after Péron.

[45] It may seem strange that Noverre does not figure in the contemporary accounts of David; but this could very well be because they were written by David's own disciples and reflect the mature views of the master, which can hardly have been favourable to Noverre (who had fled from France during the Revolution).

Salieri's Opera—Gluck and Beaumarchais

Given the close association of the *Serment* with operatic pantomime, it is hardly an accident that the sensational success of David's *Horaces*, exhibited in Paris in 1785, was followed by the production of a new opera, *Les Horaces*, in 1786. The text was by Guillard, who had written the libretto of Gluck's *Iphigénie en Tauride*. The music was by Salieri, Gluck's disciple and friend who had made his début in Paris two years before with *Les Danaïdes*, to which Gluck had even signed his name as co-author.[46] *Les Horaces* proved to be a failure, and Gluck wrote to console his friend:

Mon cher ami, *Vienne, le 29 Décembre* 1786.

Je vous remercie infiniment des notes que vous m'avez données sur la représentation de votre opéra des Horaces, qui sera toujours estimé & apprécié du Public, nonobstant ce qui paroît lui avoir déplu dans l'ensemble. Attendez qu'il s'en fasse encore quelques représentations, & les sourds recouvriront l'ouïe; car la manière dont vous avez été critiqué, n'a aucun fondement.

Je suis, &c.,

(Signé) le Chev. GLUCK.

The letter was published in the *Courier de l'Europe* on 9 February 1787[47] by 'un amateur' who added a comment of his own: 'Si la froideur patriotique a fait proscrire le poème, ce n'est pas la faute du musicien, qui n'a pas pu mettre une gavotte dans un opéra où il s'agissoit du salut de Rome.' He further adds that Gluck's letter had been transmitted to him by Beaumarchais.[48] In spite of its failure, the performance of *Les Horaces* had the considerable effect of enlisting Beaumarchais, that extravagant precursor of the Revolution, on the side of Gluck's party. He asked Salieri to write the musical score of his own lyrical drama *Tarare*, a mystery play which he produced with a display of pageantry on the pretext of reviving the theatrical art of the Greeks. Where *Les Horaces* had failed, *Tarare* succeeded. In the preface to the printed edition, dedicated to Salieri, Beaumarchais prides himself on having discovered the dramatic potentialities of his music:

Mon plus grand mérite en ceci est d'avoir deviné l'Opéra de *Tarare* dans *les Danaïdes* & *les Horaces*, malgré la prévention qui nuisit à ce dernier, lequel est un fort bel ouvrage, mais un peu sévère pour Paris.

Vous m'avez aidé, mon ami, à donner aux Français une idée du Spectacle des Grecs, tel que je l'ai toujours conçu.[49]

[46] The transaction was later regarded as very questionable, and dictated by financial interests. Salieri himself has an unfortunate reputation in the history of music. Beethoven acknowledged him as his teacher and he directed the school in which Schubert was trained, but he is chiefly remembered as the musician who, solidly entrenched at the court of Vienna, stood in the way of Mozart's preferment. For more independent estimates, see I. F. von Mosel, *Über das Leben und die Werke des Anton Salieri* (1827); and Albert von Hermann, *Antonio Salieri* (1897), which although confined to a study of his early period gives a chronological list of all his dramatic works. The opera *Les Horaces* is listed by Picot, op. cit., no. 1123.

[47] XXI, 12, p. 99.

[48] Was Beaumarchais perhaps himself the 'amateur'?

[49] *Tarare, opéra en cinq actes, avec un prologue. Représenté pour la premiere fois . . . 8 Juin 1787*, 'A monsieur Salieri', p. [3].

This declamation was, of course, indirectly addressed to Gluck, for whom Salieri served as a deputy. Writing to Salieri after his departure for Vienna (September 1787), Beaumarchais ends his letter with the words: 'Rappelez-moi au souvenir de ce géant qu'on nomme Gluck.'[50]

Even so, Guillard's text of *Les Horaces* was to outlive Beaumarchais's *Tarare*. A revised version served as the libretto for a new opera, first produced in 1800 in the presence of Bonaparte.[51] The choice of subject was undoubtedly meant to honour the First Consul as a son of the Revolution. But many among David's former co-revolutionaries refused to share his idolatry of the new hero. They suspected in the growing authority of Bonaparte another version of *le tyran*. A plot was formed to assassinate him at the opera. But the police had been forewarned, and they removed the conspirators from the house without disturbing the play. The signal chosen for the assassination was the moment when the oath was about to be sworn on stage. Contrary to David's intentions, the image had retained its revolutionary meaning.

Further Traces of Gluck and Garrick in David

Some three years after the exhibition of the *Horaces*, David painted the *Paris and Helen* (1788), which has often puzzled its interpreters (Fig. 109). How, they asked, did the artist who had painted the *Horaces* with such an emphatic display of moral intransigence here relapse into the *genre agréable* and produce a group which one critic did not hesitate to compare even to Boucher?[52]

David's interest in this type of theme cannot be dismissed as a passing mood, for it is resumed in as late a painting as the *Sappho, Phaon, and Cupid* and becomes altogether predominant in his final style. What looks at first like an anomaly is really a growing passion for the musical drama or *théâtre lyrique*, the very medium that had inspired the pantomimic transformation of the *Horaces*. David drew from the same source when he painted the love of Paris and Helen. The picture becomes at once intelligible, and appears far removed from the *genre agréable*, when seen as a transposition into painting of the dramatic idea of Gluck's *Paride ed Elena*, that is, the love of a Spartan woman for a Phrygian man.[53]

The Spartan character, associated in music with the Doric mode, is severe and harsh, and suggests a martial, puritan, and inflexible temperament—the very opposite of the pliant beauty and voluptuous indulgence of the classical Helen. To make his character of Helen consistent, Gluck even relieved her of the charge of adultery, by letting her meet and succumb to Paris when she is promised, but not yet wedded, to Menelaus. In resisting and

[50] Louis de Loménie, *Beaumarchais et son temps*, II (1880), p. 409.

[51] Picot, op. cit., no. 1125.

[52] Saunier, op. cit., p. 36.

[53] Cf. Cooper, op. cit., pp. 138 ff.

revealing her passion, she is as firm—but also as hard and, when roused, as fierce—as a true Spartan maiden. Her tragedy, and the dramatic motif of Gluck's composition, is that she is won by the tender warmth and luxurious affection of a Phrygian youth whose temperament, true to the Phrygian mode in music, is complementary to hers, yet incompatible.[54]

In David's painting, Paris appears as an accomplished devotee of Aphrodite, while Helen looks as modest as a Roman matron. She stands, statuesque, and dressed in a classical *peplos* that has slipped from one shoulder, while he is seated on a soft cushion, almost naked, wearing a Phrygian cap, and holding—as he does in Gluck's opera—a lyre with a picture of the Judgement of Paris inset in its frame. Helen leans on his shoulder, and the very contrast of standing and sitting (reversing the conventional placing of man and woman) reveals the man as the soft, effeminate character who tries to draw down to him the restrained and meditative figure of the woman bending over him with affection but without abandon. Although Paris woos her with a caressing gesture, her attitude resembles that of a Muse to whom the musician looks up for inspiration.[55]

A fitting accompaniment to this group is provided by the caryatids in the background, which have been identified—with much surprise at this intrusion of a pseudo-antique piece of French Renaissance decoration—as those designed by Goujon for the Salle des Caryatides in the Louvre. David may have known that they were intended as a setting for musicians: 'C'est pour la tribune des musiciens, "des hautbois et joueurs d'instruments", placée au-dessus de la porte d'entrée, que Lescot et Goujon réalisèrent ce portique d'allure antique.'[56]

In the very same year (1788) in which he painted the *Paris and Helen*, David also painted the double portrait of Lavoisier and his wife (Fig. 112). We do not know which picture he designed first. It is probable—in spite (perhaps even because) of the striking contrast—that he worked at them simultaneously. What appears certain is that they were associated in his imagination, and that the theme of each picture is 'parodied' (in the musical sense of the word) by the other. In the Lavoisier portrait, as in *Paris and Helen*, the man is seated and looks up over his shoulder at the woman leaning over him as his Muse. Madame Lavoisier was a pupil of David, but she placed her gifts as a draughtswoman—a folder of drawings or prints is seen in the left background—in the service of her husband's research, illustrating his treatises by graphic designs and even helping in the laboratory, where she took notes and made scientific sketches.[57] Drawings by her own hand (Figs. 110–11) show Lavoisier and his assistants carrying out experiments on respiration, while she sits at a table recording their

[54] It is he who sings the famous aria, 'Oh, del mio dolce ardor'.

[55] Hellenistic reliefs of 'Paris inspired by Love' (Palazzo Spada, Rome; further examples in T. Schreiber, *Die hellenistischen Reliefbilder*, 1894) appear to have served as David's models. [And see now E. Coche de la Ferté and J. Guey, 'Analyse archéologique et psychologique d'un tableau de David: "Les Amours de Pâris et d'Hélène"', *Revue archéologique*, XL (1952), pp. 129 ff.]

[56] P. Vitry, *Jean Goujon, Biographie critique* (1908), p. 100.

[57] Cf. É. Grimaux, *Lavoisier, 1743–1794, d'après sa correspondance, ses manuscrits, ses papiers de famille et d'autres documents inédits*, 2nd edn. (1896).

results. David's friend Ducis might have addressed to her (with only a slight fault in the metre of the third line) the verses that he wrote for Madame de Boufflers:

> Épouse et cousine à-la-fois,
> Sûre d'aimer comme de plaire,
> Pour Boufflers, soumis à vos lois,
> Vous remplissez les deux emplois
> Et de muse et de secrétaire.[58]

While the attitudes of Paris and Helen derive from classical models—Helen standing in the posture of a classic Muse, Paris imitating the *contrapposto* of an antique singer who turns his head to receive inspiration—the modern configuration adopted in the Lavoisier portrait makes it equally certain that David intended to produce the image of an 'author with his Muse'. It has been suspected (I think justly) that the composition derives from an English model. In any case, it is easy to name English pictures of this type which precede David's painting by several decades. The most important of them, and perhaps the one that impressed David most forcefully, is Hogarth's portrait of Garrick with his wife (Fig. 116), known on the continent through engravings. It is a travesty of the solemn portrait of Colley Cibber, Garrick's predecessor and unequal rival, which Van Loo had painted in England (Fig. 113); in it the old comedian and poet laureate is seen preoccupied with his papers, holding a pen which is guided by his young daughter. The painting, known as *Colley Cibber with his Muse*, must have struck both Hogarth's and Garrick's sense of the ridiculous. Hogarth reversed the situation and represented Garrick meditating the composition of a 'Prologue on Taste', but about to be disturbed by the intrusion of his Muse (in the character of his pretty wife), who approaches secretly behind his back and instead of guiding his pen tries to take it away.[59] Both men, old Cibber and Garrick, are seated holding pens, while their Muses stand over them as good—or impish—spirits. The type remains prevalent into the nineteenth century and bears witness to the continued exchange of artistic formulae between England and France; for after it had been adopted by David and endowed with an elegant nobility, it reappeared in Romney (Fig. 115), whose version in its turn seems to have provided the pattern for an 'English portrait', *Mr and Mrs Edwin Edwards* (Fig. 114), by Fantin-Latour. Although completely undramatic in the juxtaposition of the two figures and far too sober and high-minded to permit light-hearted poetical metaphors, the last two examples still stress the subordination of the man to the woman and make her so conspicuously the guiding spirit (metaphorically speaking, the Muse) that they even disregard the polite convention that a man should not sit while a woman stands.

As the imagery of David's *Paris and Helen* belongs to the sphere of the *théâtre lyrique* and

[58] *Œuvres posthumes de J.-F. Ducis, précédées d'une Notice sur sa Vie ses Ecrits, par M. Campenon* (1826), p. 239.

[59] Hogarth may have seen the Netherlandish type of double portrait in which the woman enters the room with a message (a letter) and interrupts the man in his writing. [See also F. Antal, *Hogarth and his Place in European Art* (1962), both on Hogarth's portrait *Garrick and his Wife*, pp. 69 f., and on his impact on David, pp. 199 f.]

relates the painting to the music of Gluck, so the Lavoisier portrait has its place in the tradition of Anglo-French portraiture and embodies an elegant dramatic type that had been sponsored by Garrick. The union of Gluck and Garrick that we observed in Noverre as a presupposition of David's *Serment des Horaces* now returns in David himself, who polarizes it into two separate pictures: the one lyrical and antique, the other fashionable and English, but both treating a similar theme and both painted in the same year.[60]

[60] [Although for several decades art historians have discounted the thesis that *The Oath of the Horatii* was a revolutionary painting with a political message, this thesis has recently been revived in a stimulating article by T. Crowe. Surprisingly, Crowe bases his new interpretation on the art criticism of Louis Carrogis, writing as Louis de Carmontelle, who, he argues, 'could describe art as an instrument of reaction and oppression' (see above, p. 37 and note 93, and Fig. 30). The difficulties and contradictions of Crowe's position are analysed by F. Haskell, 'Recent Literature on Jacques-Louis David', *The Burlington Magazine*, CXXIII (1981), pp. 749 f.]

XII · In Defence of Composite Portraits

It seems natural for a modern spectator to laugh when he sees portraits in the style of Louis XIV, where men with perukes or moustaches pose as Apollo, Endymion, or Hercules; where ladies appearing as ancient goddesses—Diana, Juno, Thetis, or Hebe—might seem less ludicrous if the fur of the huntress or the crescent of the moon better suited their build or their complexion. Alas! too many resemble the painting by Largillière in which a classical nymph couched gracefully near a river exhibits the podgy features of Charlotte von der Pfalz (Fig. 38).[1]

Laughter at seeing the unexpected should signal danger to an ingenuous spectator. Young anthropologists are said to have laughed when they first saw the Pueblo Indians performing the dance of the antelopes.[2] It struck them as comic that men should studiously imitate the appearance and behaviour of leaping quadrupeds. But once the meaning of the ritual was understood there was little room for laughter. Primitive men, being genuinely humble, believe that animals are superior to themselves. They also believe that, by taking

[1] [*The Princesse Palatine as a Water-Nymph*, portrait of Élisabeth-Charlotte de Bavière, second wife of Louis XIV's brother, Philippe d'Orléans. In a famous letter to her sister (22 April 1698), she disarmingly described herself in even less flattering terms than the painter has done with his brush: '. . . ma taille est monstrueuse de grosseur, je suis aussi carrée qu'un cube; ma peau est d'un rouge tacheté de jaune; mes cheveux deviennent tout gris; mon nez a été tout bariolé par la petite-vérole, ainsi que mes deux joues; j'ai la bouche grande, les dents gâtées. . . .' (*Correspondance complète de Madame*, trans. by M. G. Brunet, I, 1857, pp. 32 f.) See A. Châtelet, F.-G. Pariset, and R. de Broglie, *Chantilly—Musée Condé. Peintures de l'École française, XVe–XVIIe siècles* (1970), no. 104; also N. M. Rosenfeld, *Largillière and the Eighteenth-Century Portrait*, Montreal Museum of Fine Arts (1981), no. 63b; and especially A. Schnapper's essay 'The Position of the Portrait in France at the End of the Reign of Louis XIV (1680–1715)', ibid., pp. 60 ff.]

[2] 'Als ich den Antilopen-Tanz in San Ildefonso zu sehen bekam, machte er auf mich zunächst einen sehr harmlosen und beinahe komischen Eindruck. Für den Folkloristen, der die Wurzeln der menschlichen Kulturäußerungen biologisch erforschen will, gibt es aber keinen gefährlicheren Augenblick, als wenn er bei volkstümlich-komisch erscheinenden Gebräuchen lacht. Wer über das Komische in der Volkskunde lacht, hat Unrecht, dem verschüttet sich im selben Augenblick die Einsicht in das tragische Element.' [A passage from the unpublished German text of A. Warburg's lecture on serpent ritual, 'Bilder aus dem Gebiet der Pueblo-Indianer in Nord Amerika', delivered on 21 April 1923 in a nursing-home at Kreuzlingen (see Wind, *The Eloquence of Symbols*, p. 111) and now in the archives of the Warburg Institute, London (Unpublished Sources, no. 16, p. 9); see the English trans. by W. F. Mainland, *Journal of the Warburg Institute*, II (1939), pp. 277–92, here p. 282. Extracts from the text of the lecture and from Warburg's notes are quoted by E. H. Gombrich, *Aby Warburg. An Intellectual Biography* (1970), pp. 20, 88 f., and 216 ff. Warburg's anthropological approach to European history is discussed by F. Saxl, 'Warburg's Visit to New Mexico', *Lectures*, I (1937), pp. 325 ff. On Wind's application of Warburg's comment to the study of English portraiture, see B. Buschendorf's forthcoming paper in *Idea*, IV (1985), '"War ein sehr tüchtiges gegenseitiges Fördern": Edgar Wind und Aby Warburg'.]

Journal of the Warburg Institute, I (1937), pp. 138–42, the introduction to Wind's 'Studies in Allegorical Portraiture. Albrecht von Brandenburg as St. Erasmus', loc. cit.; see *The Eloquence of Symbols* (1983), pp. 58 ff.

the shape of superior beings, men can acquire their powers, and that those beneficial gifts can be preserved and transmitted by the force of imitation.

From animal-worship to hero-worship the distance is not very far; nor is the ritual very different. The Roman emperor Commodus had his portrait sculptured with lifelike accuracy, yet with a lion's skin around his head and a heavy club in his hand (Fig. 35). Being thus vested with the emblems of Hercules, he presented himself as possessed of his virtues, and the honours due to the ancient demi-god were transferred to the emperor in whom he was embodied.[3]

Such a ritual will survive with the force of an atavism. From the ceremonies attached to the Roman emperor-cult, it descended to the court masques of the modern dynasts: the Virgin Queen assumed the role of Diana, the *Roi Soleil* invested himself with the symbols of Phoebus Apollo.

The days were past when man acquired superior power by transforming himself into another being; yet transformation remained the method of expressing one's power through a visible sign. The gods of Olympus no longer bestowed the gifts of strength, leisure, and wealth, but powerful, idle, and rich men would play the Olympian in order to exhibit them. And such is the secret wisdom of social habits that the symbolic acts which display these things tend to work as means of securing them. They preserve in the actor the sense of his dignity and produce in the beholder a feeling of awe. What was real and literal in primitive ritual has now become fiction and metaphor. But the theatrical effect reverses the process: what began in fiction terminates in reality,[4] so that the human beings who played at being gods were venerated as actual god-like beings.

'Though this be madness, yet there is method in't'. The joining of incompatibles has ever been the secret of witchcraft. The painters of the court of Louis XIV made it very clear that the men and women they portrayed were masters in the blending of contraries: while they had themselves adorned with the traits of the gods, they stressed the retention of their own (Fig. 37).[5] What they aspired to was the union of the two, and they found their most powerful instrument in those composite forms which, seeming to modern eyes incongruous, now make us laugh.

Distorted Gods, Composite Beings—all very true. But why not carry this idea to its ultimate conclusion? These beings are, to put it quite plainly, artificial monsters. Dissolve

[3] The cult of Commodus is not, of course, an isolated instance. According to H. Jeanmaire, the fight between Octavian and Mark Antony at the Battle of Actium took the form of a mythical combat between Apollo and Dionysus; see his article 'La Politique religieuse d'Antoine et de Cléopâtre', *Revue archéologique*, XII (1924), pp. 241 ff. See also A. D. Nock, 'Notes on Ruler-Cult', *The Journal of Hellenic Studies*, XLVIII (1928), pp. 21 ff. (or now his *Essays on Religion and the Ancient World*, ed. Z. Stewart, I, 1972, pp. 134 ff.); Lily R. Taylor, *The Divinity of the Roman Emperor* (1931); A. Alföldi, 'Insignien und Tracht der römischen Kaiser', *Mitteilungen des deutschen Archaeologischen Instituts, Römische Abteilung*, XLIX (1934), pp. 1–156, and 'Die Ausgestaltung des monarchischen Zeremoniells am römischen Kaiserhofe', ibid., L (1935), pp. 1–231; K. Scott, *The Imperial Cult under the Flavians* (1936).

[4] On Reynolds's quotation from James Harris, see above, pp. 21, 71.

[5] See, for example, Largillière's portrait of Madame de Gueidan as Flora, in Honoré d'Urfé's novel *L'Astrée* (Musée Granet, Aix-en-Provence), and his portrait of the actress Mademoiselle Duclos as Ariadne, in Thomas Corneille's *Ariane* (Comédie-Française, Paris); both now illustrated in Rosenfeld, op. cit., nos. 63 and 49. [See also Châtelet, Pariset, and de Broglie, op. cit., no. 105, for the Chantilly version of the portrait of Mademoiselle Duclos.]

them into their component parts—and we will be left with the idea of a god and a specimen of human flesh, neither of them very surprising. As with the Sphinx and the Minotaur, it is the *union* that works the spell.

Those who, on looking at these portraits, find it difficult to believe that the spell ever worked should try to picture to themselves (as a sort of preliminary exercise) an actual performance of a tragedy by Racine. There the ancient heroes would appear speaking French, giving vent to their wild pagan grief in that apostrophizing style of diction and mime which French actors play to perfection. Is the emotion of Phèdre, this truly Graeco-Gallic madness, any less composite than her garment? Could we imagine Hippolyte inspiring that passion if he were formed in the image of an archaeological cast? It is curious that our modern purists who concede grandeur to a heathen passion expressed in Racine's French should laugh at seeing Roman armour combined with a moustache and a wig.

Look at Molière playing Julius Caesar (Fig. 36). Does he not deserve the title of 'a mind naturalized in antiquity'?[6] Naturalized indeed. This is not a native Roman; nor is it a genuine hero. The laurel has descended upon a peruke. The Roman has been grafted on to a Frenchman. But as a tree after grafting develops new strength and produces new fruit, so this composite being possesses the power to move his audience, to raise their emotion until the remote past and the present melt into one and the concerns of the Roman become our own.

The passion for the grandeur of the past would have no strength if it were not deeply rooted in the taste for the contemporary. The actor, by retaining in his make-up as Julius Caesar the essentials of the fashion of his own day, can address his audience as if there were no distance of style or period between them. He can *tease* them into that feeling of the sublime which, without this familiar appeal to their imagination, might well remain intangible and abstract.

This powerful union of opposites exasperated Sir Joshua Reynolds. On the well-pruned plant of his Academy, every branch had its proper place, and was not to intertwine with the others. Modern fashions were not to be mingled with the genuine 'habits of nature'. Hence his stern rejection of 'that ridiculous style which has been practised by some painters, who have given the Grecian Heroes the airs and graces practised in the court of Lewis the Fourteenth'.[7]

This verdict was a death-blow to the composite portrait. But the pliant manner of the President of the Academy gave it the form of a mitigated sentence: 'It may be asserted that the great style is always more or less contaminated by any meaner mixture. But it happens in a few instances, that the lower may be improved by borrowing from the grand. Thus if a portrait-painter is desirous to raise and improve his subject, he has no other means than by approaching it to the general idea.'[8]

In this diluted form the composite portrait was readmitted. 'Borrowing from the grand',

[6] *Discourses on Art* (as in Wark), XIV, l. 297.

[7] Discourse III, ll. 242–5.

[8] Discourse IV, ll. 459–63.

'improving the subject', 'ennobling the character of a countenance'—this was Reynolds's guarded advice, and it meant leaving out 'all the minute breaks and peculiarities in the face' and changing 'the dress from a temporary fashion to one more permanent, which has annexed to it no ideas of meanness from its being familiar to us'.[9] Clearly, this was not a blending or union of opposites. It was a fading out of contrasts.

The results may not strike one as ludicrous, as do some of the classical French examples, but nor will they strike one as very meaningful. Not for one moment does Reynolds assume that in depicting a woman as Hebe or Juno (Fig. 39) he is investing her with some rightful sign of divinity. 'Hebe' is merely the most fitting compliment for a young beauty, and 'Juno' for a more mature one. And if it so happens that three pretty sisters have received the flattering nickname of 'The Irish Graces' he will gladly seize on that compliment and derive from it his pictorial formula (Fig. 8). He also varies his epithets in depicting the same person: he paints Kitty Fisher both as Danaë[10] and as Cleopatra (Fig. 81). The first is a very daring, the second a very flattering, title for a courtesan; and both of them, in being witty, are entirely non-commital. There is nothing left of that magical power of words which forced Diane de Poitiers to have herself worshipped in the style of the goddess after whom she was named.

These are the necessary concomitants of Enlightenment. Metaphors have become mere embellishments, and to choose them properly is a function of the artist's 'taste'. An incongruity or superabundance of symbols will be avoided—not because it is ominous, but because it is ugly; a fitting symbol will be applauded—not because of its magical force, but because of its wit.

Once wit and taste have become autonomous, arbitrariness will be the constant danger. This was the Vicar of Wakefield's grim experience when he saw his family 'using art':

> As for our neighbour's family, there were seven of them, and they were drawn with seven oranges, a thing quite out of taste, no variety in life, no composition in the world. We desired to have something in a brighter style, and, after many debates, at length came to an unanimous resolution of being drawn together, in one large historical family piece. . . . As we did not immediately recollect an historical subject to hit us, we were contented each with being drawn as independent historical figures. My wife desired to be represented as Venus, and the painter was desired not to be too frugal of his diamonds in her stomacher and hair. Her two little ones were to be as Cupids by her side, while I, in my gown and band, was to present her with my books on the Whistonian controversy. Olivia would be drawn as an Amazon, sitting upon a bank of flowers, dressed in a green joseph, richly laced with gold, and a whip in her hand. Sophia was to be a shepherdess, with as many sheep as the painter could put in for nothing; and Moses was to be drest out with an hat and white feather. Our taste so much pleased the 'Squire, that he insisted on being put in as one of the family in the character of Alexander the great, at Olivia's feet.[11]

Goldsmith could hardly be expected to realize that the object of his ridicule—the

[9] Ibid, ll. 463–6.
[10] See above, p. 42 note 106.
[11] *The Vicar of Wakefield*, xvi (Friedman, iv, pp. 82 f.). See above, pp. 14 f.

arbitrariness in the choice of heroic characters—had a common root with his own sobriety of judgement. Both were symptoms of the fundamental fact that the superiority of ancient grandeur was no longer felt as a compelling force. A new self-assurance was gaining strength which rendered the appeal to higher powers superfluous and incongruous. Men were beginning to discuss 'historical truth'. Why should heroism be confined to ancient garments? Why should modern heroes not appear in modern dress? In the controversy which arose over West's *Death of General Wolfe* (Fig. 95), West himself spoke the significant words:

> . . . the event intended to be commemorated took place on the 13th September, 1758 [*sic*], in a region of the world unknown to the Greeks and Romans, and at a period of time when no such nations, nor heroes in their costume, any longer existed. The subject I have to represent is the conquest of a great province of America by the British troops. It is a topic that history will proudly record, and the same truth that guides the pen of the historian should govern the pencil of the artist.[12]

This rationality about the hero is the beginning of his destruction. One is reminded of Cervantes's description of the approaching end of Don Quixote, how when they heard him speaking so reasonably they knew that he was dying.

The thoroughly rational and sober man may gain a very clear view of his subject; but in proportion to the clarity he gains the subject will cease to affect him. No doubt Apollo was much more real to the man who sought to enhance his normal attire by adopting some of the god's traits than to the man who had acquired an archaeological view and thought of Apollo as a piece of marble. It may depress many that vivacity of approach should thus produce distortion, and that truth, though clarifying if sought in moderation, should prove deadening if surrendered to without restraint. Regret, however, cannot amend the matter. We must learn to look upon distortions as instruments rather than as mere impediments of a living mind, and we may yet find that we are none the worse for being in a world in which the limitations of our powers determine their use.

And there is consolation, even for archaeologists. No truth of theirs can be quite so deadening as not to reveal or provoke some freak of the imagination. It is not so very long ago that they shocked the world, and themselves, by discovering that Greek sculpture was coloured—whereupon one of them is said to have remarked: 'I am afraid it is true, but I hate to think of it.' These words might equally well have come from a great classical actor of the grandiloquent style, on learning that Julius Caesar was clean-shaven and bald.

Such conceits of the imagination are signs at once of our frailty and our strength. Seen *sub specie aeternitatis*, they may seem to make men behave like fools, yet it is they who, *sub specie temporis*, keep gods and heroes alive.

[12] John Galt, *The Life and Works of Benjamin West . . . composed from materials furnished by himself*, Part II (1820), pp. 48 f. See above, pp. 89, 100.

Index

When the location of a painting has not been stated in the text, it is given here after the title of the work.

PLATES

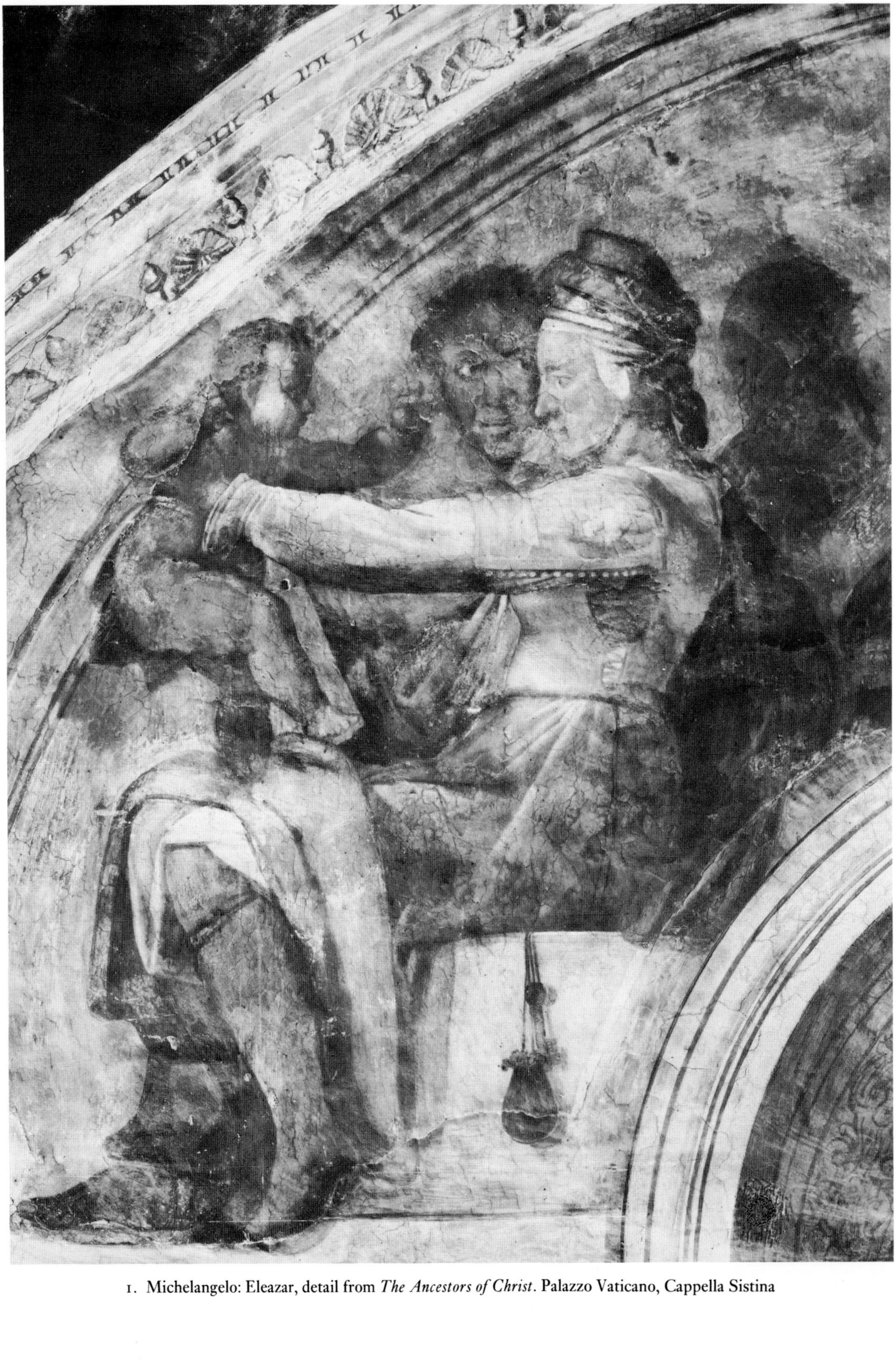

1. Michelangelo: Eleazar, detail from *The Ancestors of Christ*. Palazzo Vaticano, Cappella Sistina

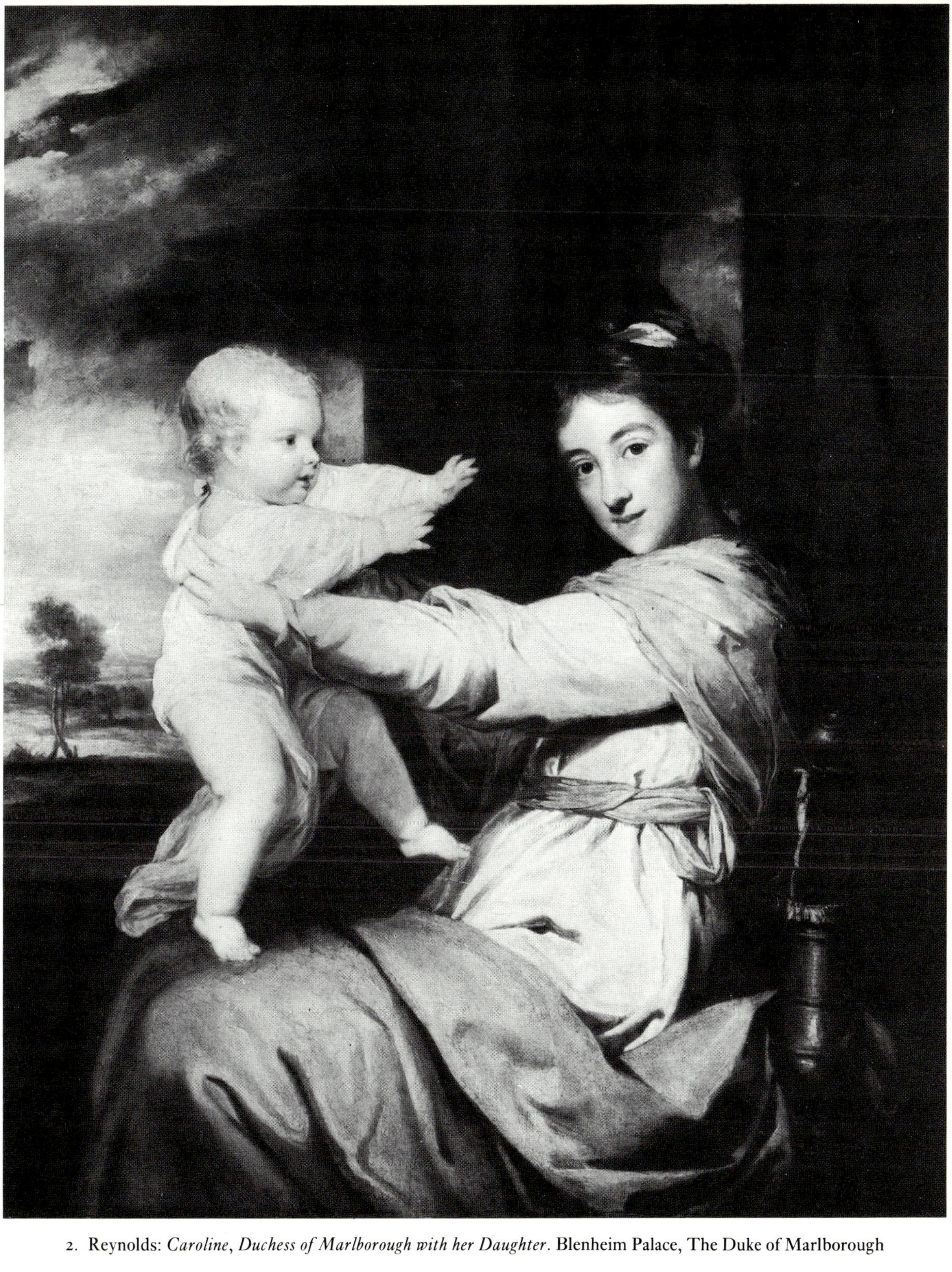

2. Reynolds: *Caroline, Duchess of Marlborough with her Daughter.* Blenheim Palace, The Duke of Marlborough

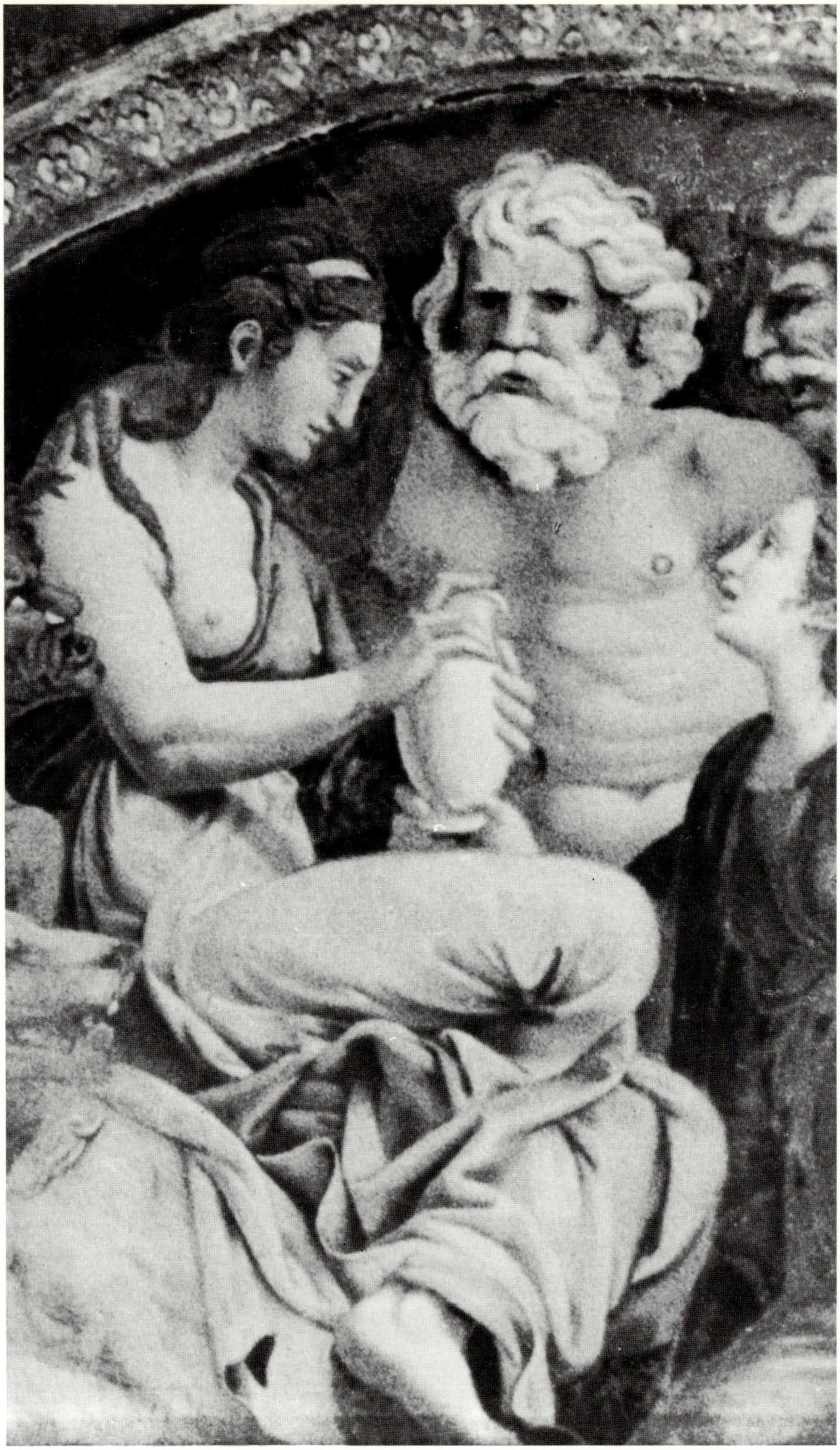

3. Giulio Romano and assistants: Psyche receiving the vase of ointment, detail from the Sala di Psiche. Mantua, Palazzo del Te

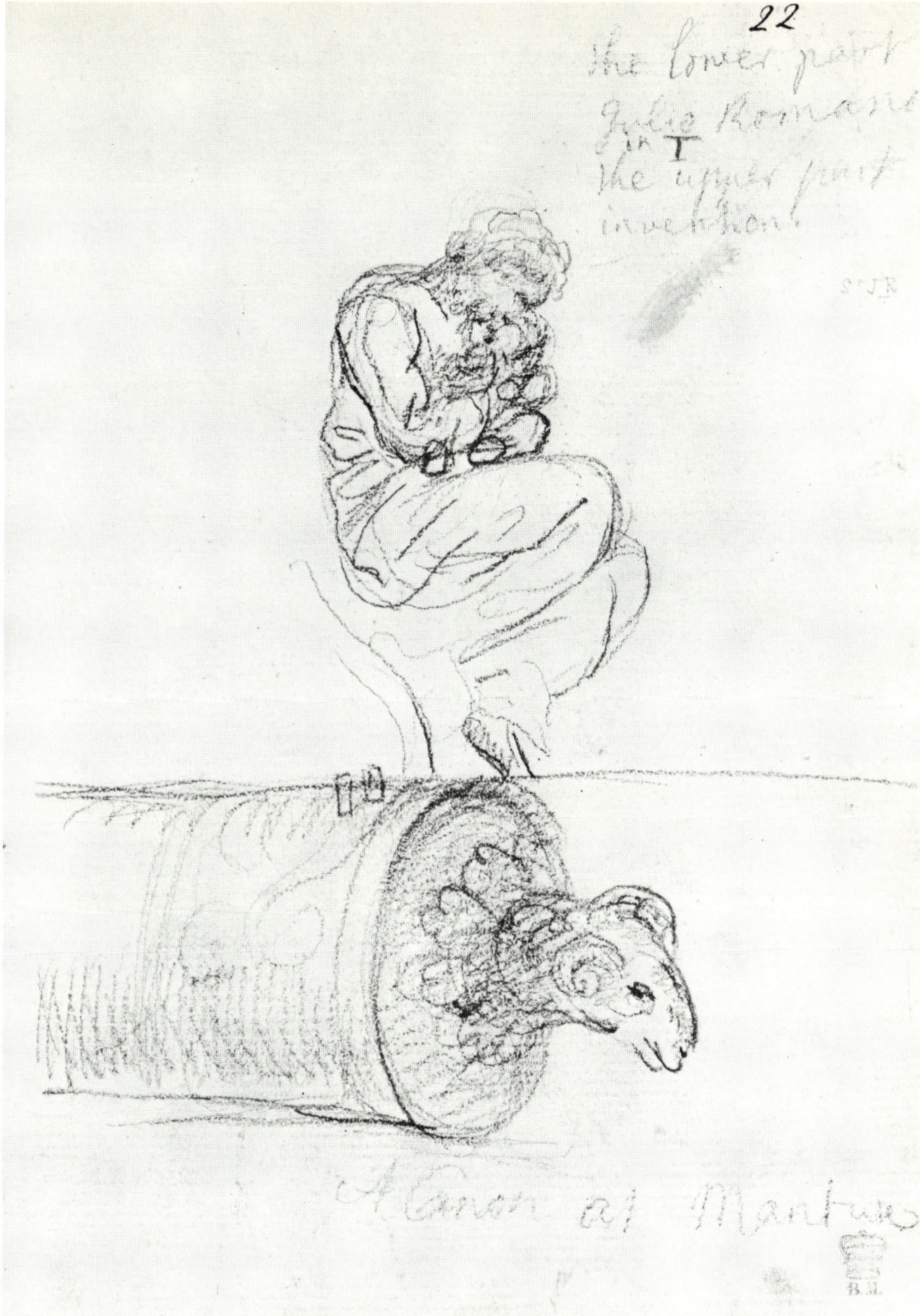

4. Reynolds: Drawing after Giulio Romano, from an Italian Sketchbook. London, British Museum

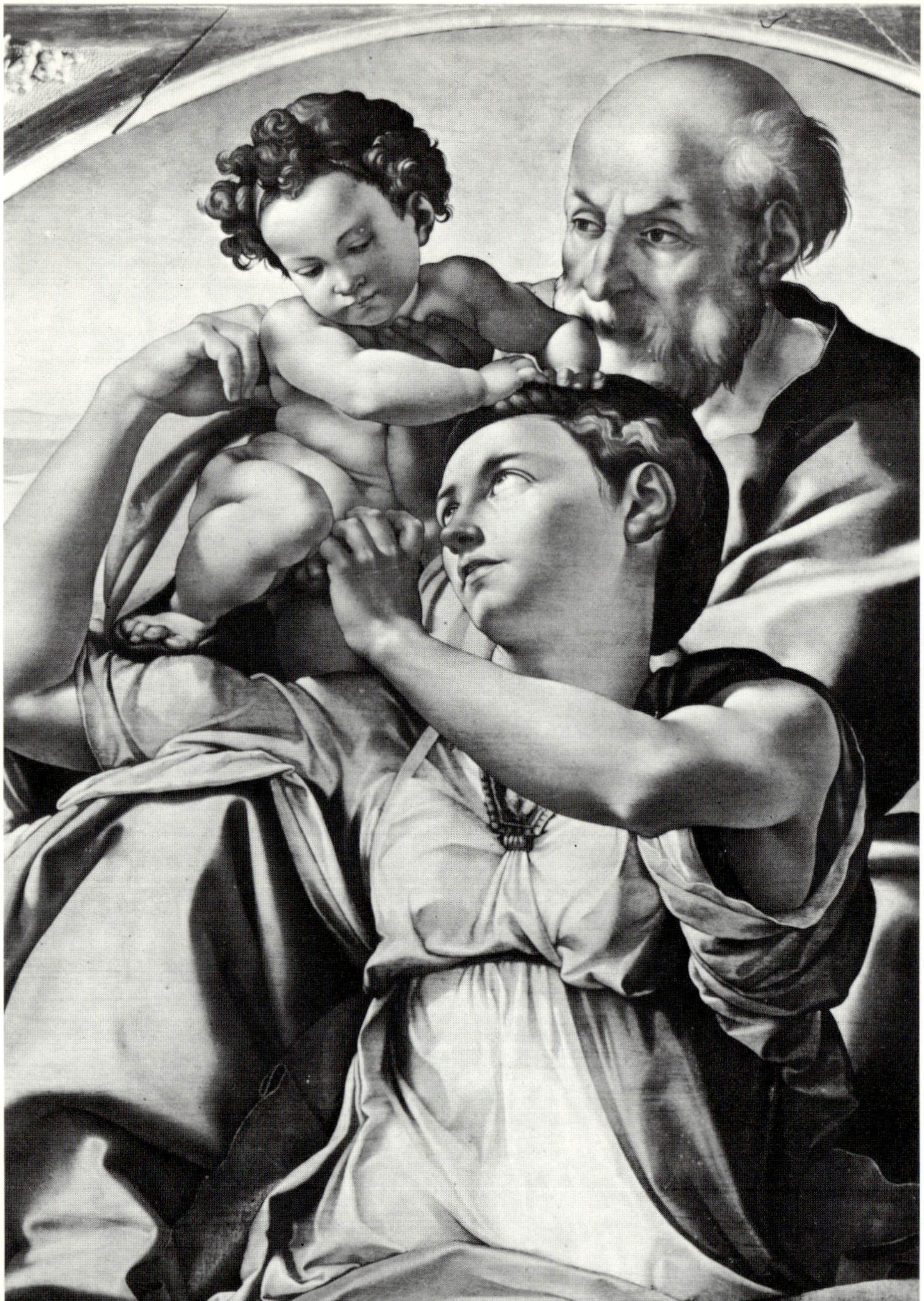

5. Michelangelo: *Madonna Doni*, detail. Florence, Galleria degli Uffizi

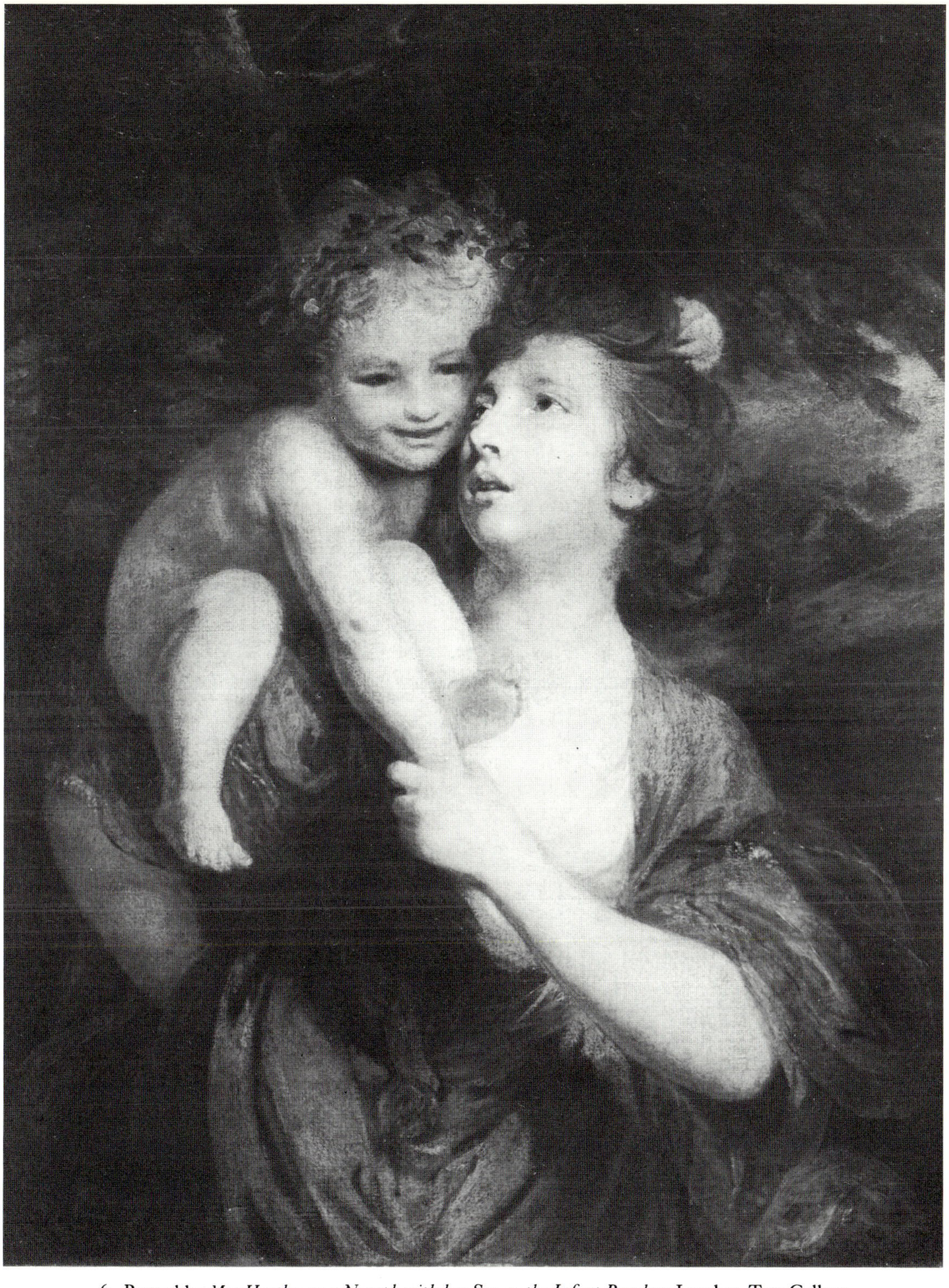

6. Reynolds: *Mrs Hartley as a Nymph with her Son as the Infant Bacchus.* London, Tate Gallery

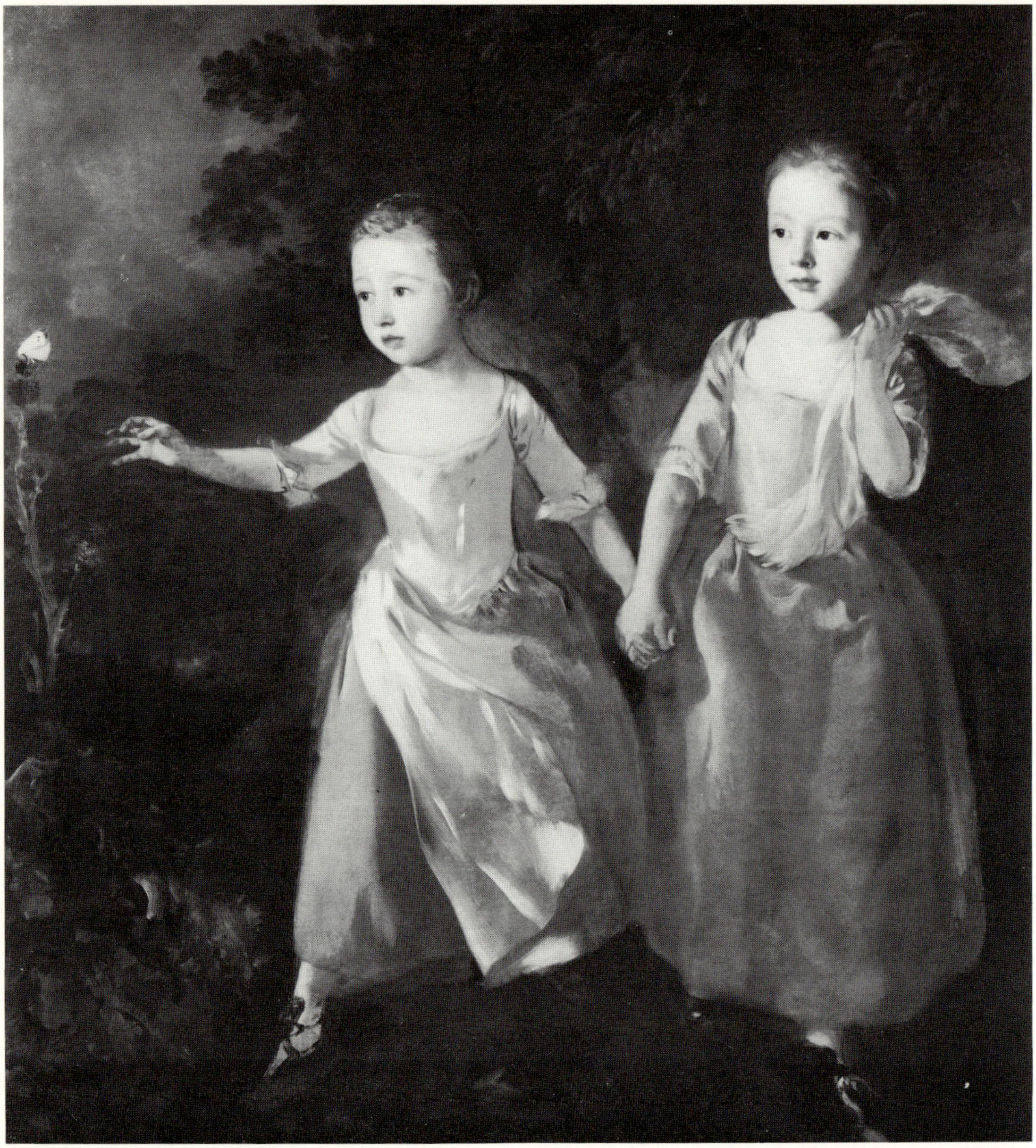

7. Gainsborough: *The Painter's Daughters chasing a Butterfly*. London, National Gallery

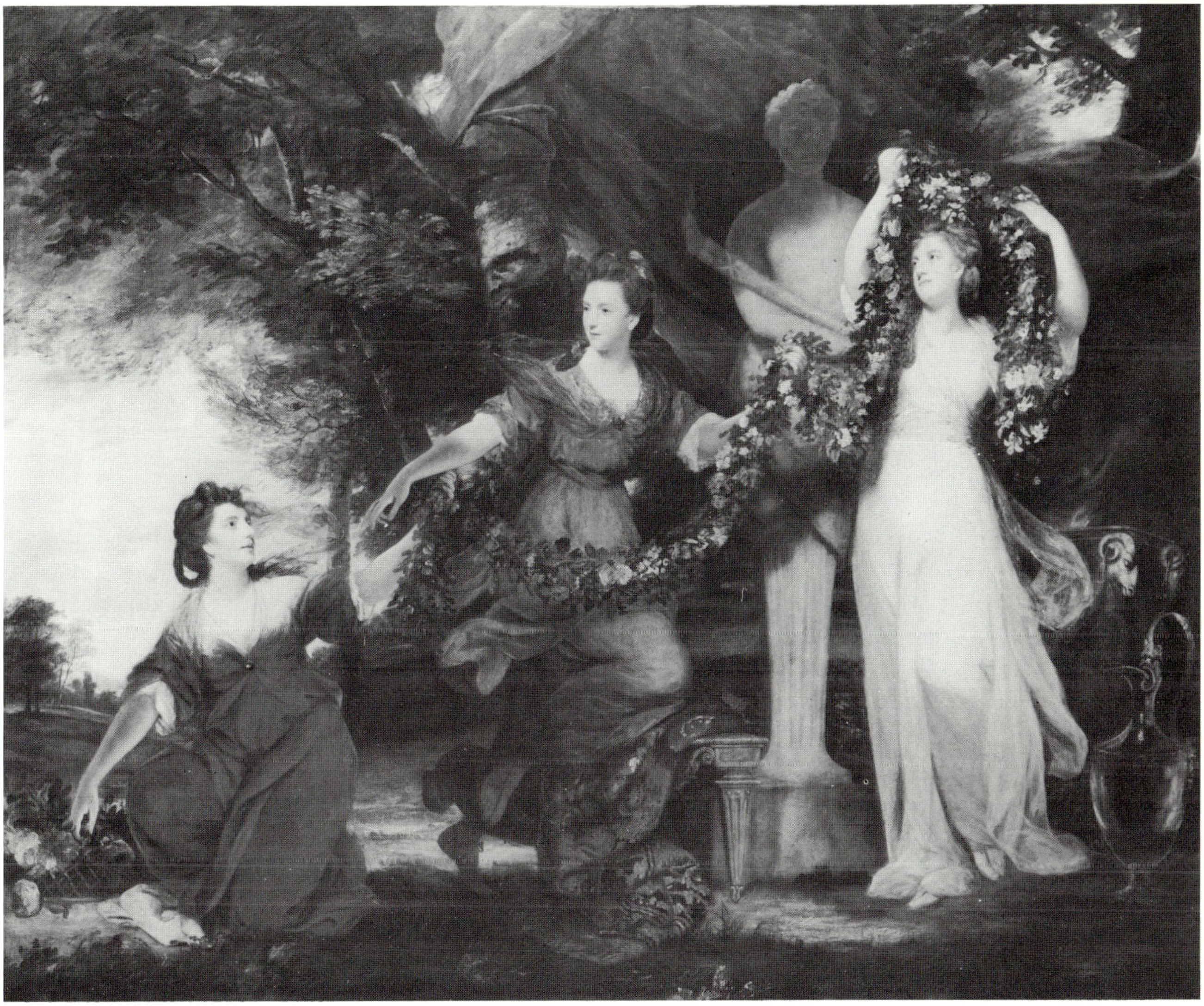

8. Reynolds: *The Daughters of Sir William Montgomery as 'The Graces adorning a term of Hymen'*. London, Tate Gallery

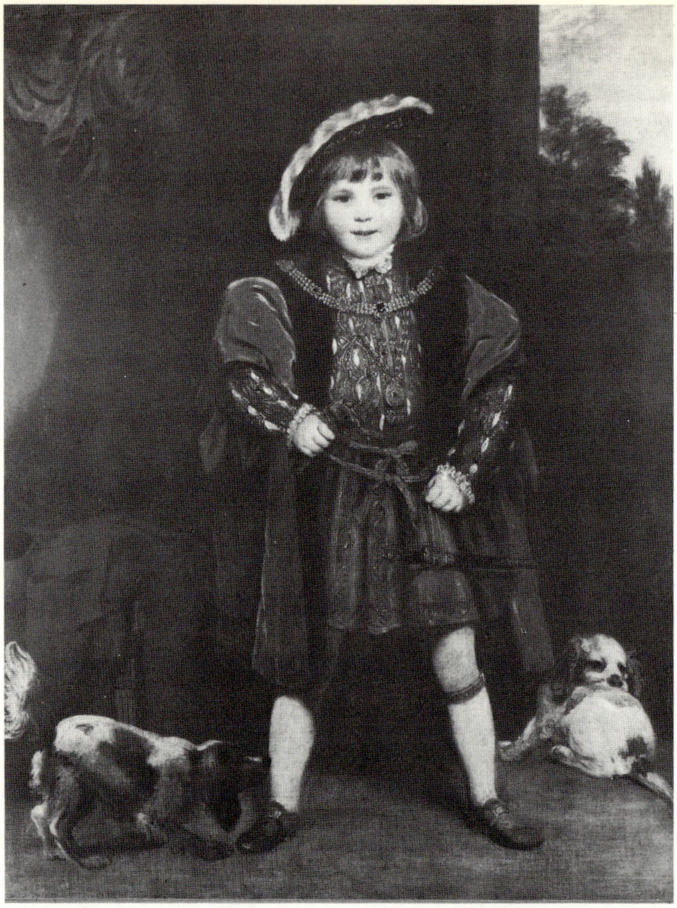

9. Reynolds: *Master Crewe as Henry VIII.*
Shane's Castle, Lord O'Neill

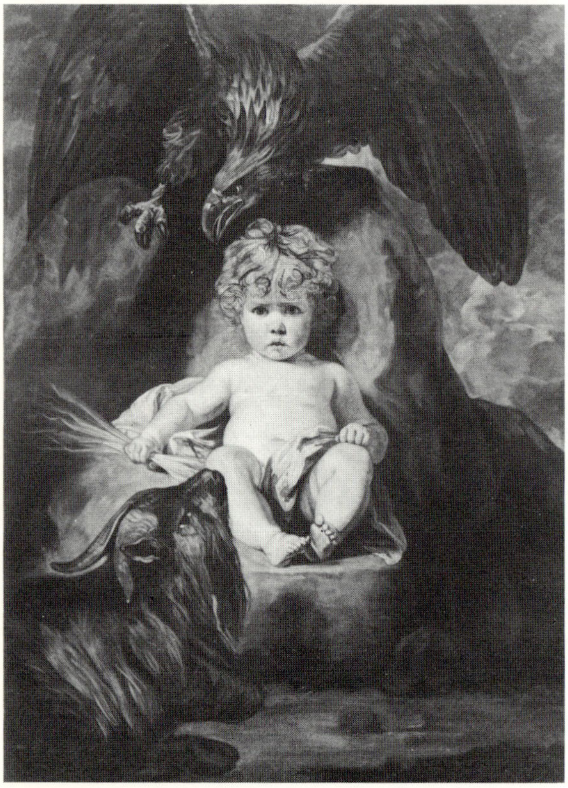

10. Reynolds: *The Infant Jupiter*. Mezzotint by J. R. Smith. London, British Museum

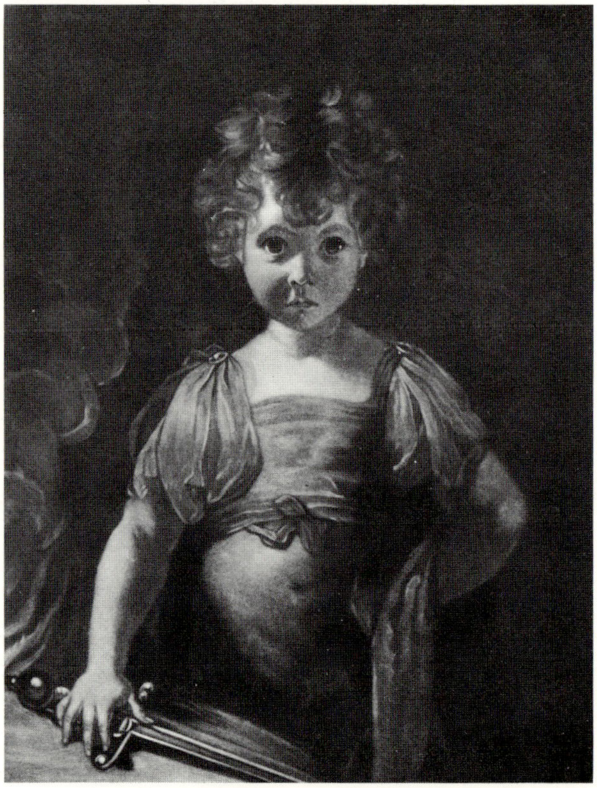

11. Reynolds: *Master Cox as Hannibal.* Mezzotint by Charles Townley. London, British Museum

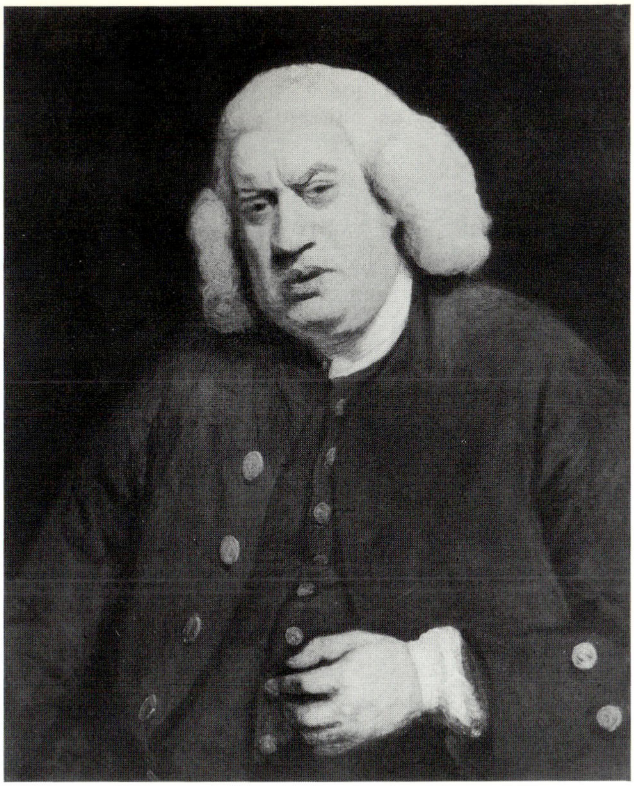

12. Reynolds: *Samuel Johnson.* London, Tate Gallery

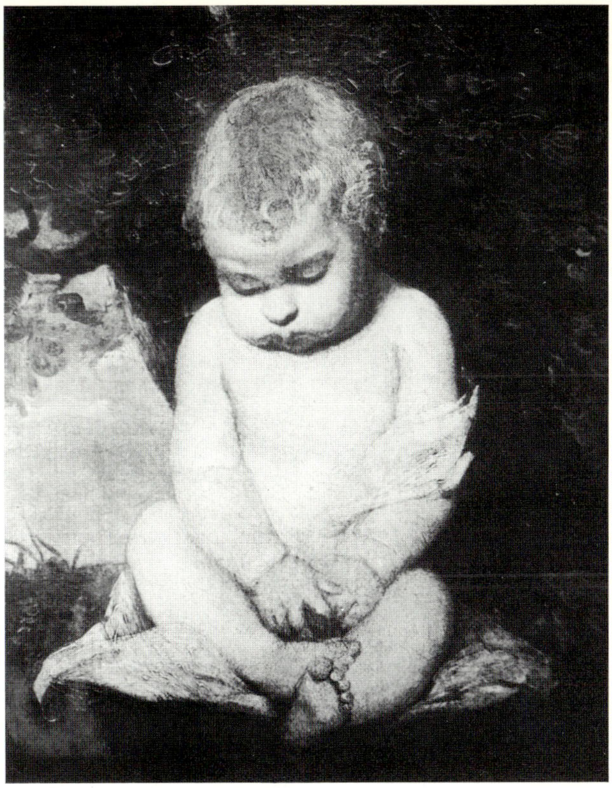

13. Reynolds: *The Infant Johnson.* Somerville (New Jersey), The Hyde Collection

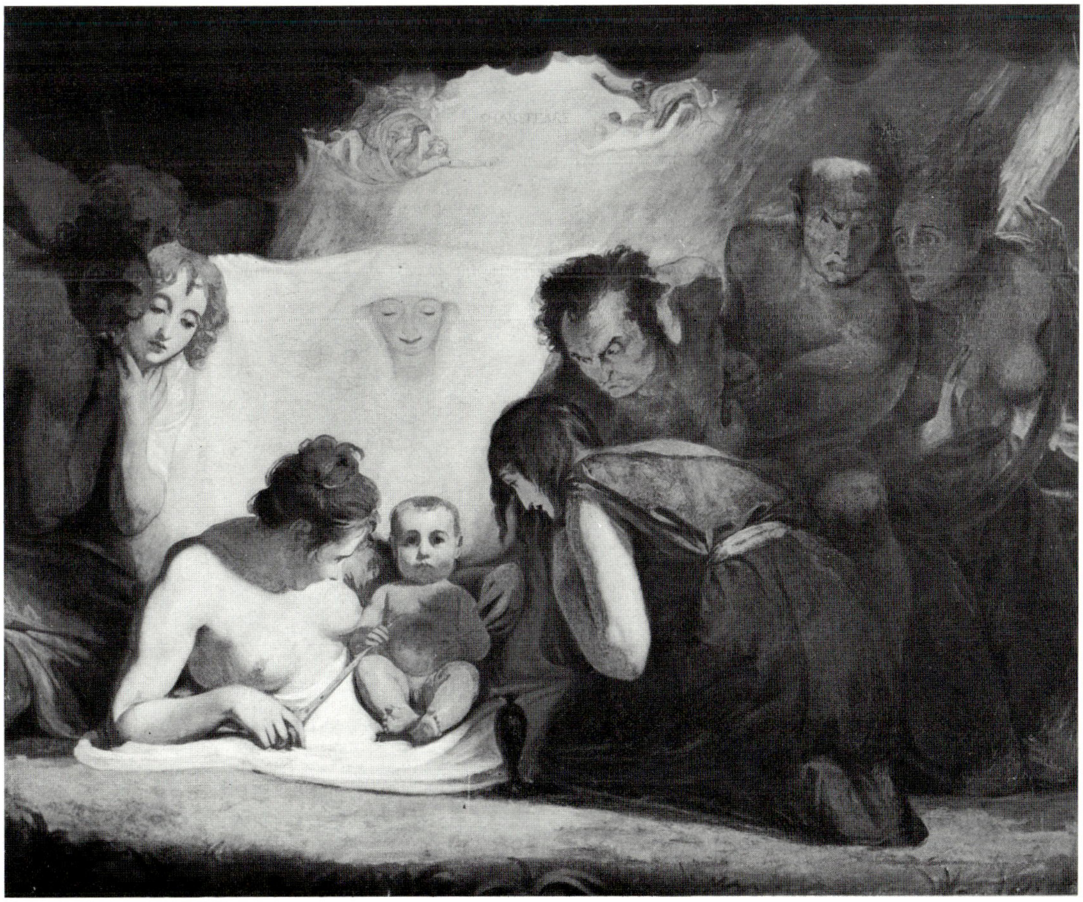

14. Romney: *The Infant Shakespeare attended by Nature and the Passions.* Unfinished oil sketch. Stratford-upon-Avon, The Shakespeare Memorial Theatre Picture Gallery and Museum

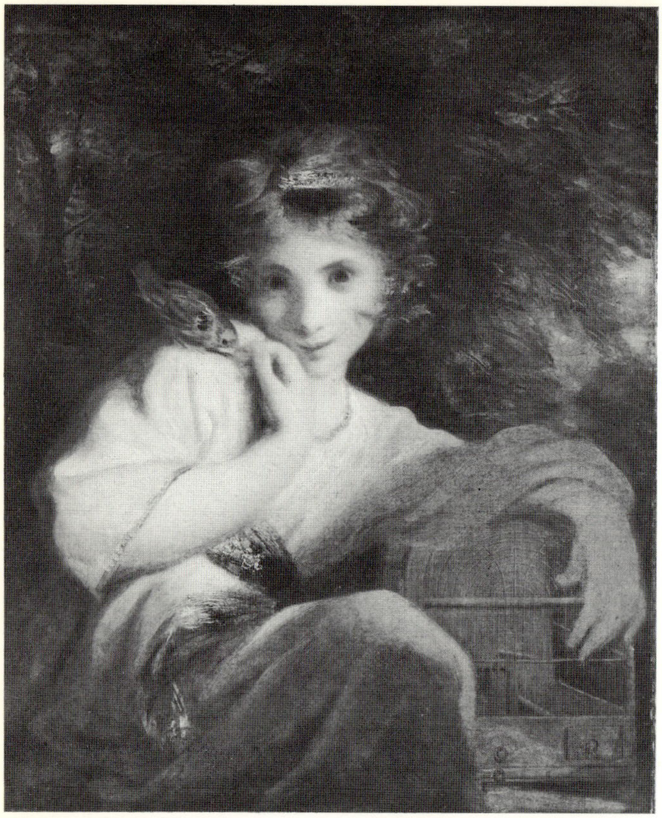

15. Reynolds: *Robinetta* (Magdalena Lewis, later the Hon. Mrs Tollemache). London, Tate Gallery version Lady Anne Cavendish-Bentinck

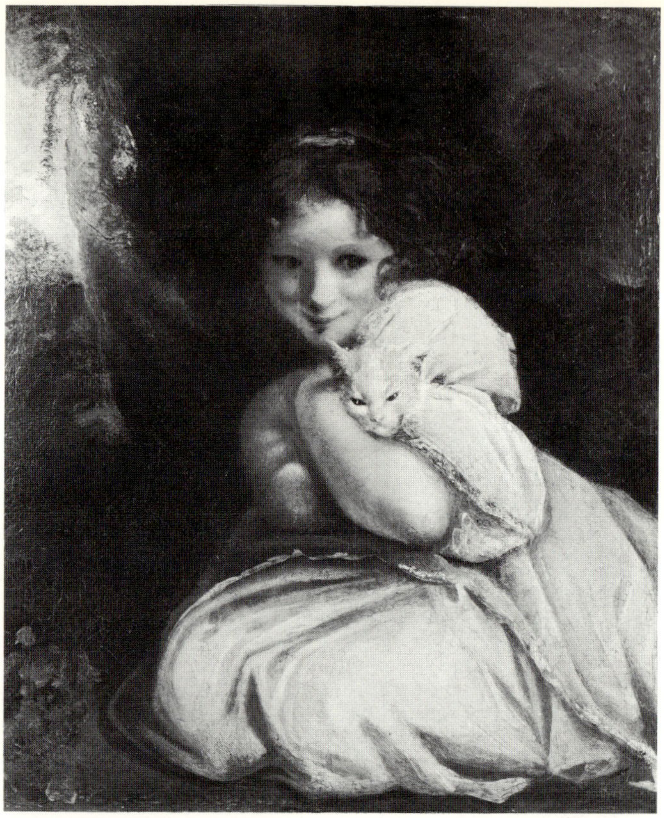

16. Reynolds: *Felina* (Theophila Palmer, later Mrs Gwatkin). Detroit, The Detroit Institute of Arts, Dodge Bequest

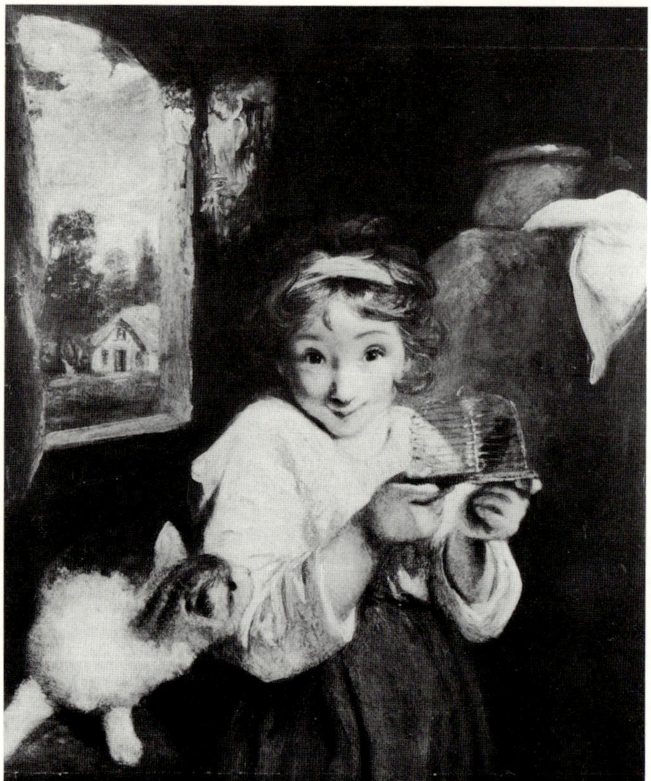

17. Reynolds: *Muscipula*. London, Lady Teresa Agnew

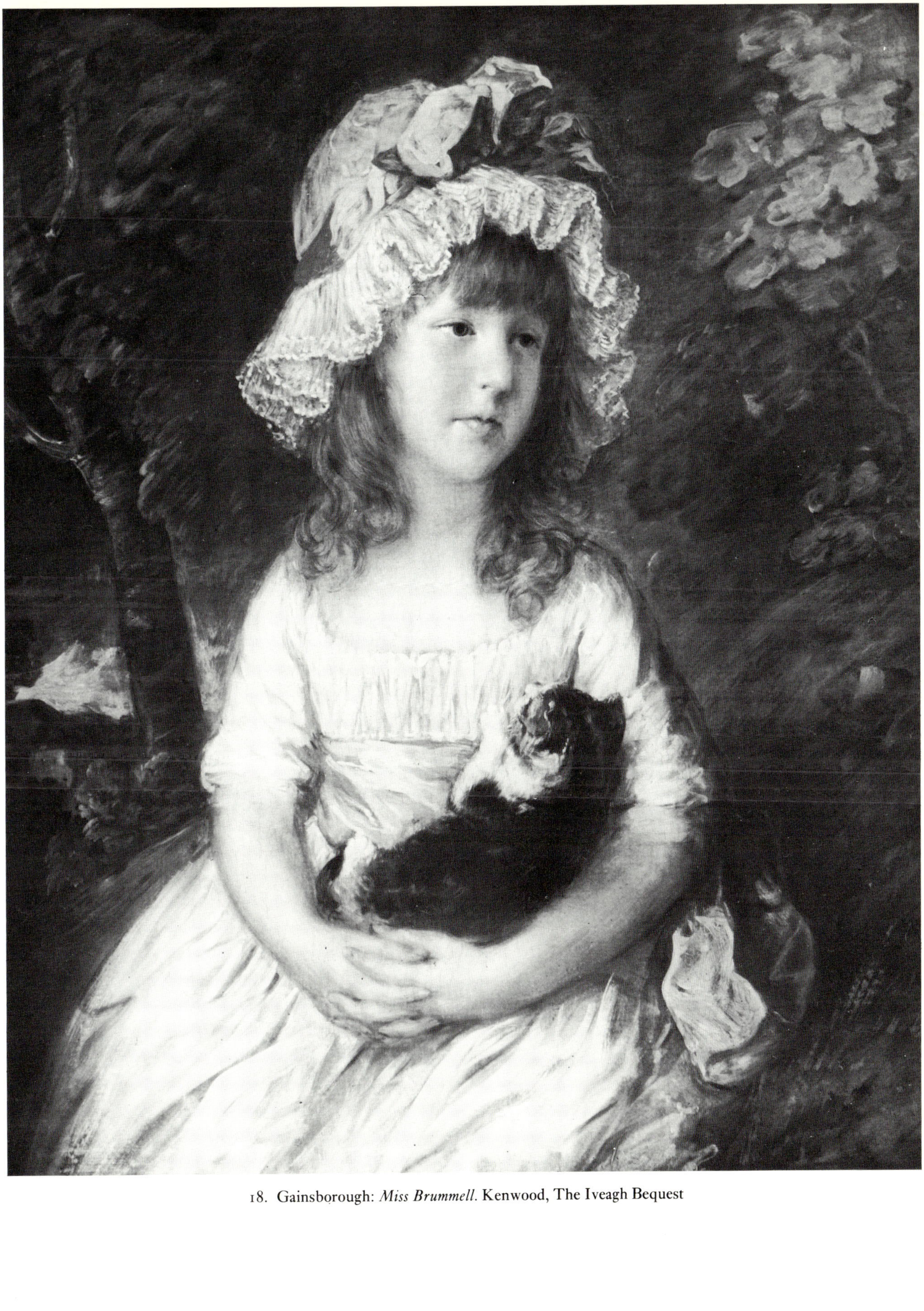

18. Gainsborough: *Miss Brummell.* Kenwood, The Iveagh Bequest

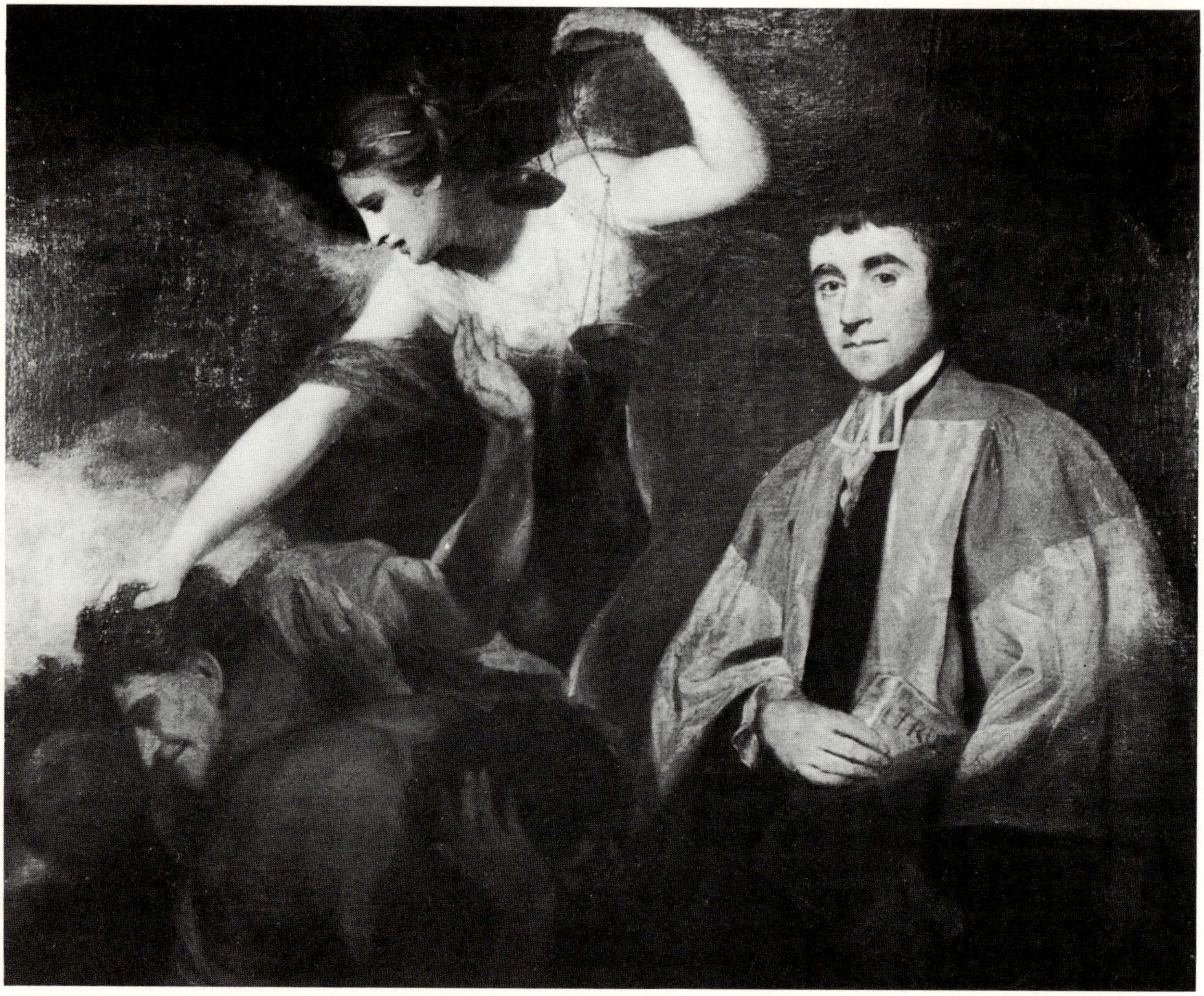

19. Reynolds: *Dr Beattie* or *The Triumph of Truth*. University of Aberdeen

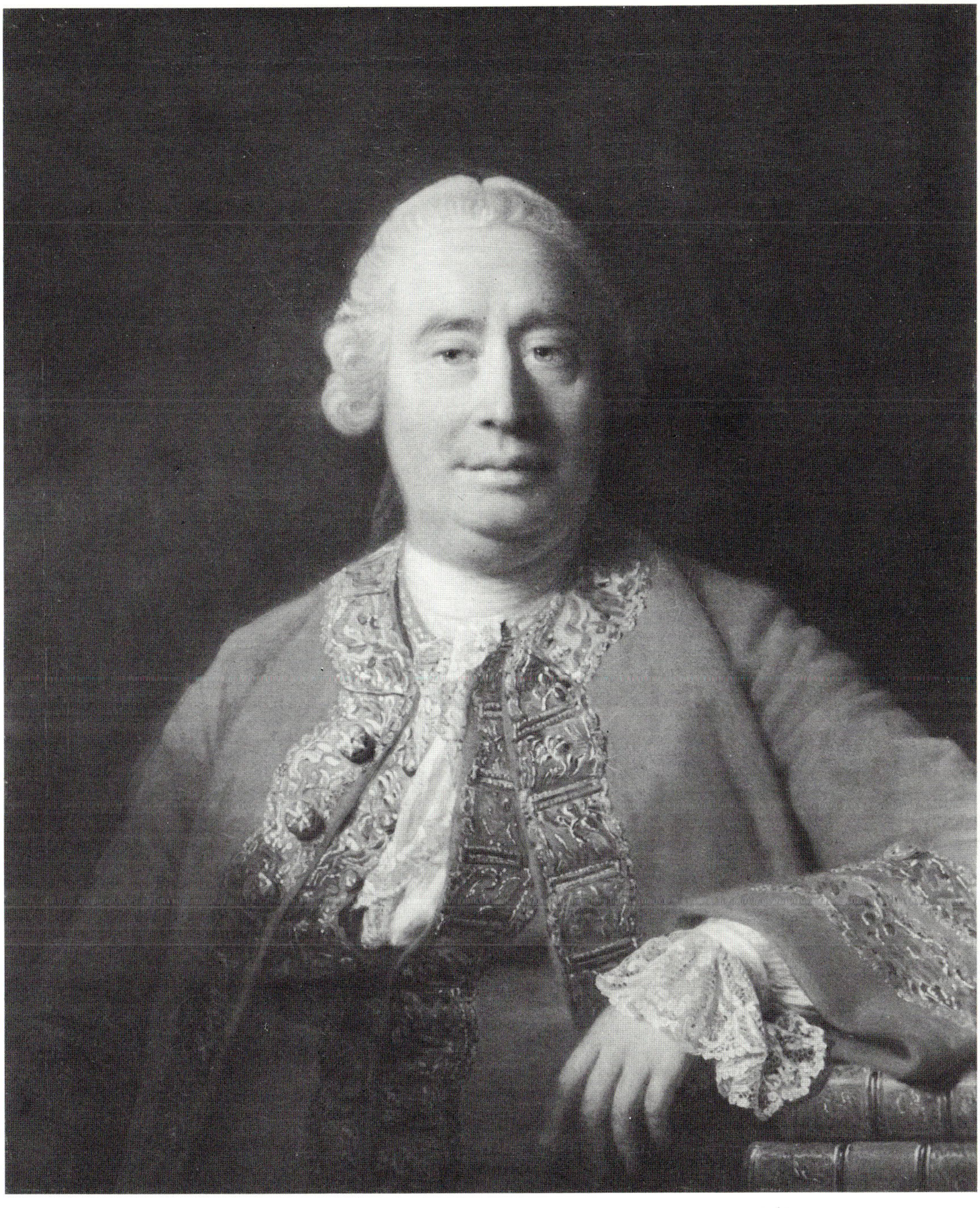

20. Allan Ramsay: *David Hume*. Edinburgh, The Scottish National Portrait Gallery

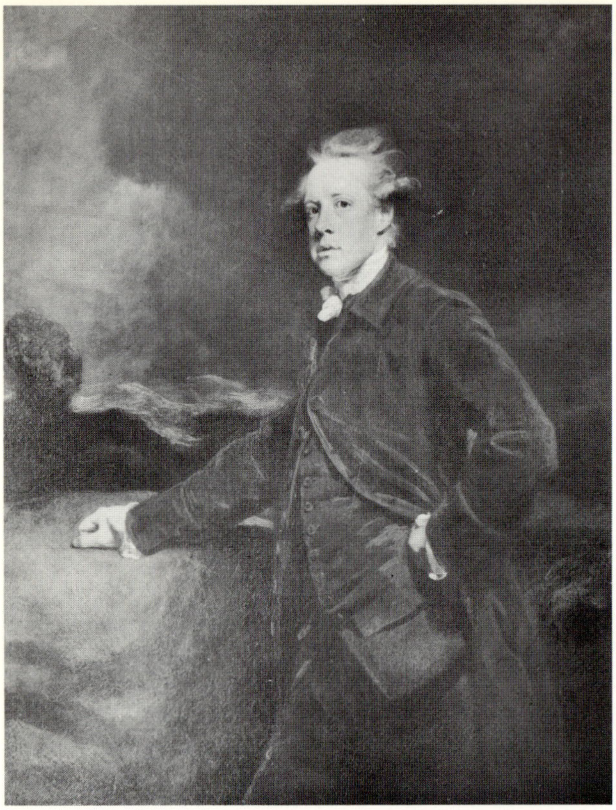

21. Reynolds: *Lord Richard Cavendish*. Welbeck Woodhouse, Lady Ann Cavendish-Bentinck

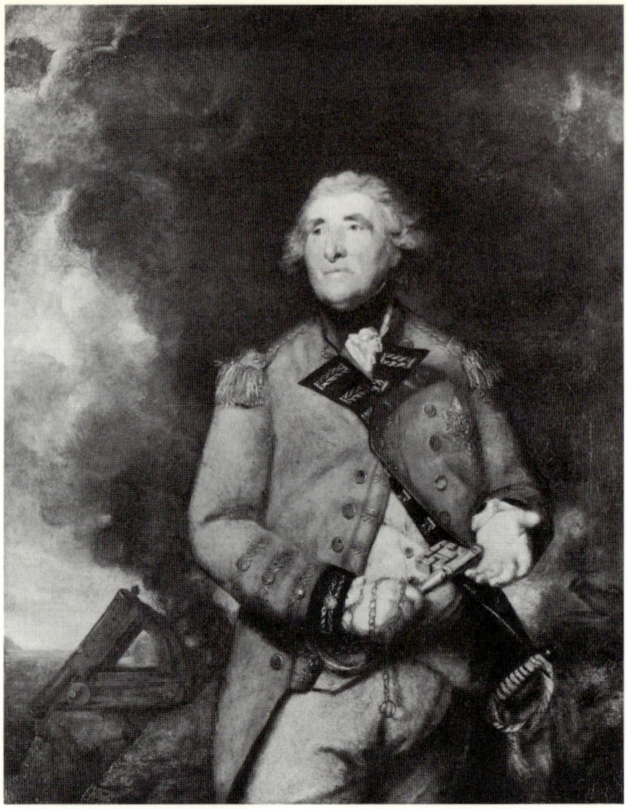

22. Reynolds: *Lord Heathfield, Governor of Gibraltar*. London, National Gallery

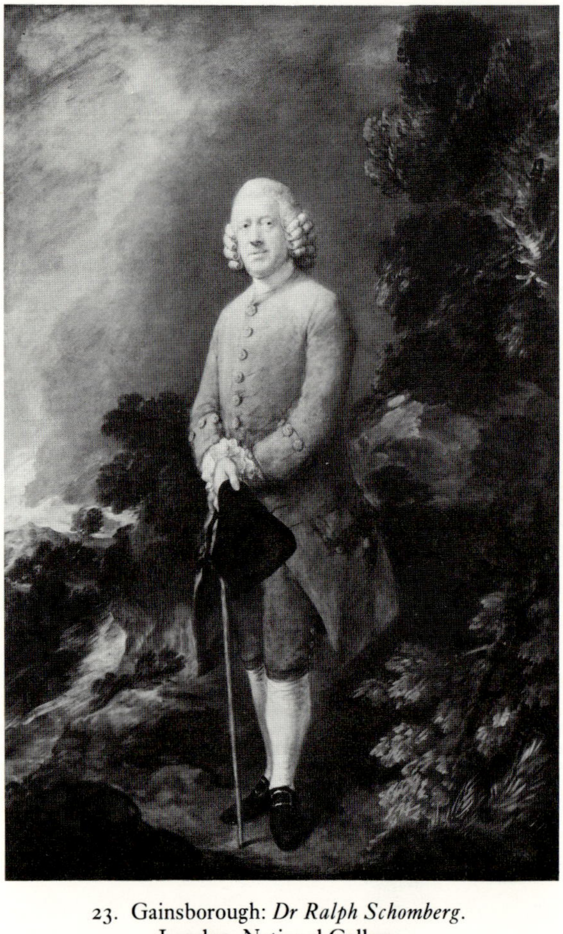

23. Gainsborough: *Dr Ralph Schomberg*. London, National Gallery

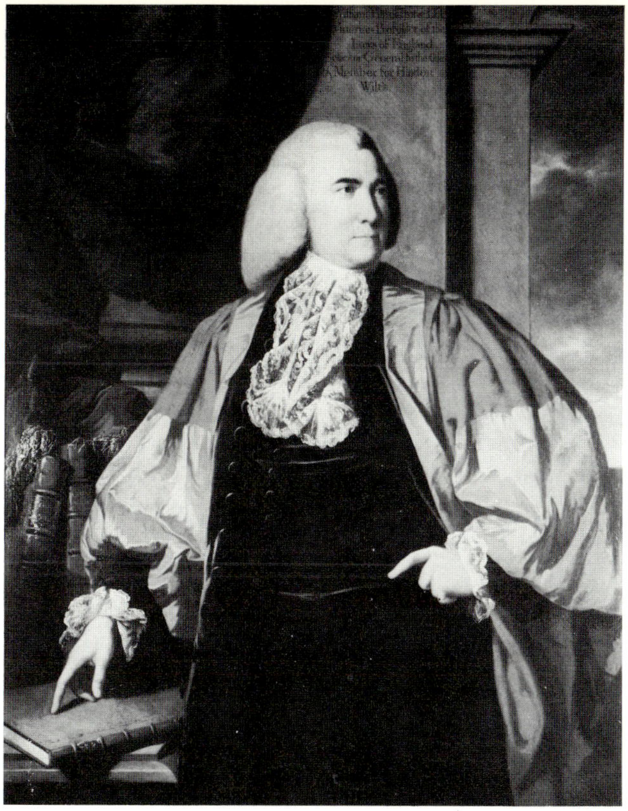

24. Tilly Kettle: *Sir William Blackstone.*
Oxford, Bodleian Library

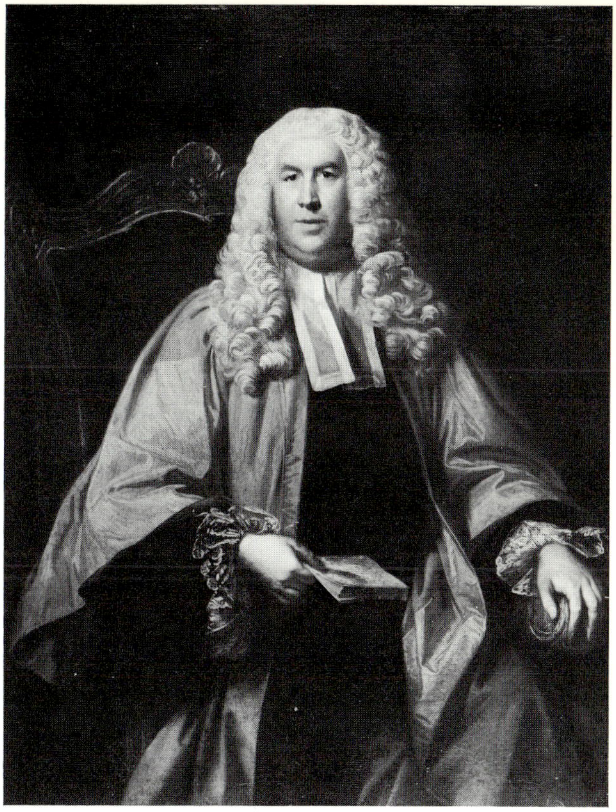

25. Reynolds [attributed to]: *Sir William Blackstone.*
London, National Portrait Gallery

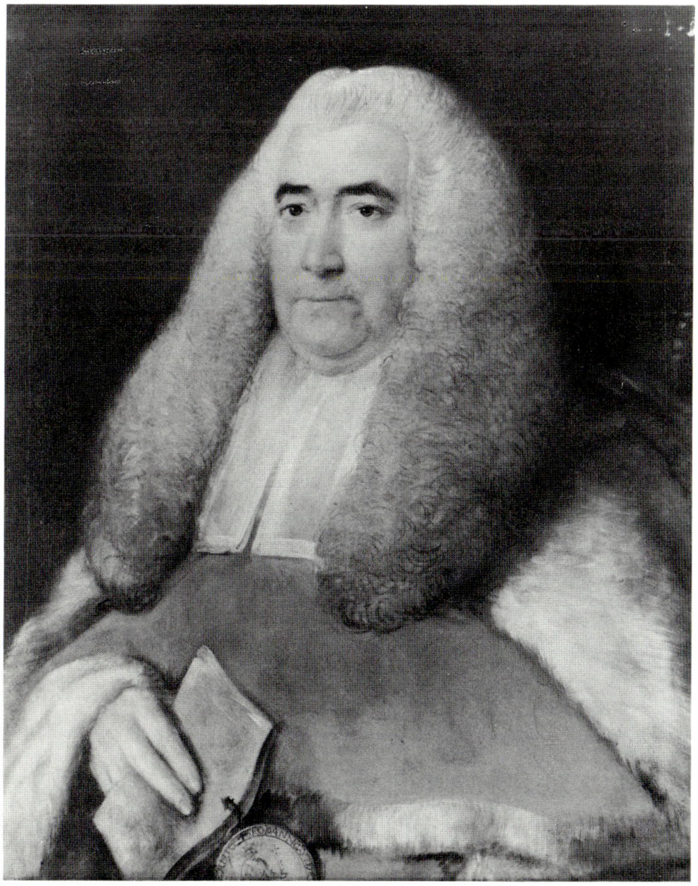

26. Gainsborough: *Sir William Blackstone.* London, Tate Gallery

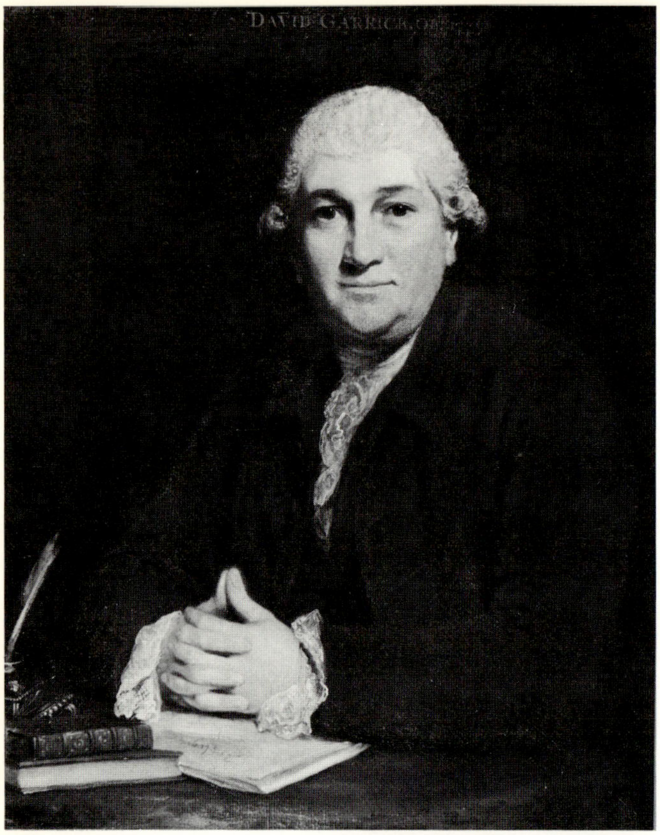

27. Reynolds: *Garrick as Author of Prologues*. Knole, Lord Sackville

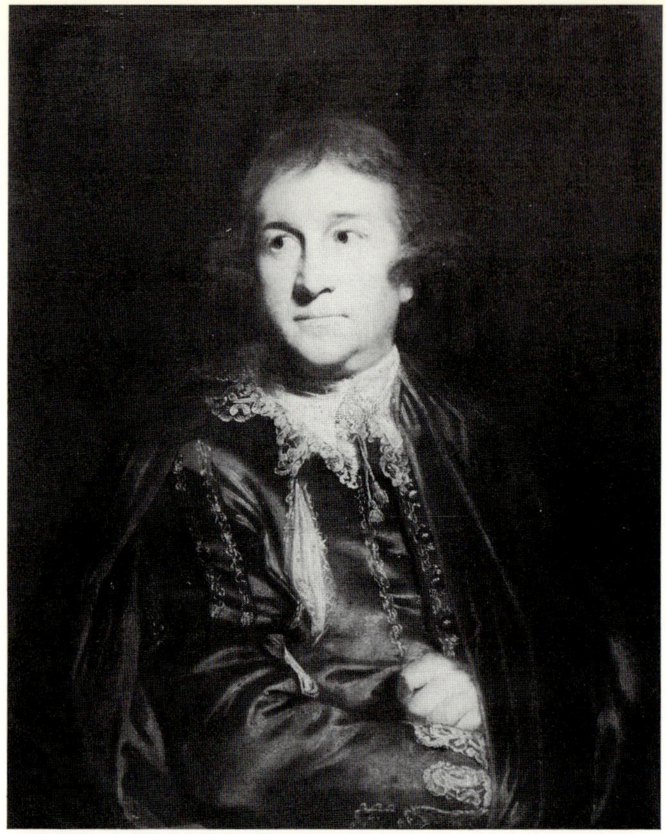

28. Reynolds: *Garrick as Kiteley in 'Every Man in his Humour'*. Windsor Castle. By gracious permission of Her Majesty the Queen

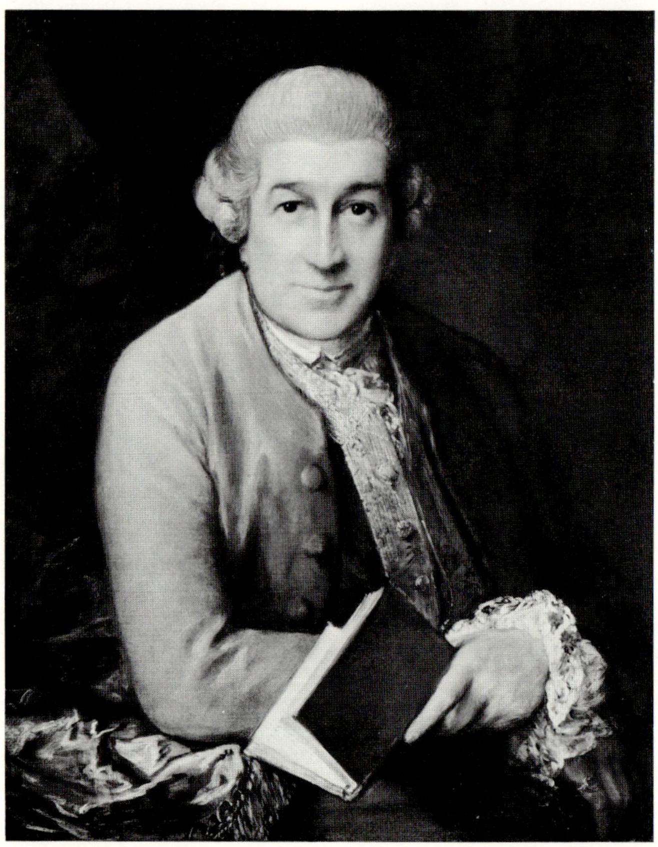

29. Gainsborough: *Garrick*. London, National Portrait Gallery

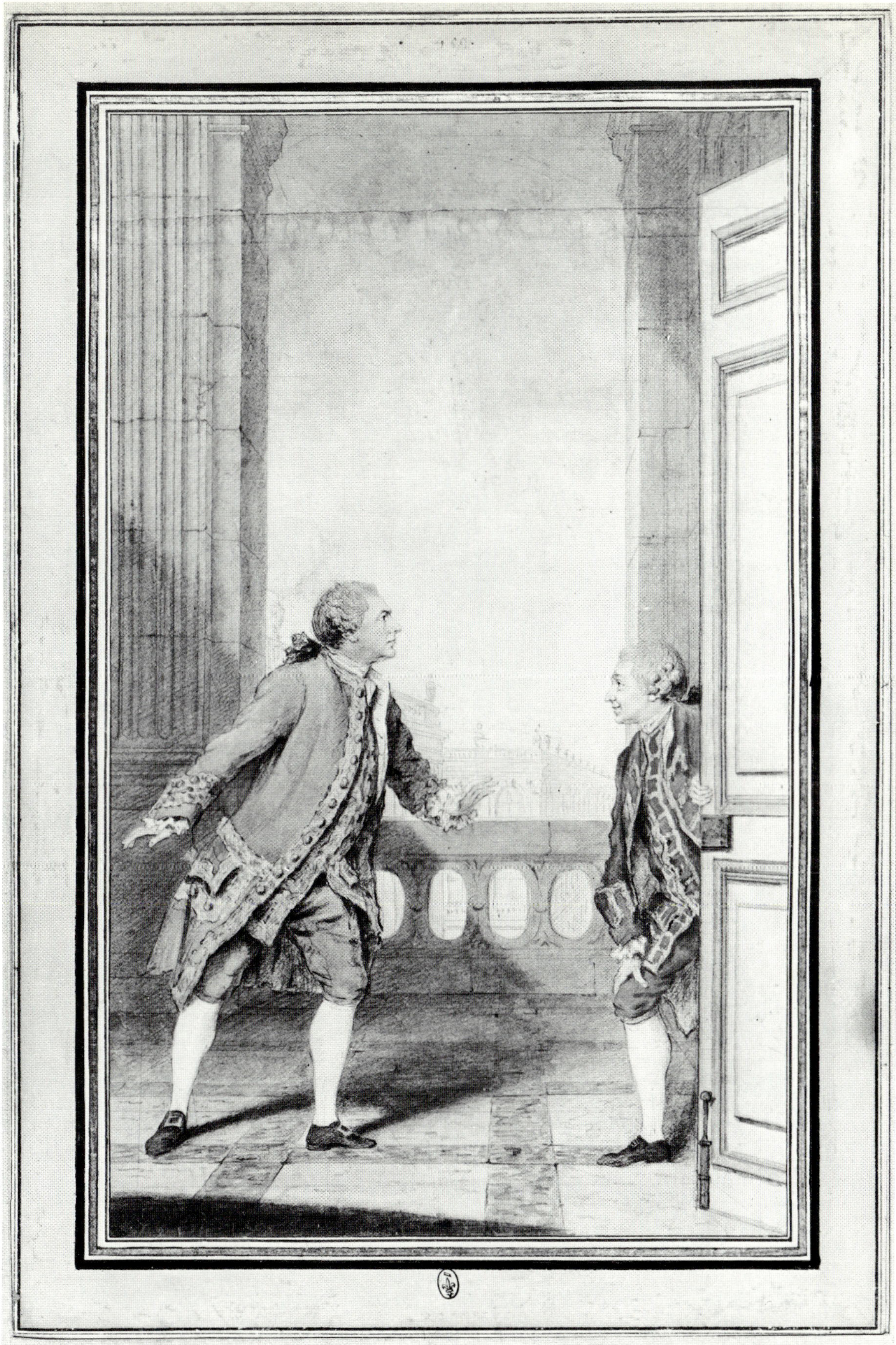

30. Louis de Carmontelle: *Le célèbre Garrick, tragique et comique . . . 1765. . . .* Drawing, inscribed by Lédans. Chantilly, Musée Condé

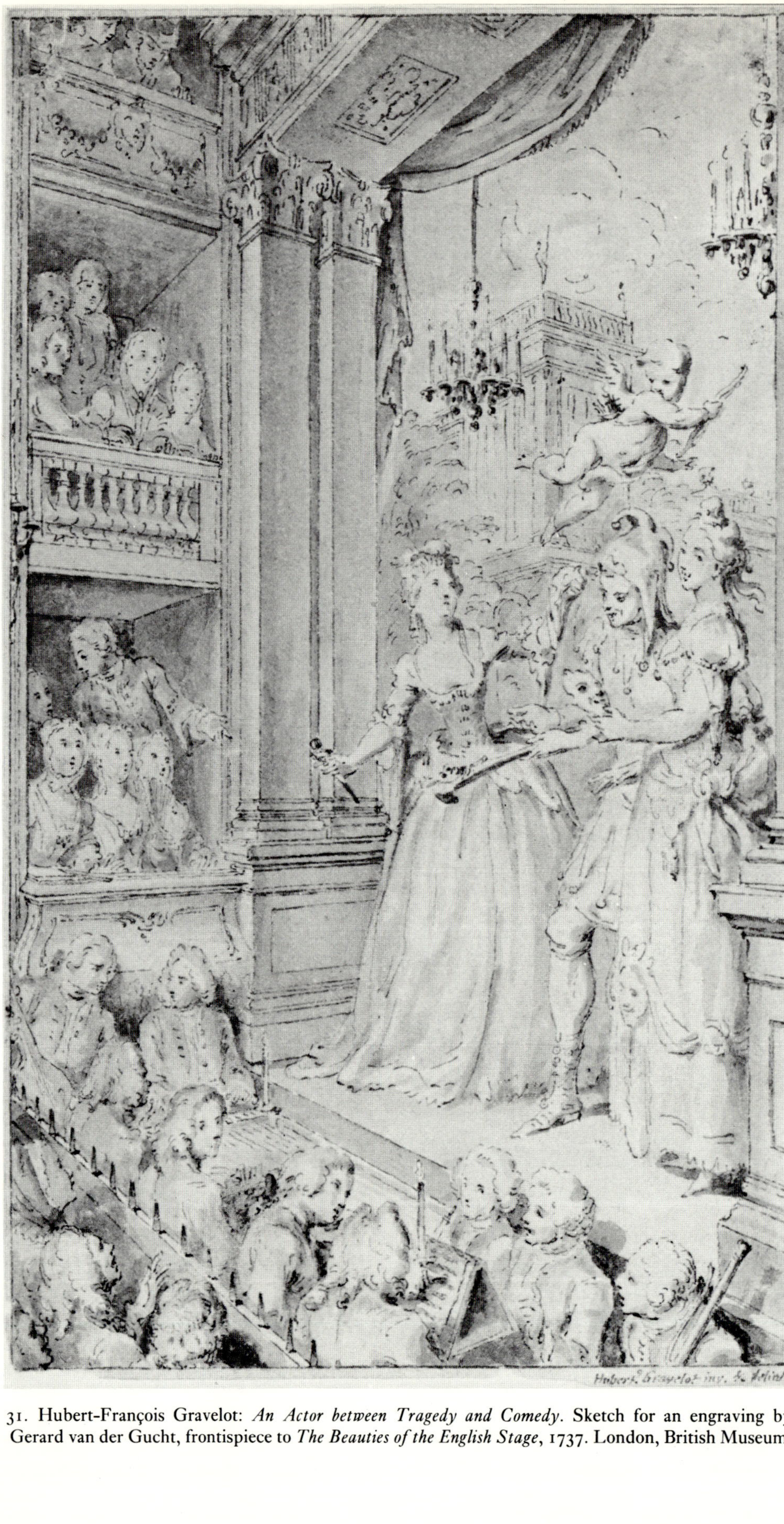

31. Hubert-François Gravelot: *An Actor between Tragedy and Comedy*. Sketch for an engraving by Gerard van der Gucht, frontispiece to *The Beauties of the English Stage*, 1737. London, British Museum

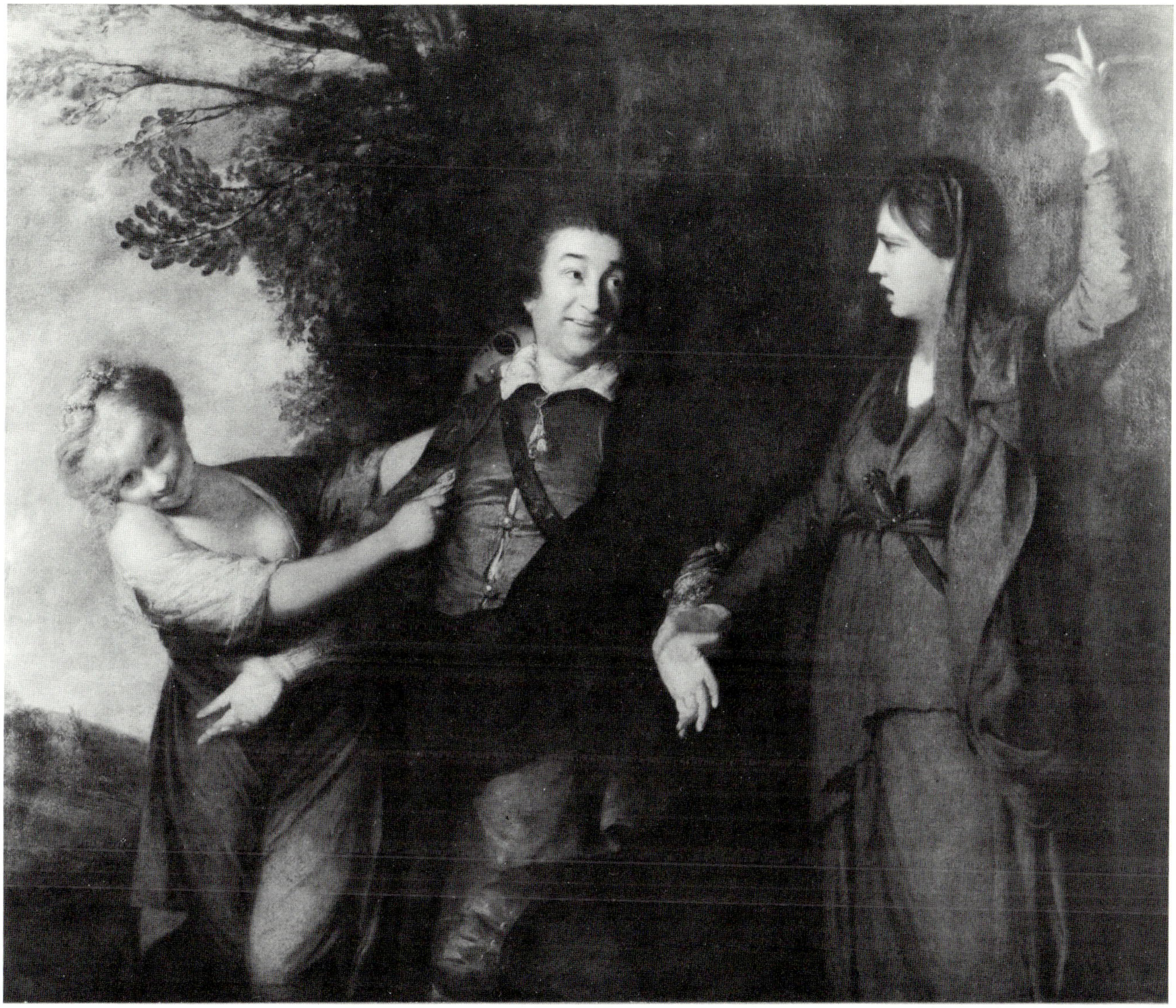

32. Reynolds: *Garrick between Tragedy and Comedy*. Private Collection

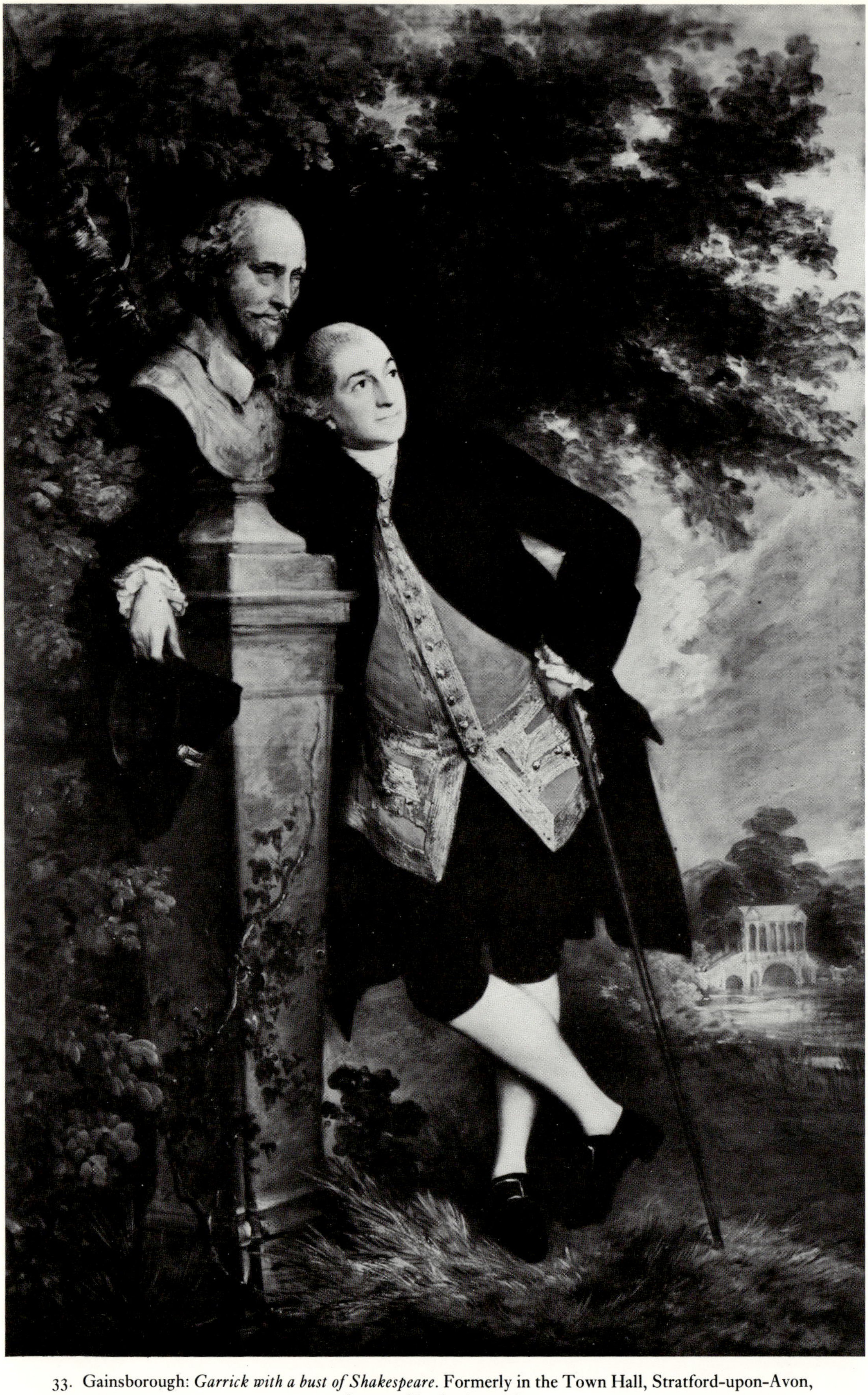

33. Gainsborough: *Garrick with a bust of Shakespeare*. Formerly in the Town Hall, Stratford-upon-Avon, destroyed by fire in 1946

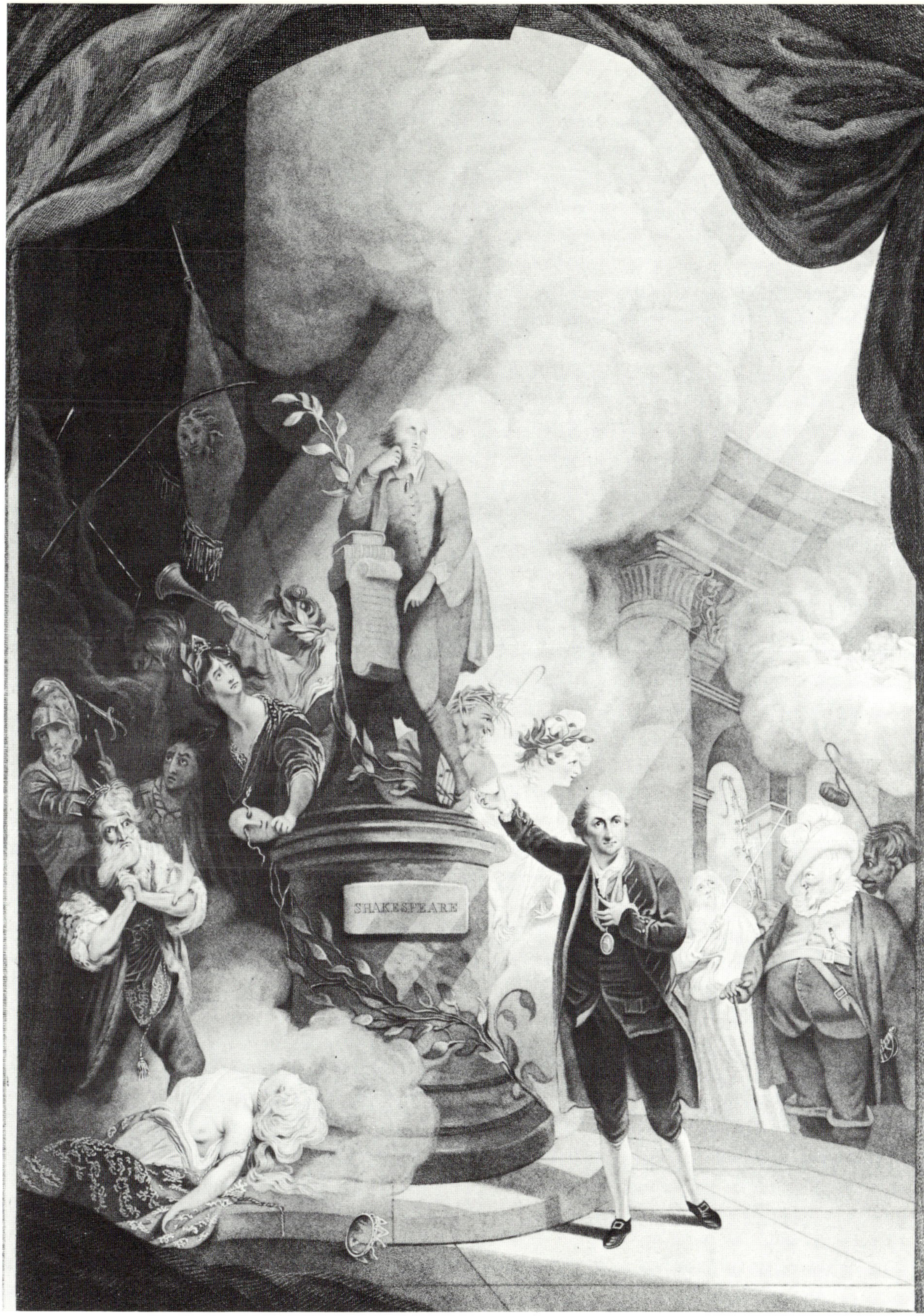

34. Robert Edge Pine: *Garrick reciting an Ode to Shakespeare*. Stipple engraving by Caroline Watson. London, British Museum

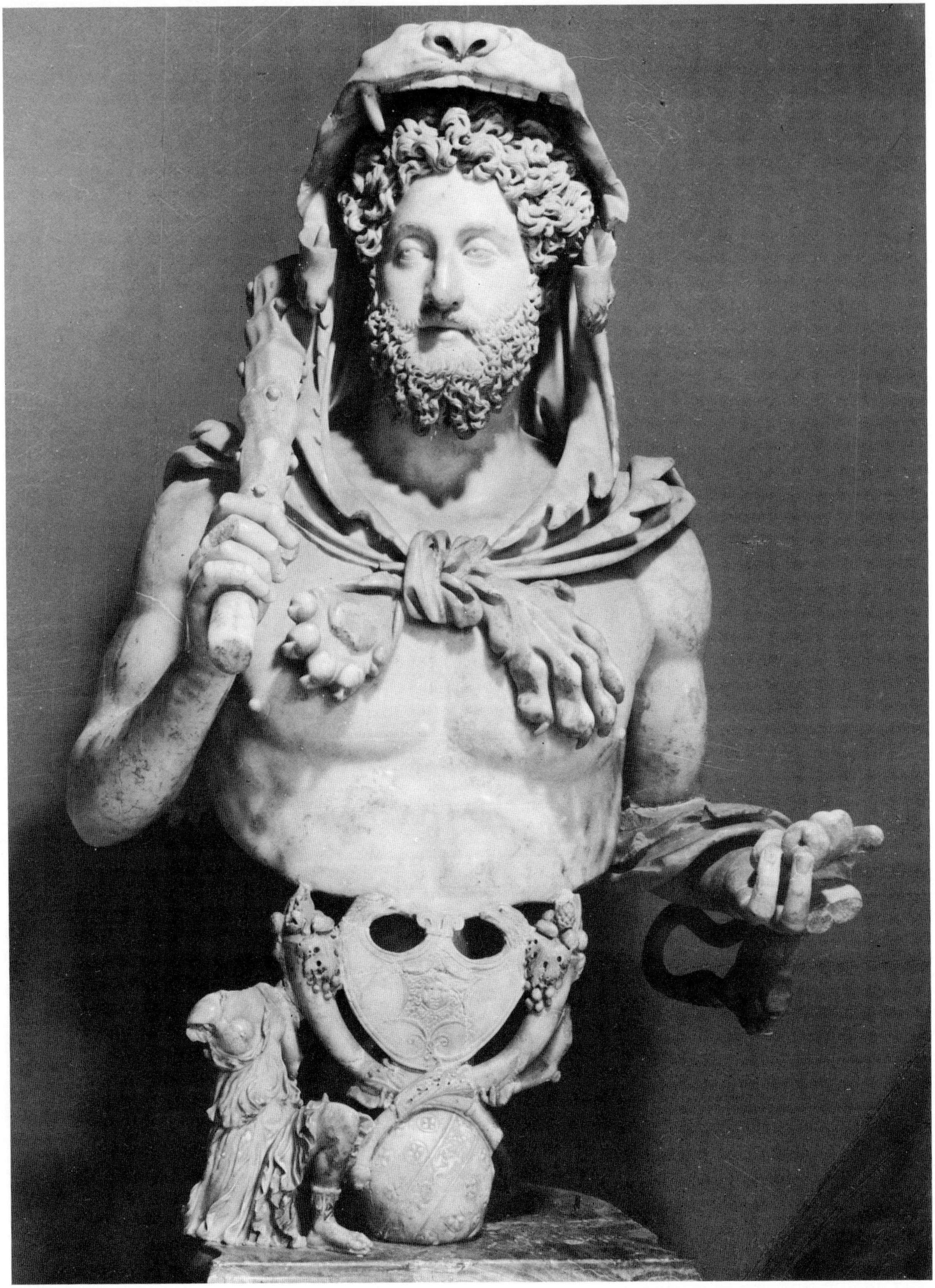

35. Second-century Roman bust: *Commodus as Hercules*. Rome, Museo Capitolino

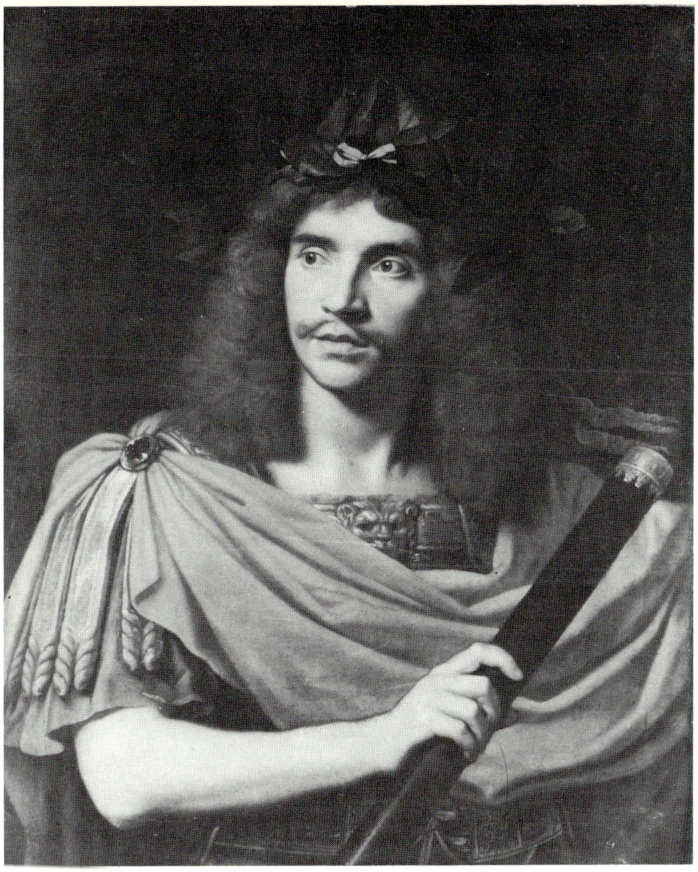

36. Mignard: *Molière as Julius Caesar*. Paris, Comédie-Française

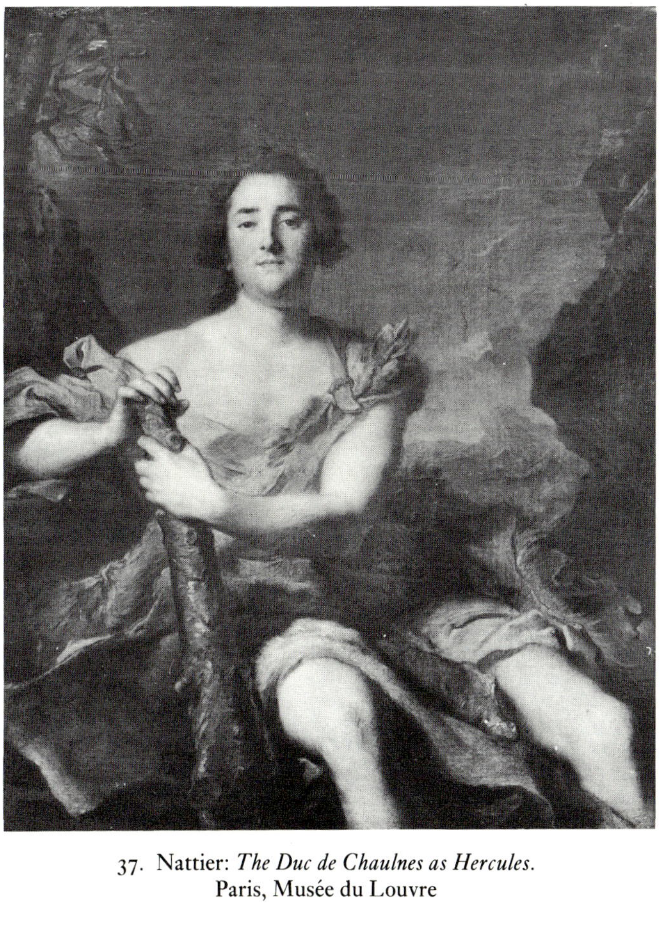

37. Nattier: *The Duc de Chaulnes as Hercules*. Paris, Musée du Louvre

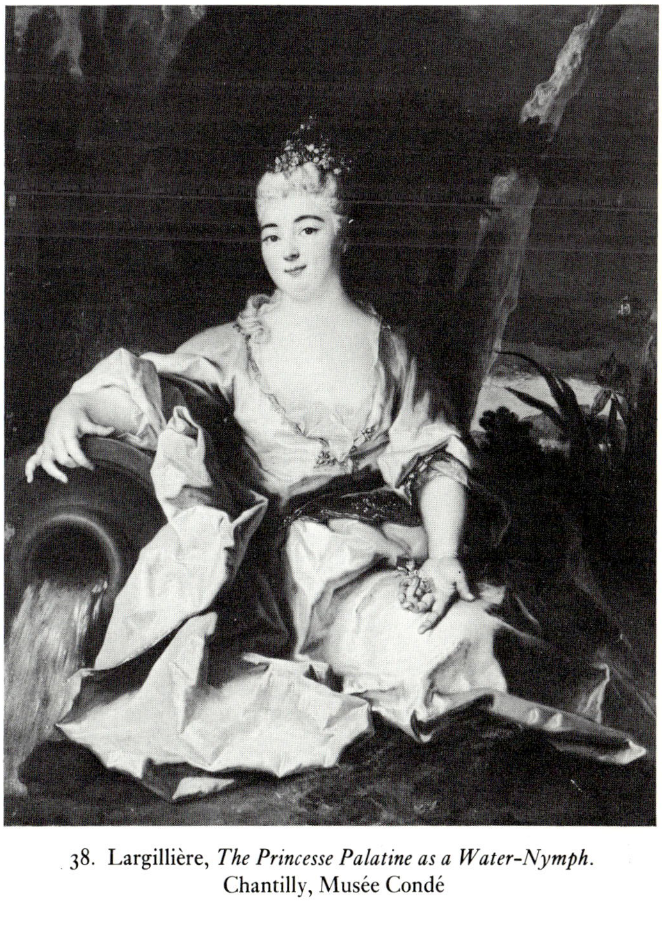

38. Largillière, *The Princesse Palatine as a Water-Nymph*. Chantilly, Musée Condé

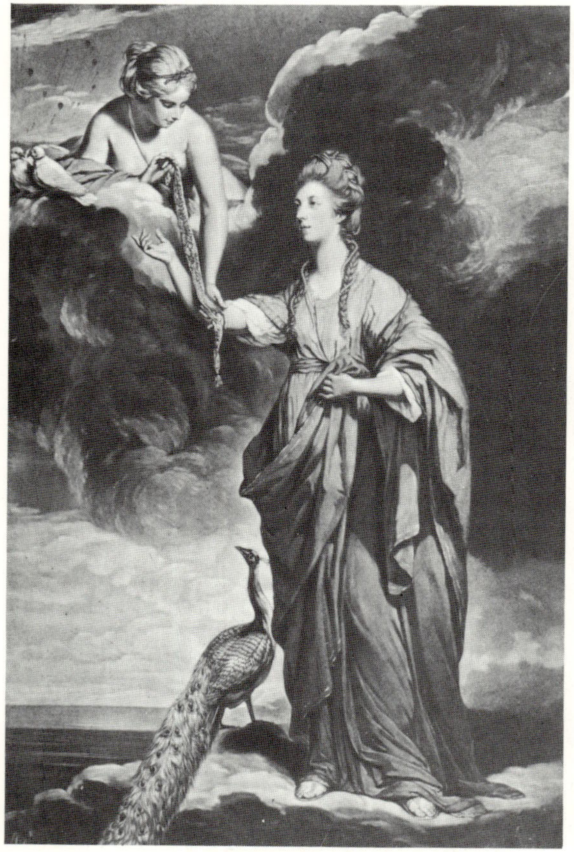

39. Reynolds: *Lady Blake as Juno receiving the cestus from Venus*. Mezzotint by John Dixon. London, British Museum

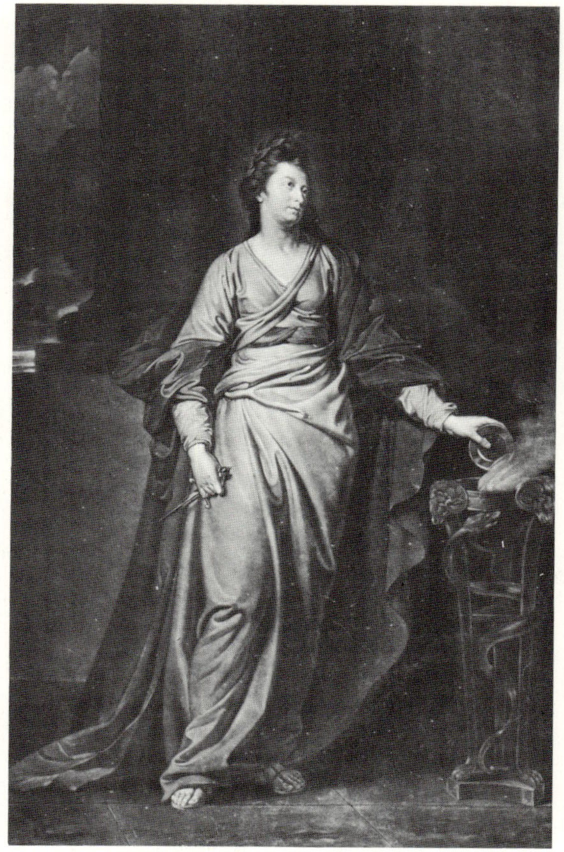

40. Romney: *Mrs Yates as the Tragic Muse*. Mezzotint by Valentine Green. London, British Museum

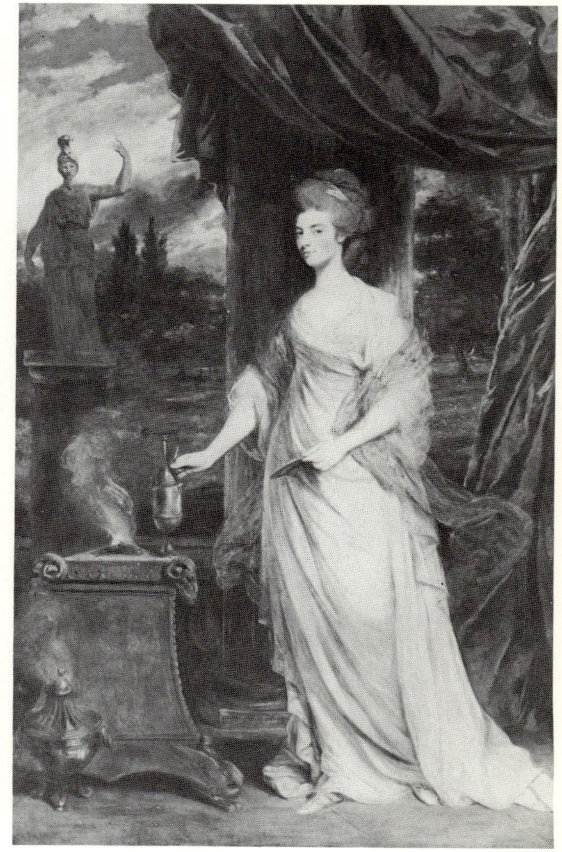

41. Reynolds: *Lady Charlotte Hill, Countess Talbot, sacrificing to Minerva*. London, Tate Gallery

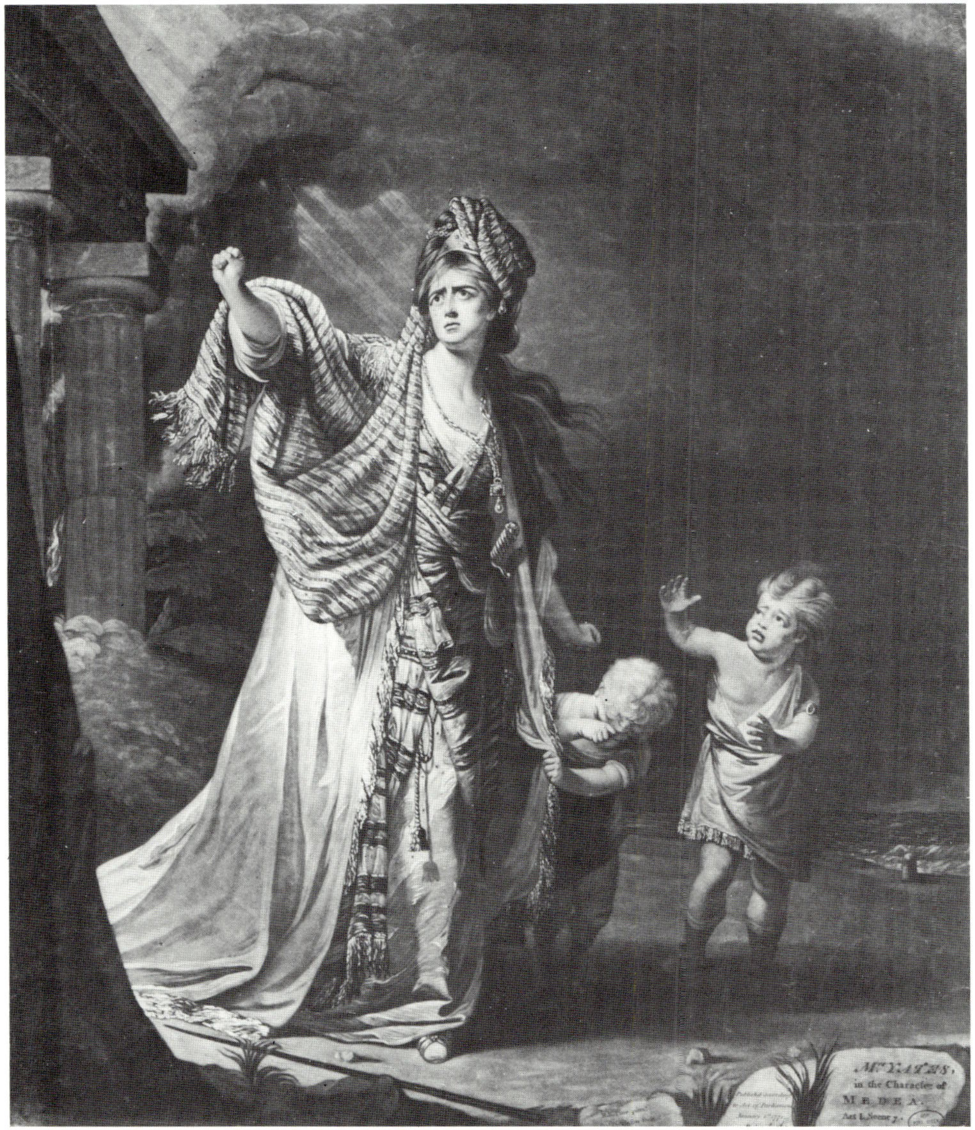

42. Ralph Edge Pine: *Mrs Yates in the character of Medea*. Mezzotint by William Dickinson. Oxford, Ashmolean Museum.

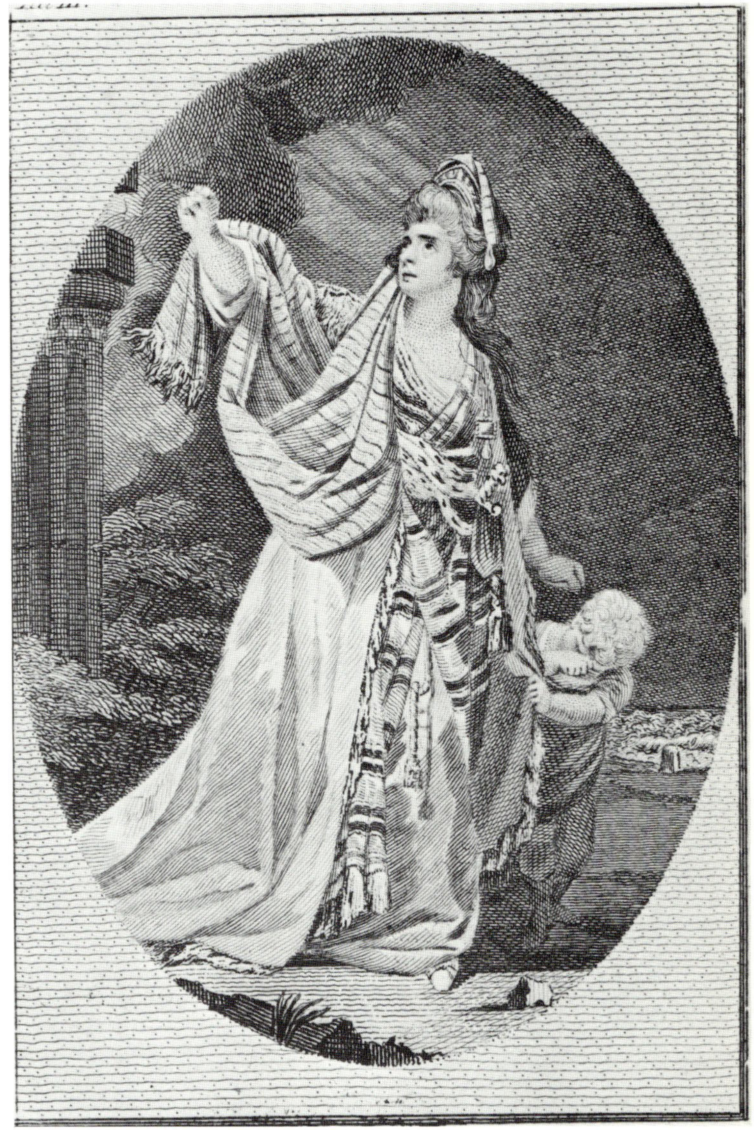

43. J. Thornthwaite: *Mrs Siddons as Medea*. Stipple and line engraving. London, British Museum

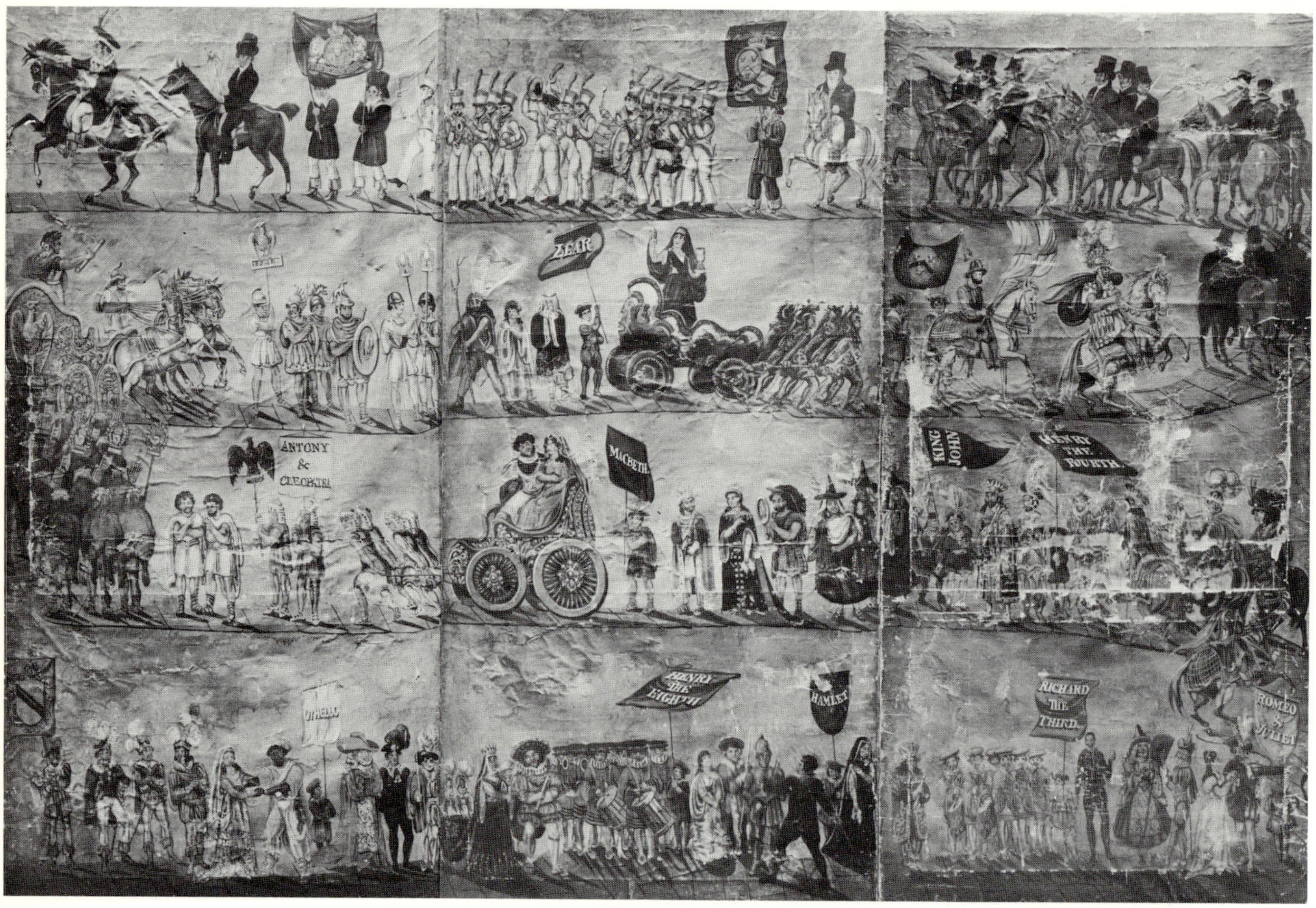

44. *Procession of the Tragic Muse*, from the Stratford Jubilee Screen, 1830. Formerly in the Town Hall, Stratford-upon-Avon, destroyed in 1955

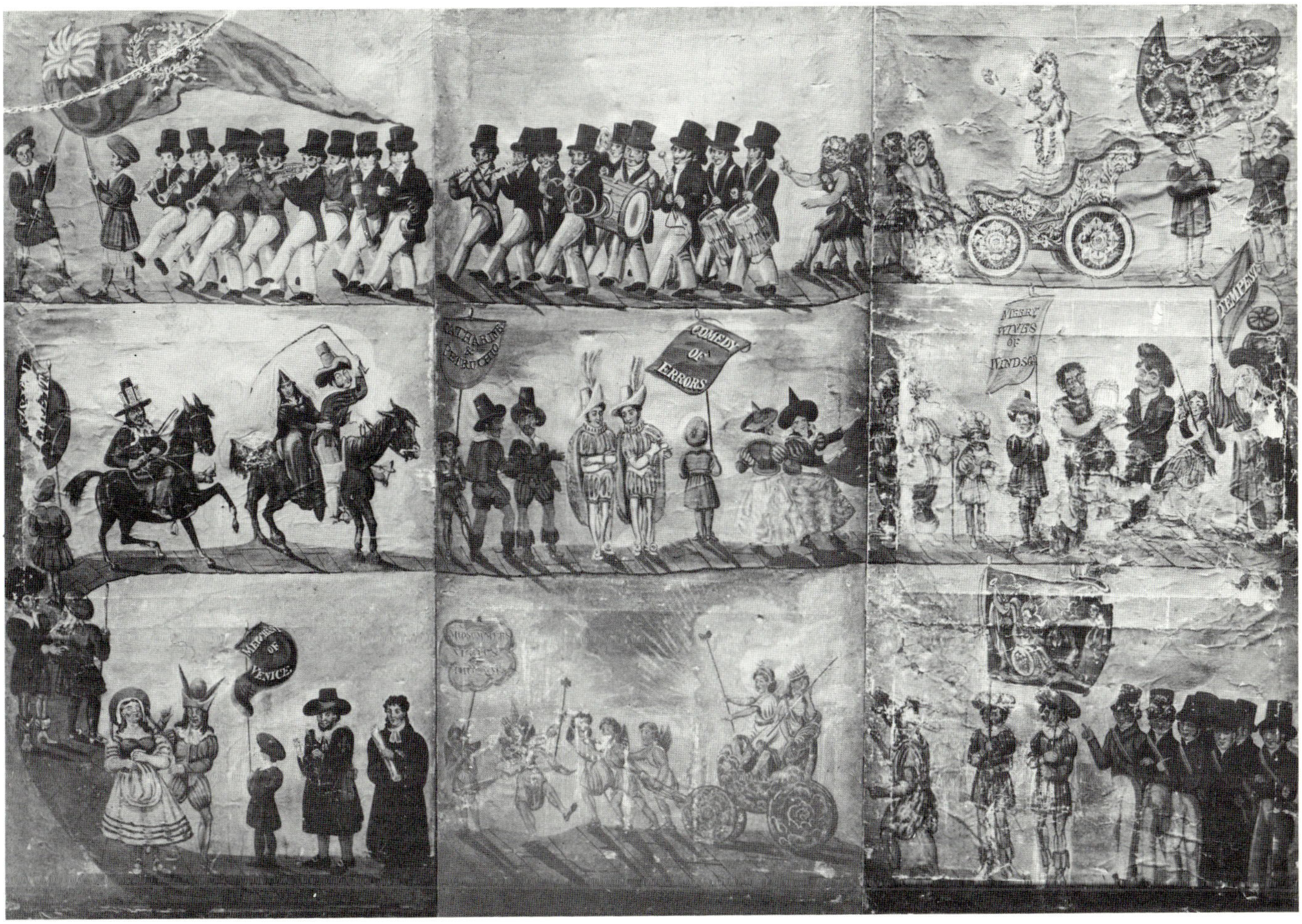

45. *Procession of the Comic Muse*, from the same Jubilee Screen

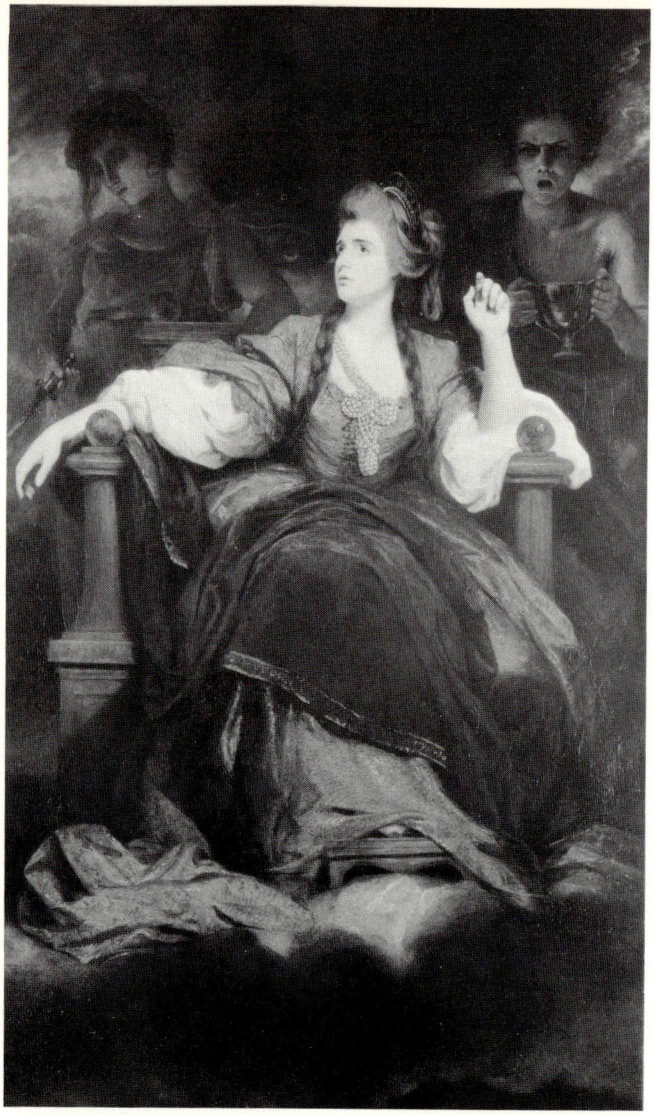

46. Reynolds: *Mrs Siddons as the Tragic Muse*. San Marino (California), The Henry E. Huntington Library and Art Gallery

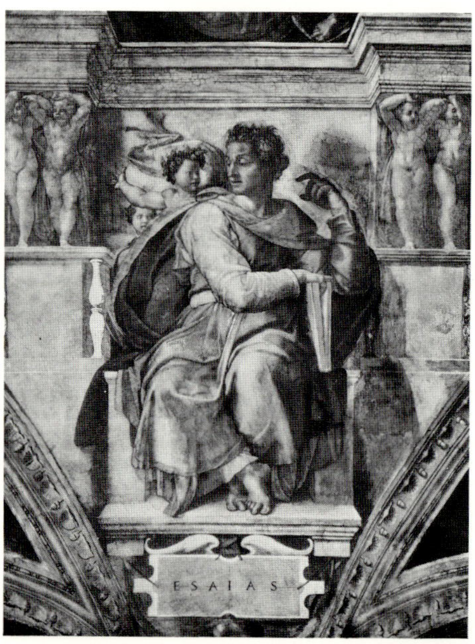

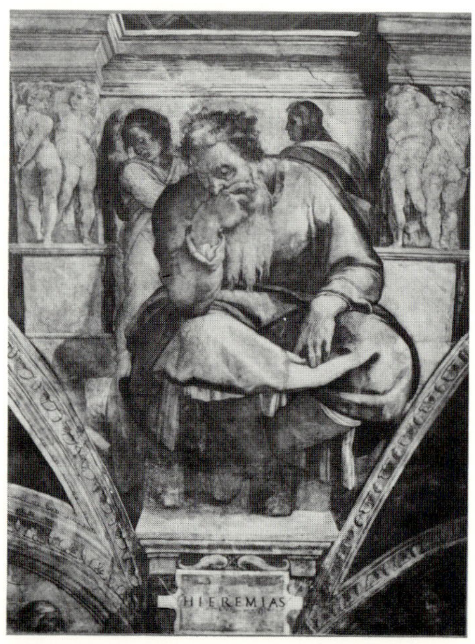

47–8. Michelangelo: *Isaiah* and *Jeremiah*. Palazzo Vaticano, Cappella Sistina

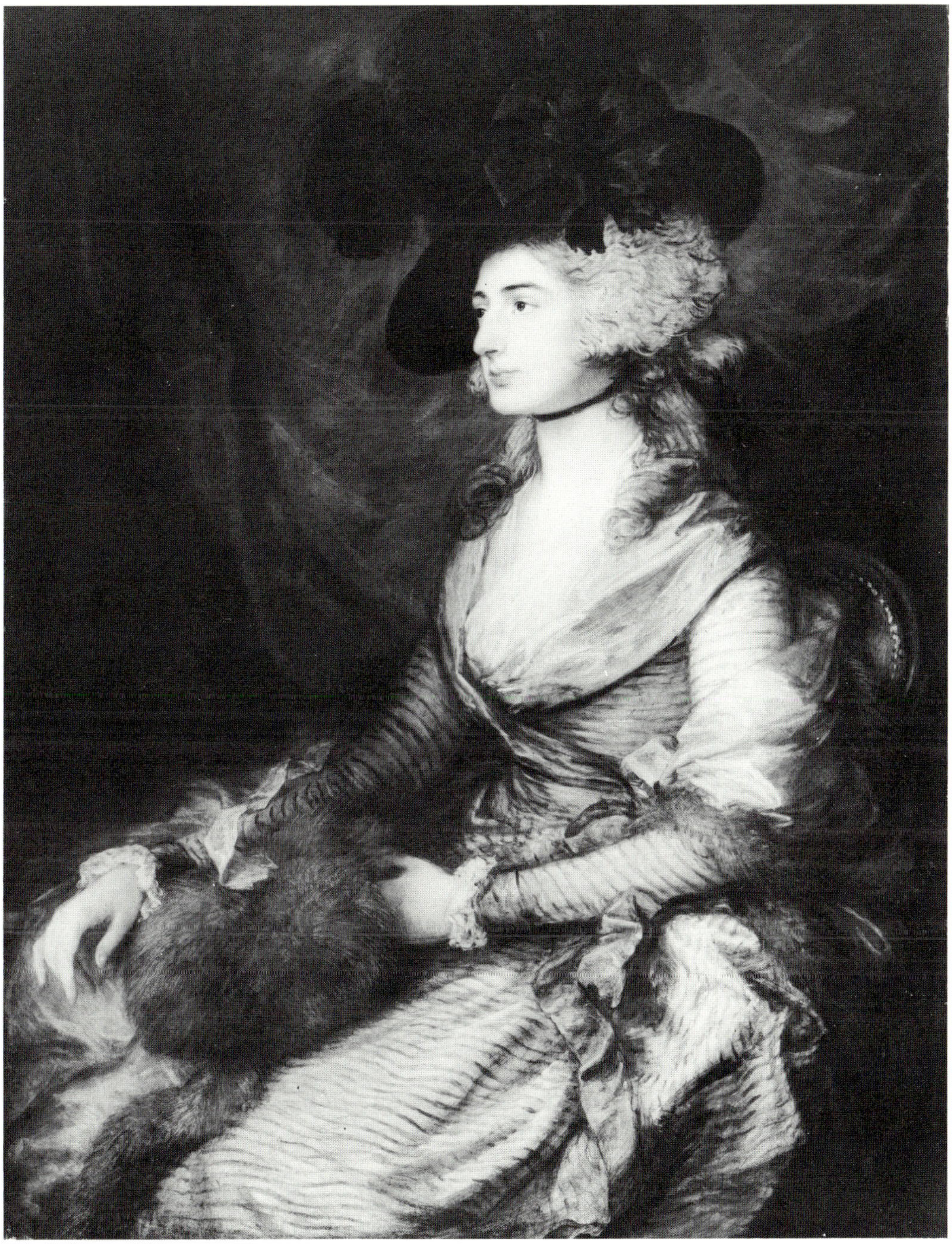

49. Gainsborough: *Mrs Siddons*. London, National Gallery

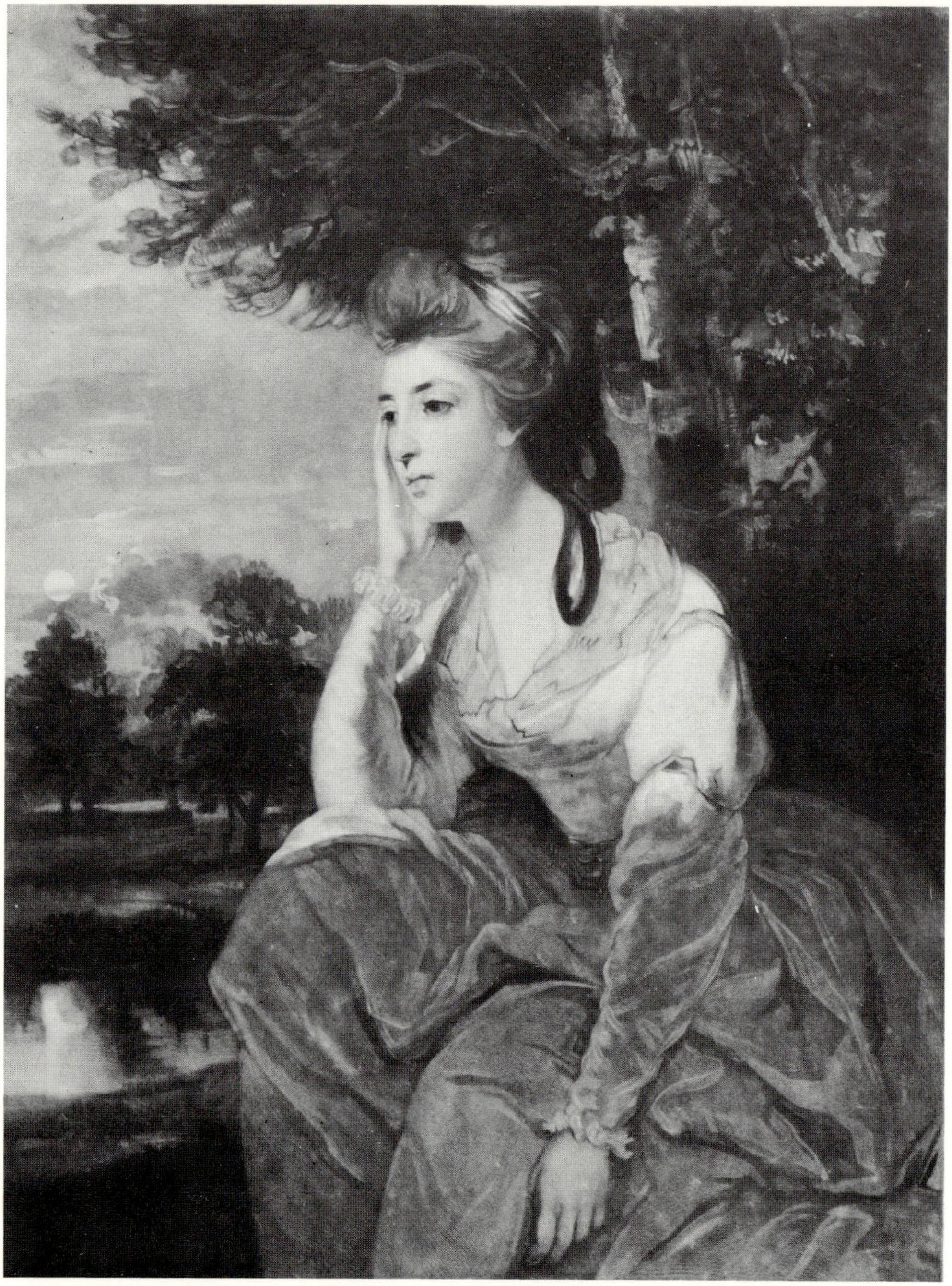

50. Reynolds: *The Hon. Mrs Stanhope* (or *Spencer?*) *as Melancholy*. Mezzotint by J. R. Smith. London, British Museum

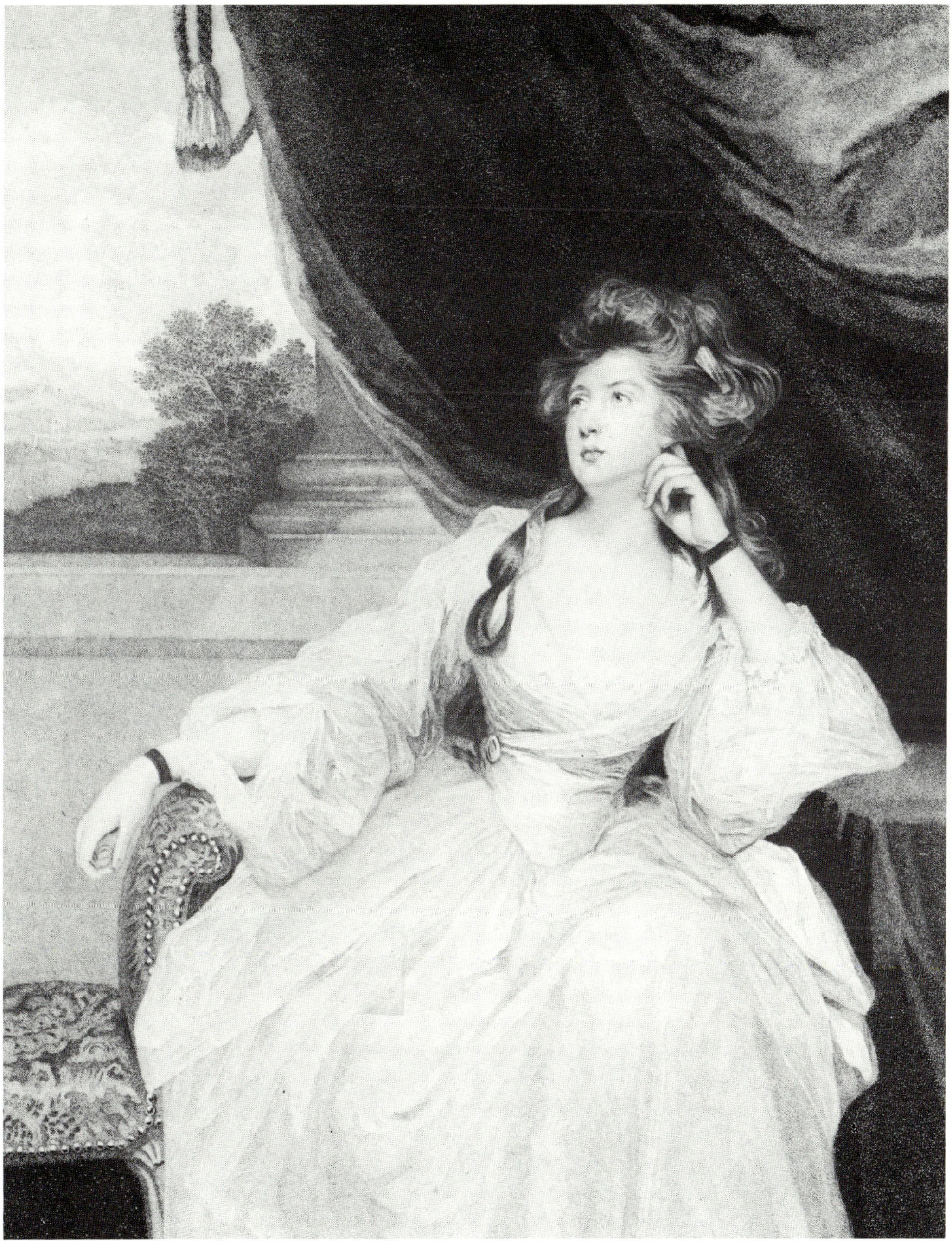

51. Reynolds: *The Hon. Mrs Stanhope as Contemplation*. Engraving by Caroline Watson. London, British Museum

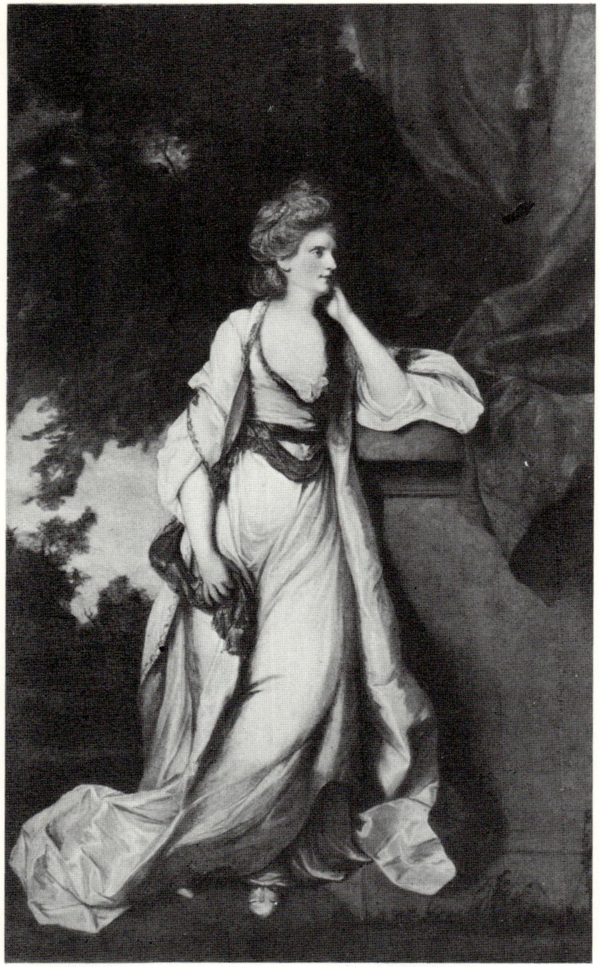

52. Reynolds: *Lady Louisa Manners, Countess of Dysart*. Kenwood, The Iveagh Bequest

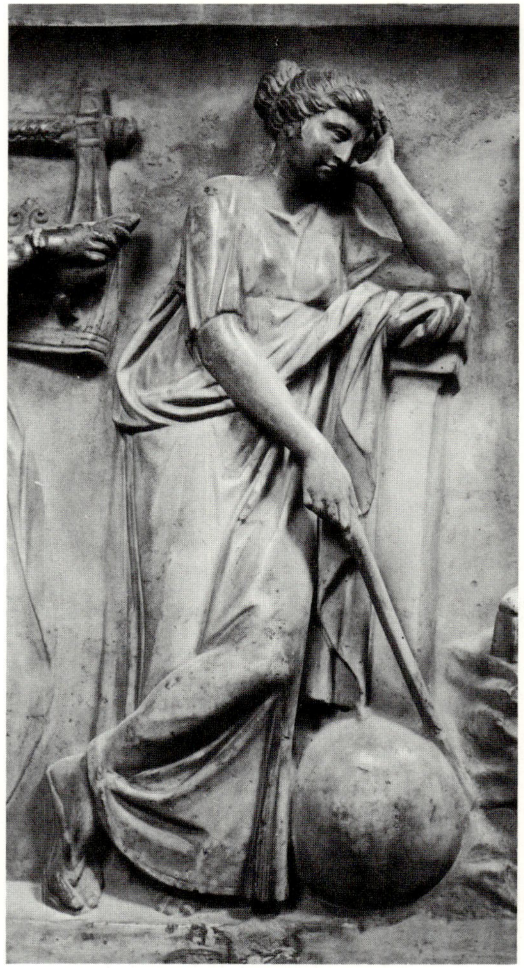

53. Second-century Roman sarcophagus: *The Nine Muses*, detail of Urania. Paris, Musée du Louvre

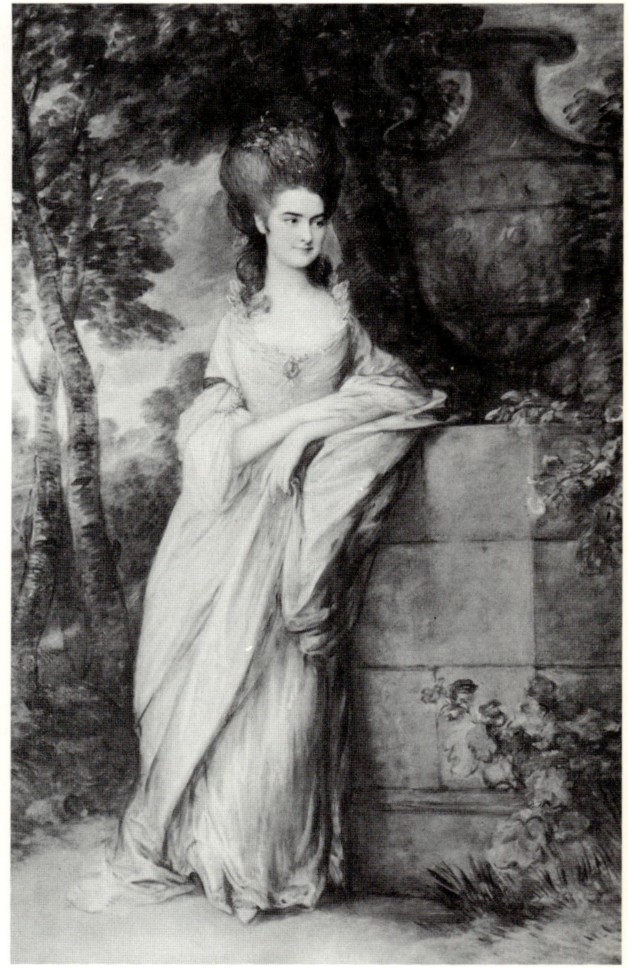

54. Gainsborough: *Mrs John Meares*. San Marino (California), The Henry E. Huntington Library and Art Gallery

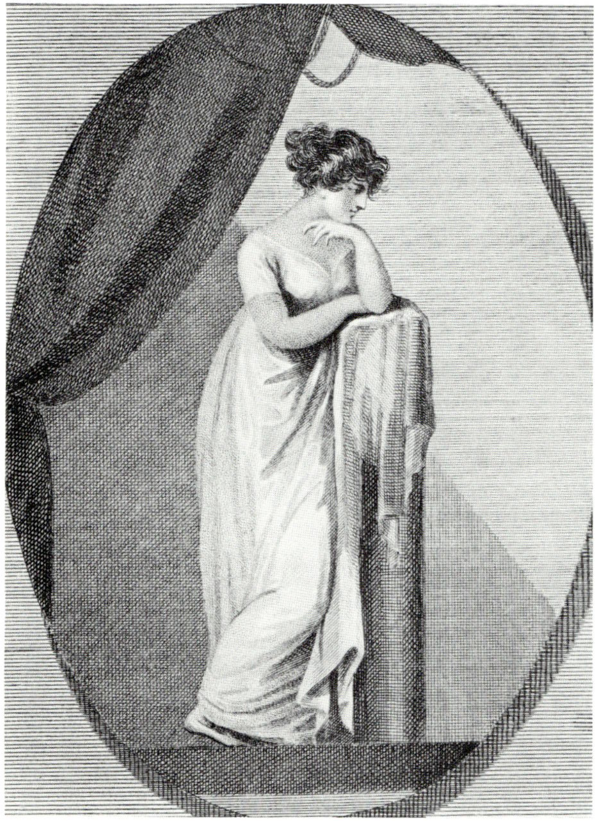

55. Adam Buck: *Mrs Siddons as Hermione*. Engraving by J. Alais, 1802. London, British Museum

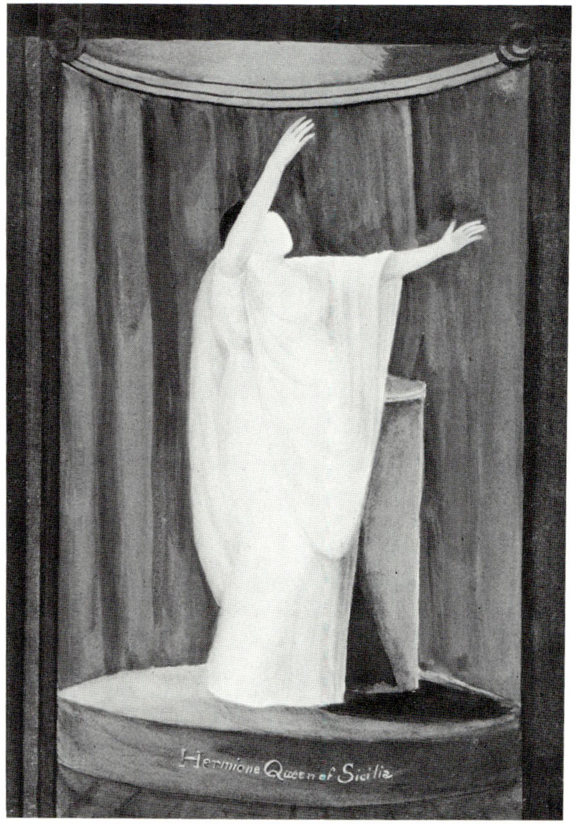

56. Mary Hamilton: *Mrs. Siddons as Hermione*. Drawing. London, British Museum

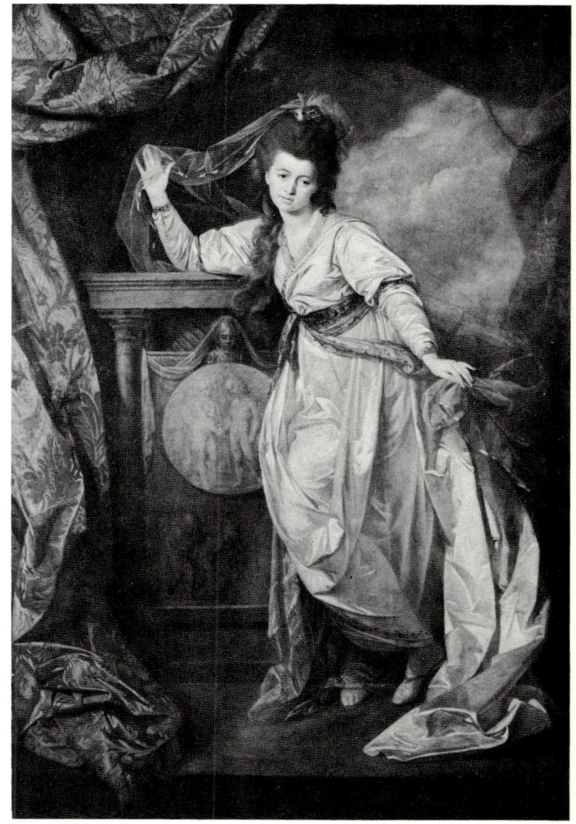

57. Zoffany: *Elizabeth Farren, Countess of Derby, as Hermione in 'The Winter's Tale'*.
Melbourne, The National Gallery of Victoria,
Everard Studley Miller Bequest

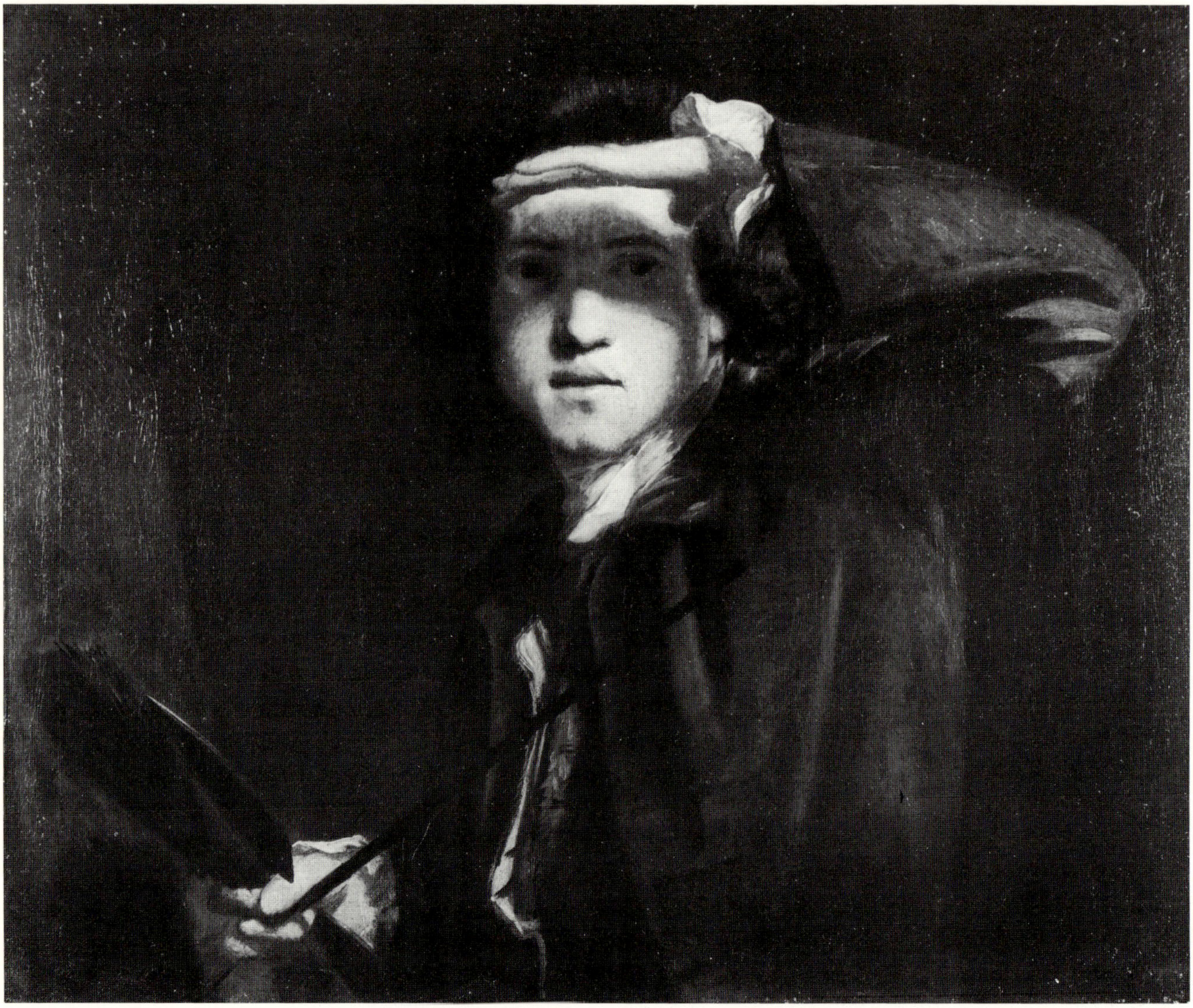

58. Reynolds: *Self-portrait*. London, National Portrait Gallery

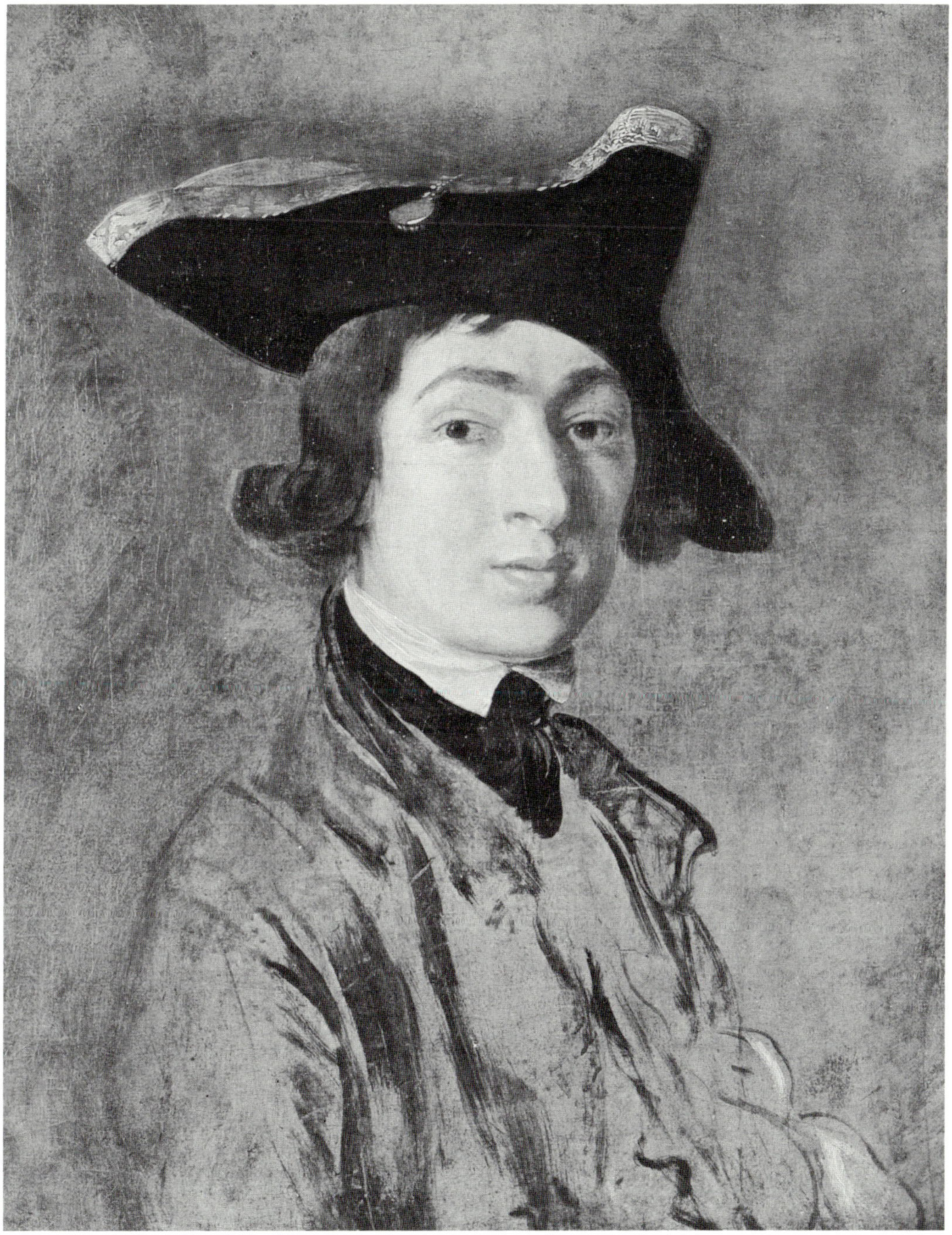

59. Gainsborough: *Self-portrait*. Private Collection

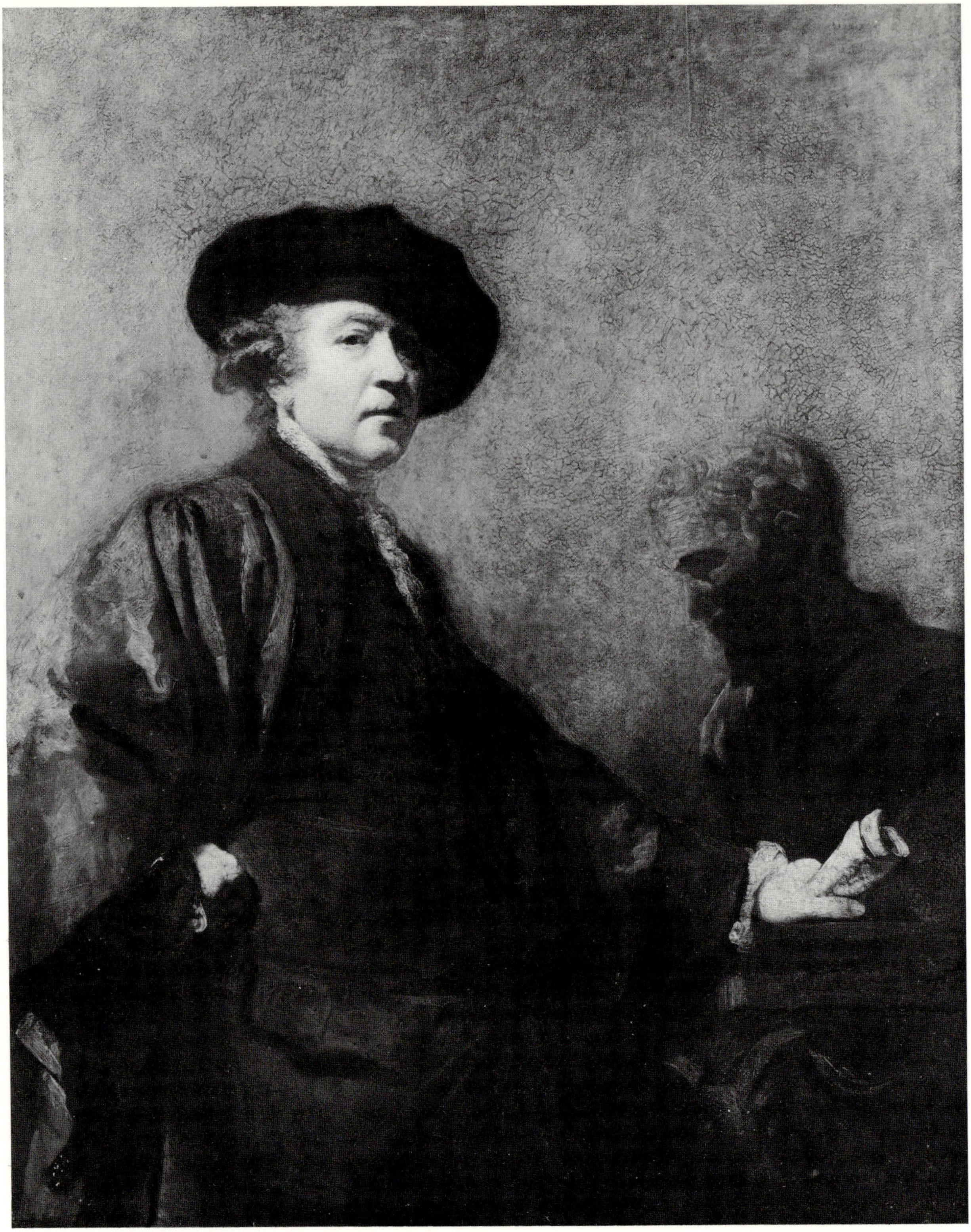

60. Reynolds: *Self-portrait with a bust of Michelangelo.* London, The Royal Academy of Arts

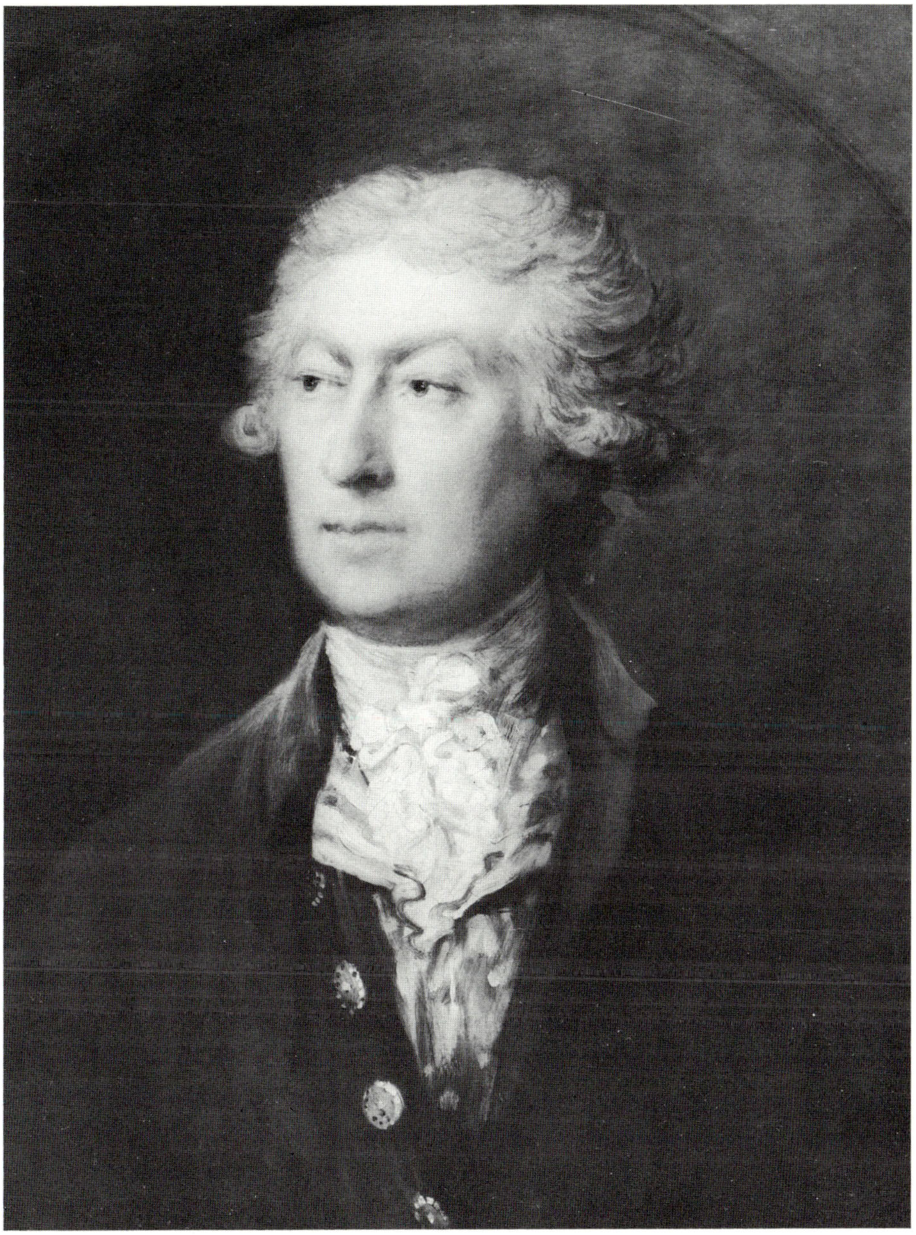

61. Gainsborough: *Self-portrait*. Bakeham House, Sir Michael Sobell

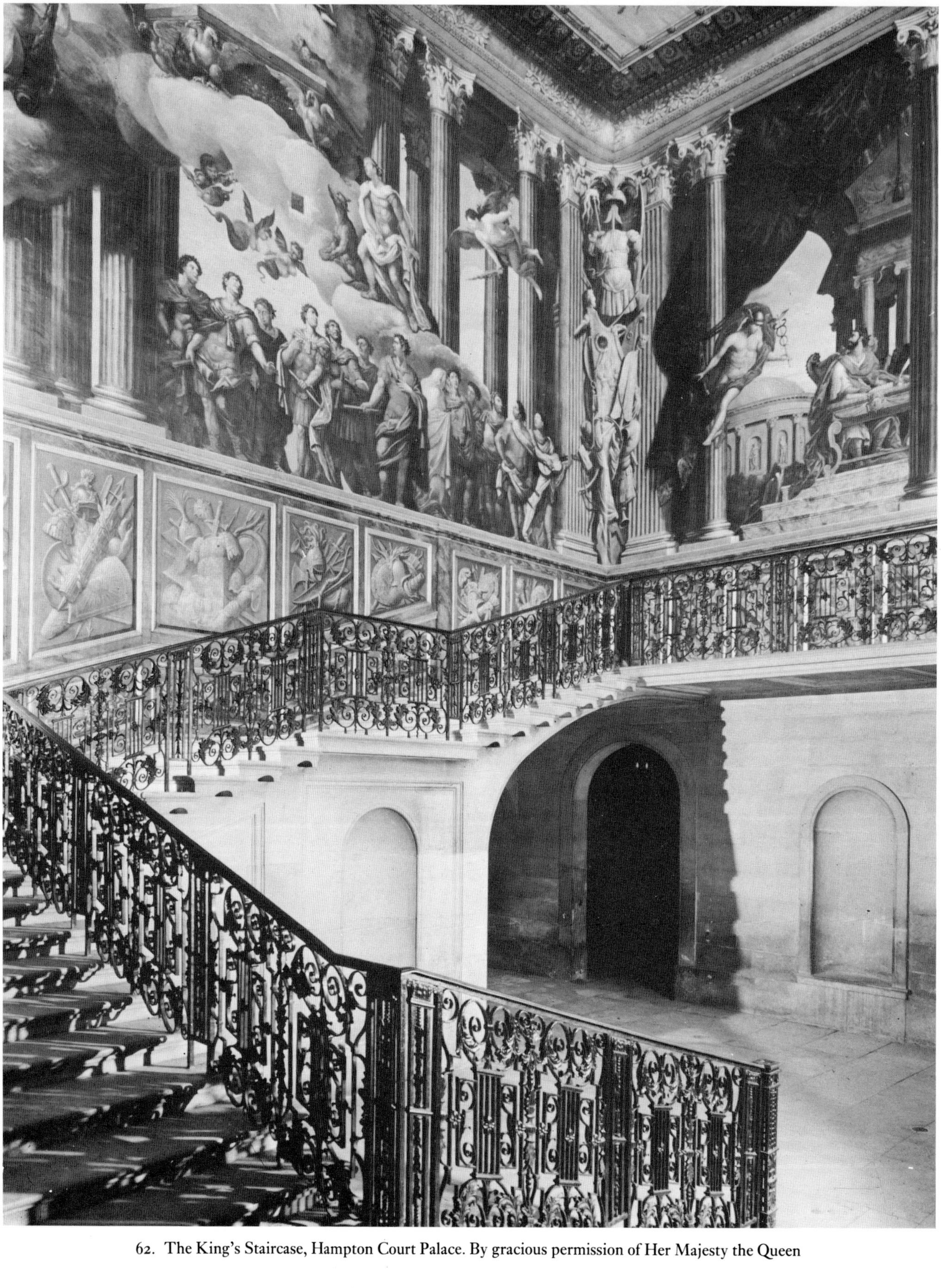

62. The King's Staircase, Hampton Court Palace. By gracious permission of Her Majesty the Queen

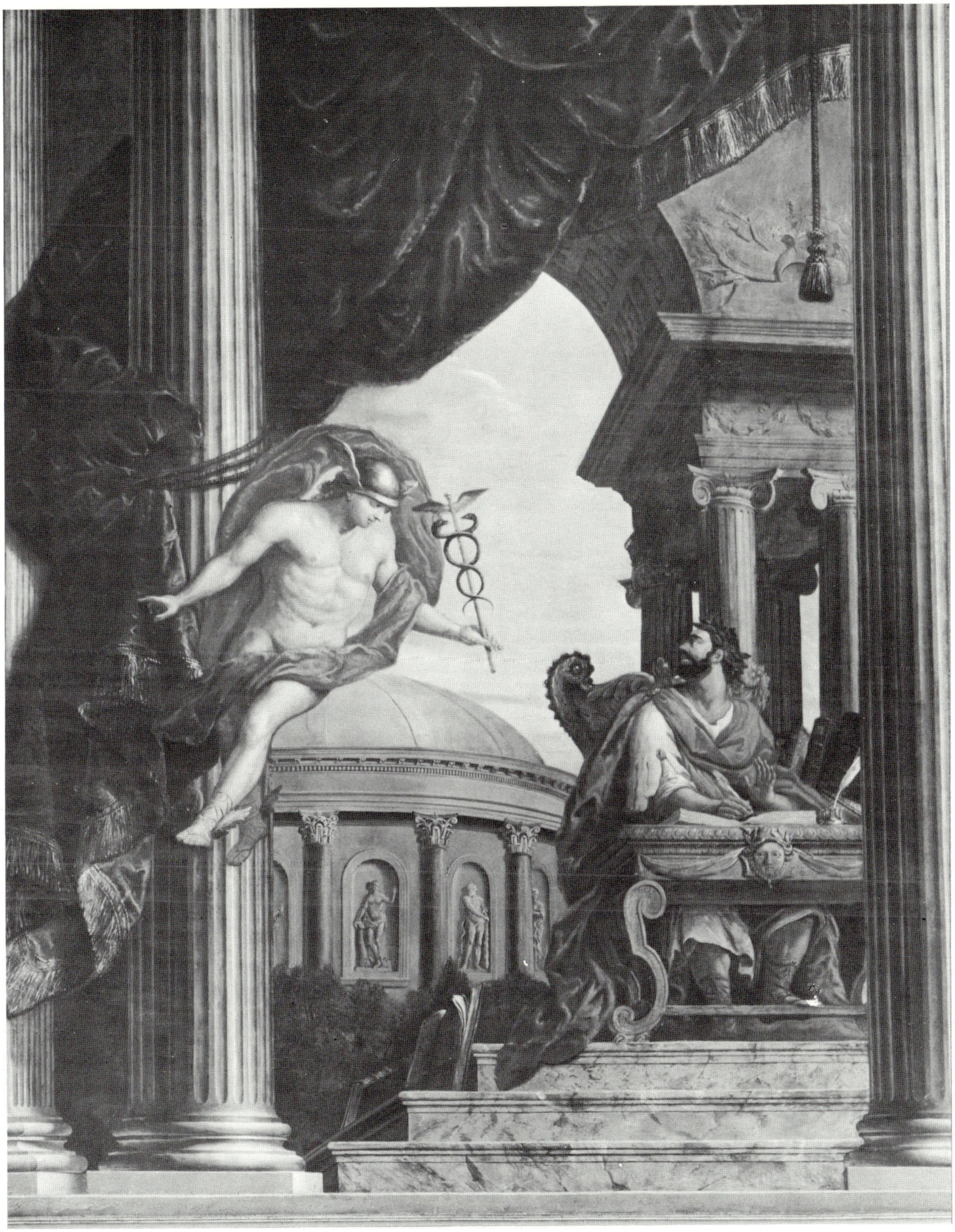

63. Antonio Verrio: *Julian the Apostate*, from the King's Staircase. By gracious permission of Her Majesty the Queen

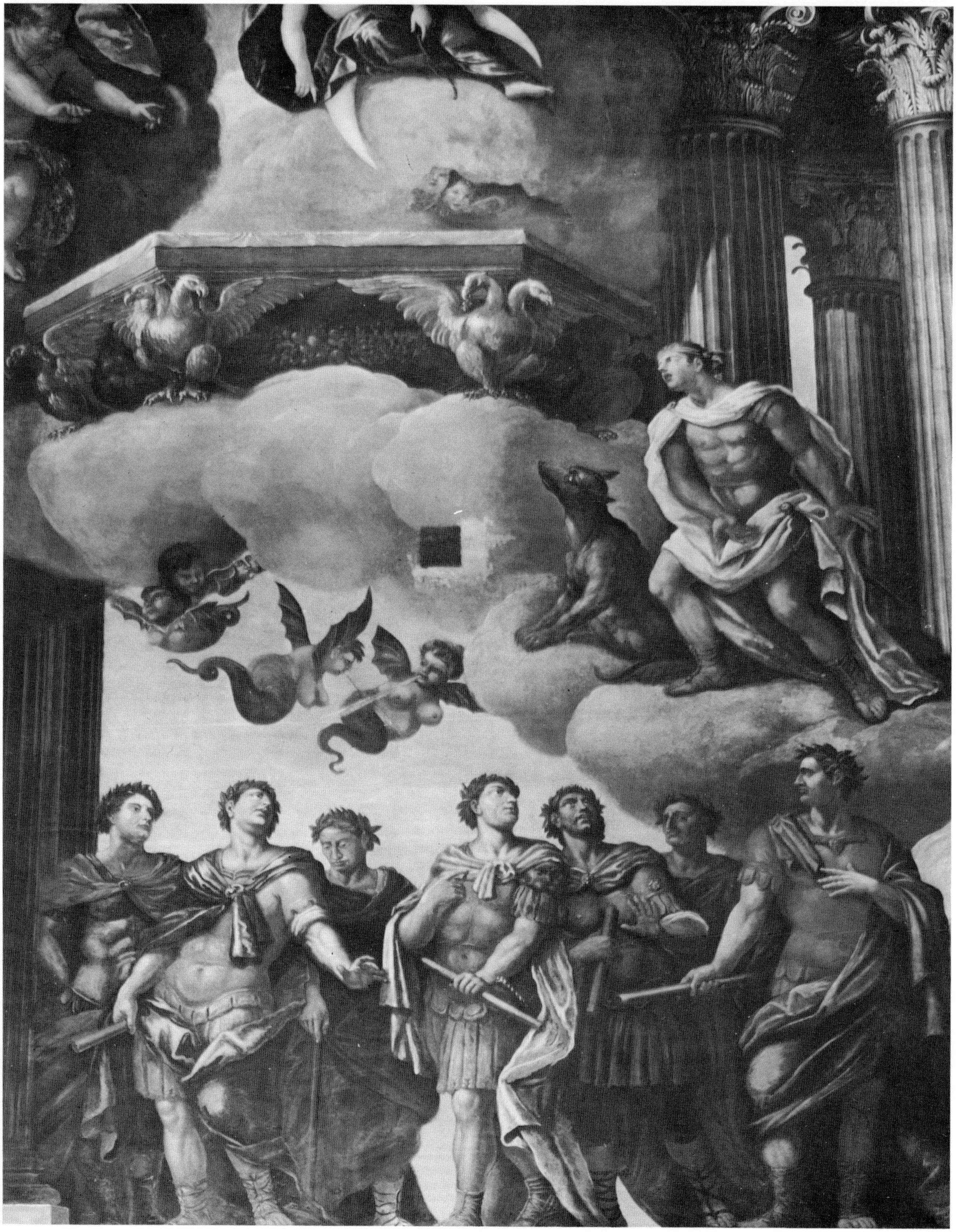

64. Antonio Verrio: Romulus, detail from Fig. 66

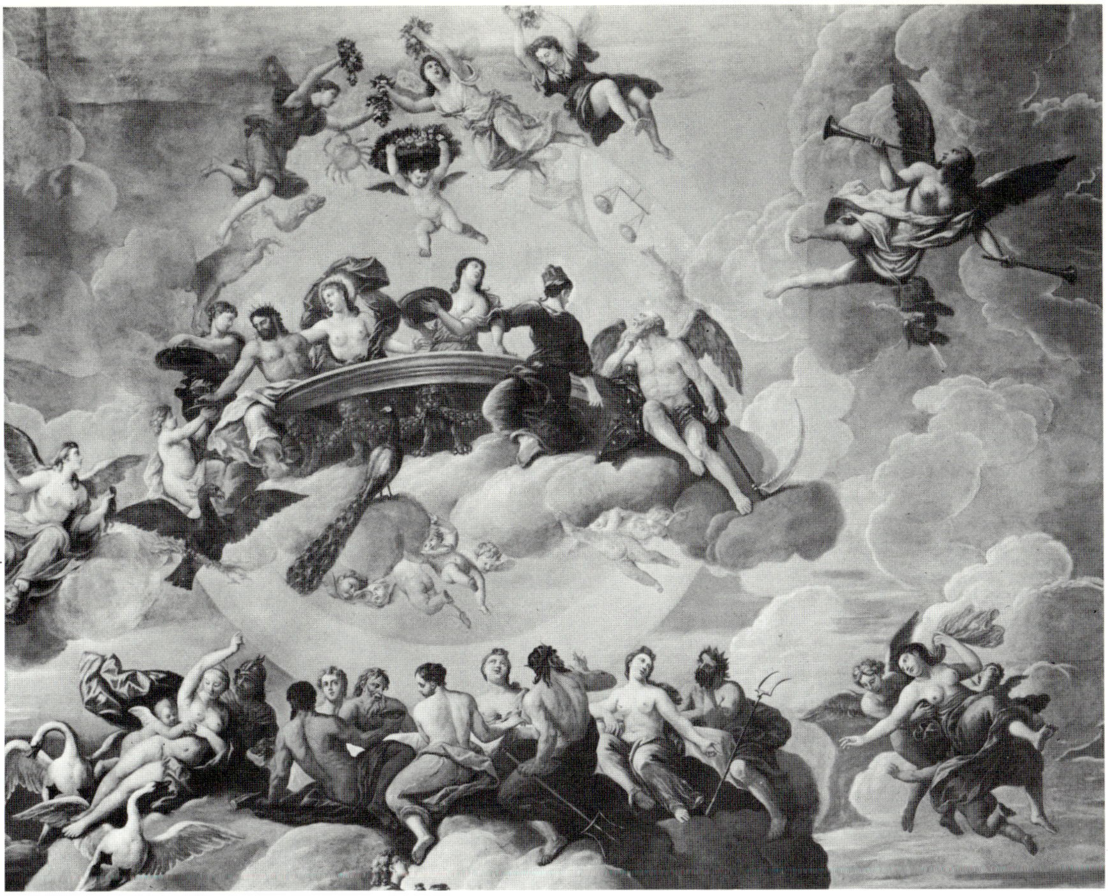

65. Antonio Verrio: *The Feast of the Gods*. Ceiling above the King's Staircase. By gracious permission of her Majesty the Queen

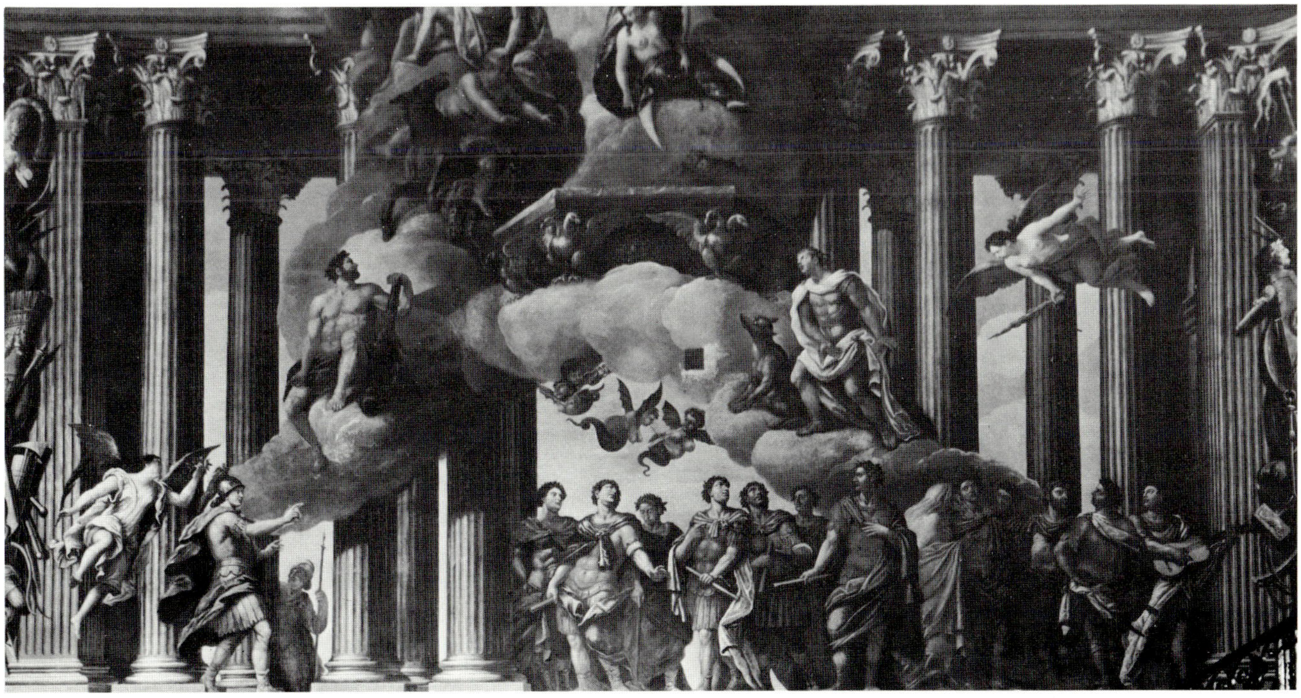

66. Antonio Verrio: *The Caesars*, from the King's Staircase. By gracious permission of Her Majesty the Queen

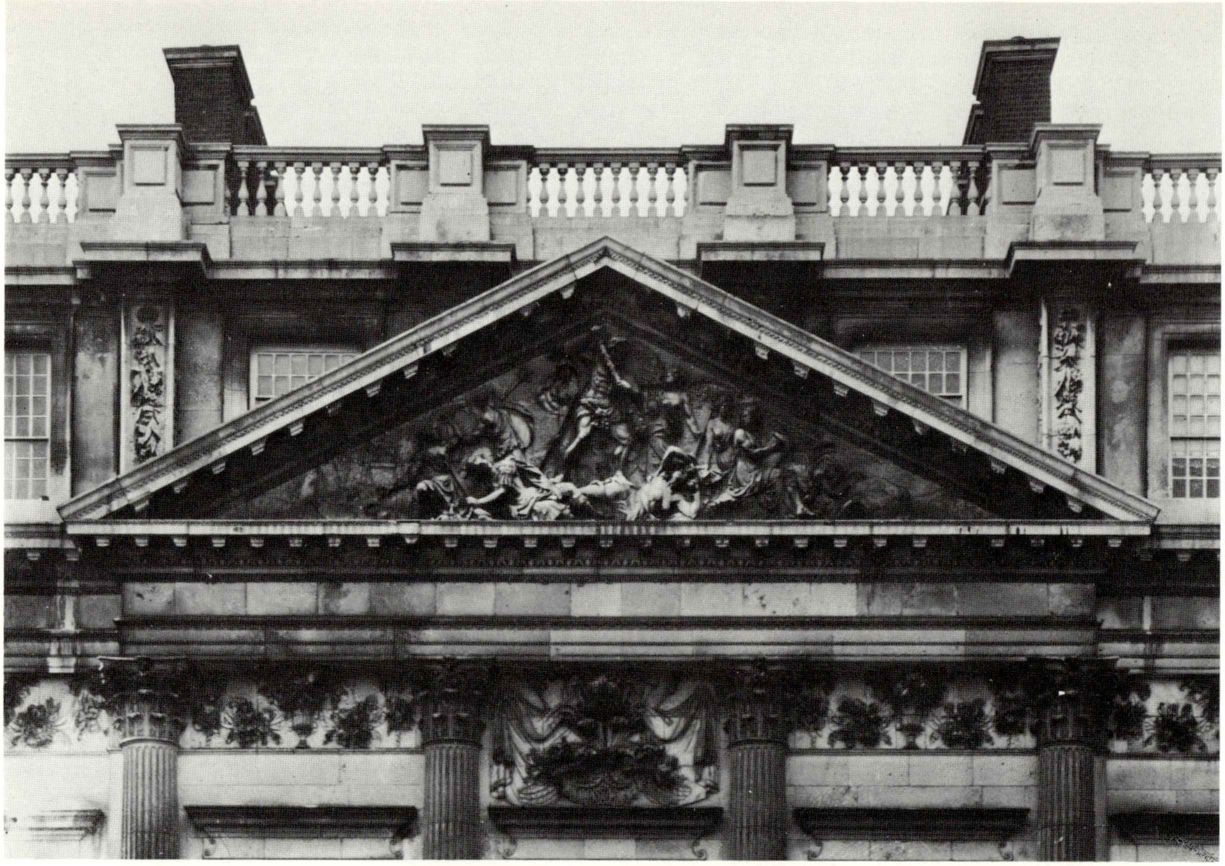

67. Caius Gabriel Cibber: *Hercules victorious over the Vices*. Pediment from the Garden Front, Hampton Court Palace. By gracious permission of Her Majesty the Queen

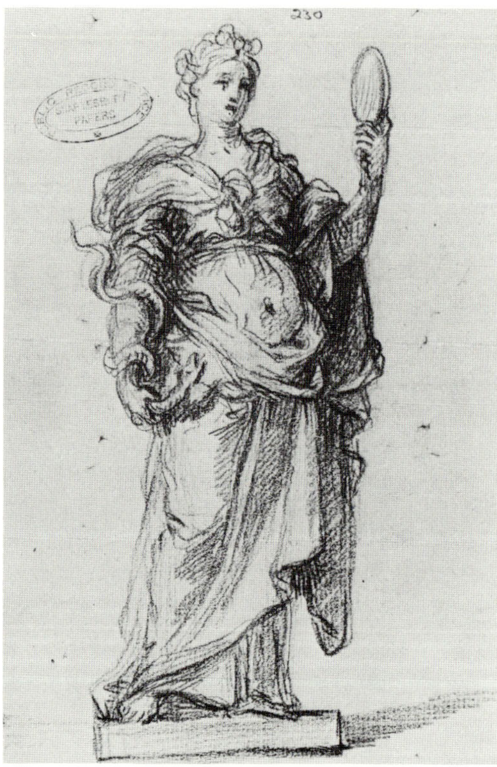

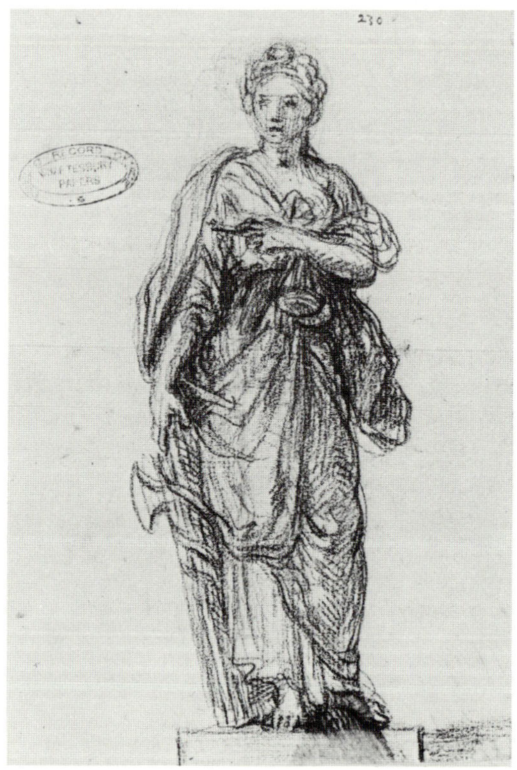

68–9. Domenico Guidi: *Prudence* and *Justice*. Drawings. London, Public Record Office, Shaftesbury Papers

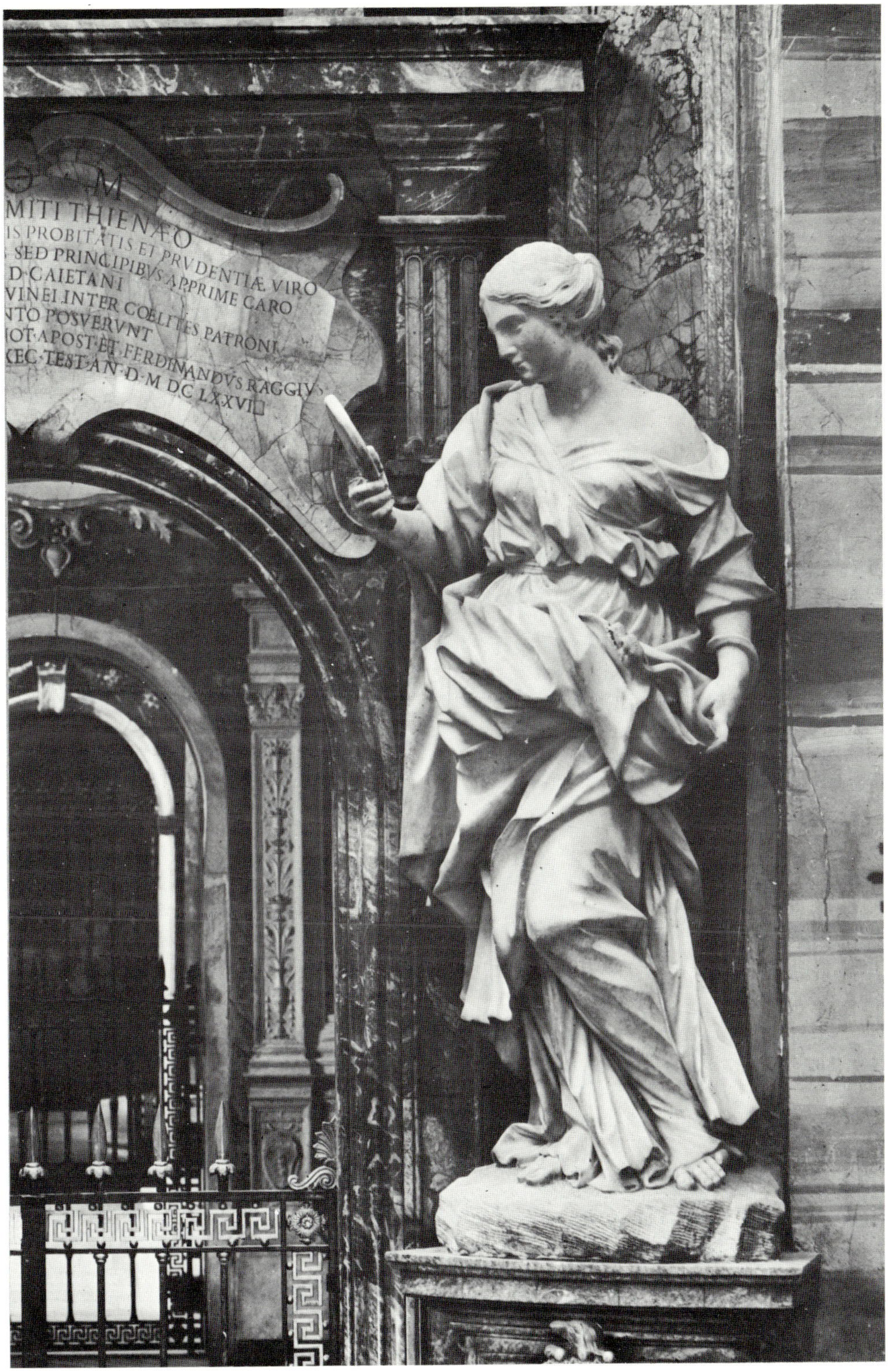

70. Domenico Guidi: *Prudence*, from the tomb of Count Gaspare Thiene.
Rome, Sant'Andrea della Valle

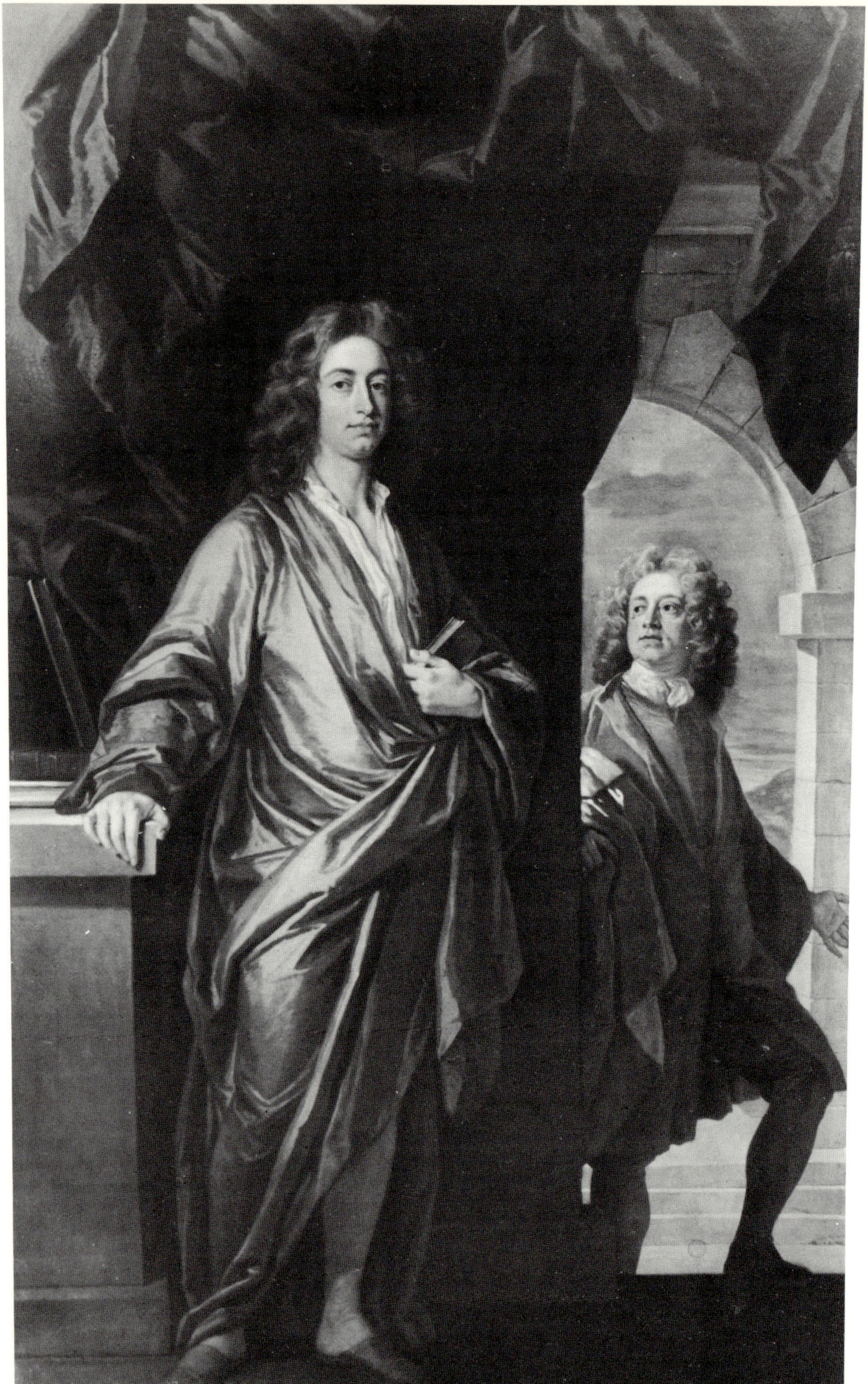

71. John Closterman: *Anthony Ashley–Cooper, 3rd Earl of Shaftesbury*. Private Collection

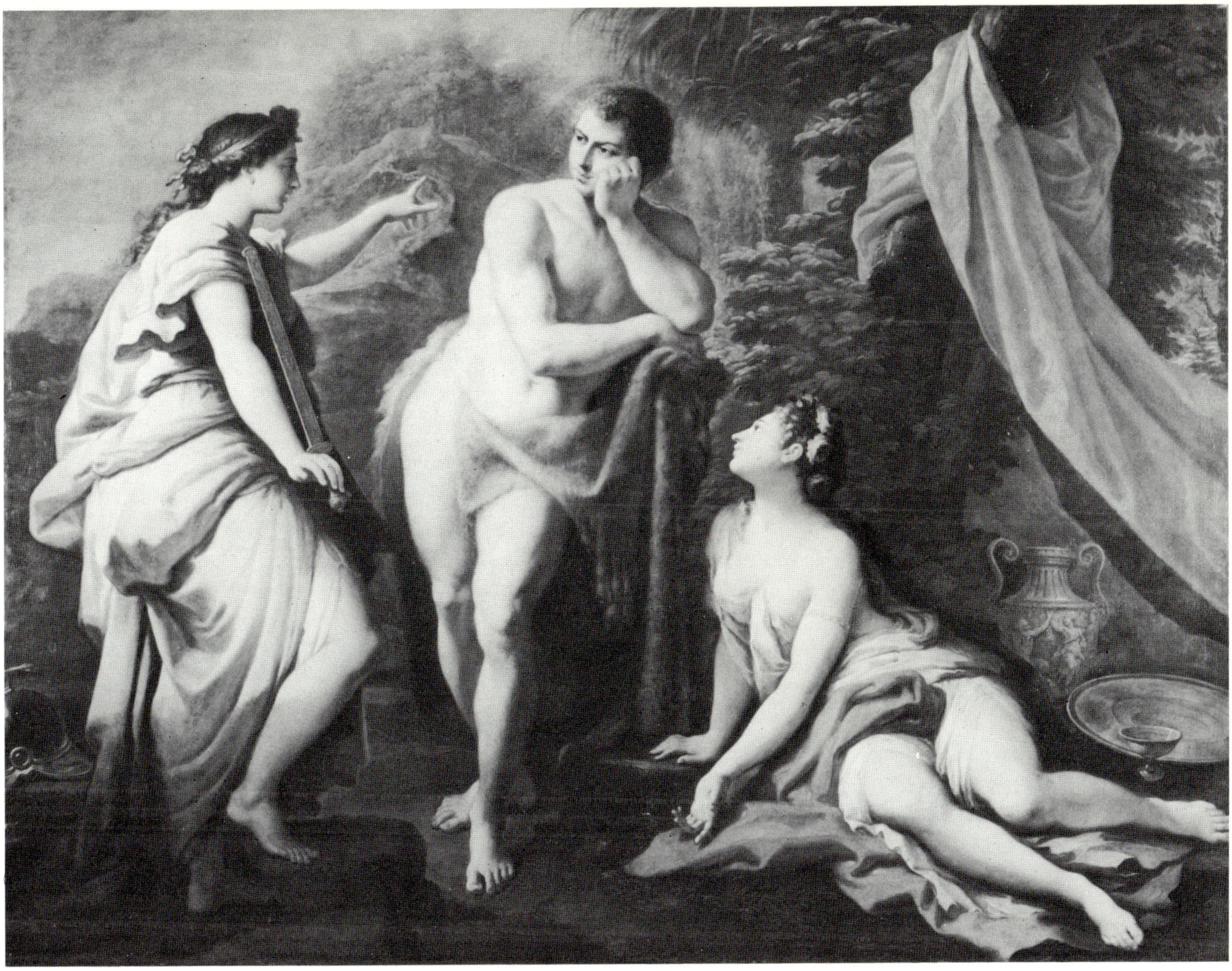

72. Paolo de Matteis: *The Choice of Hercules*. Oxford, Ashmolean Museum

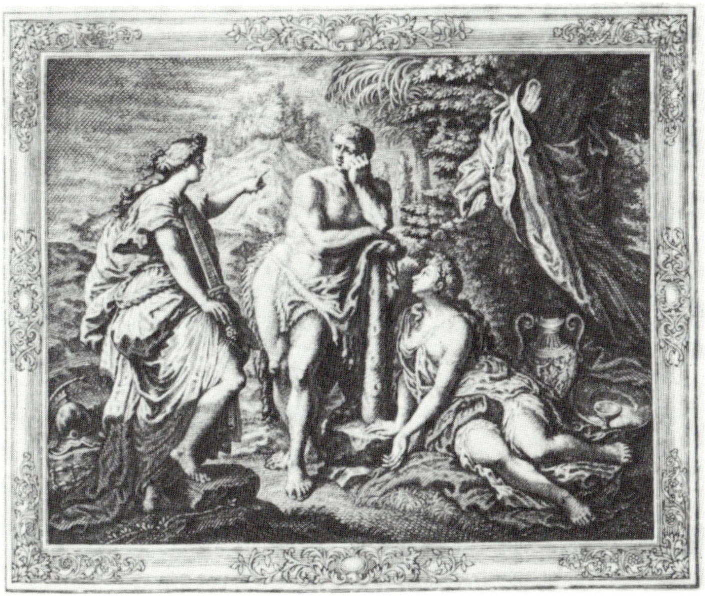

73. Paolo de Matteis: *The Choice of Hercules*. Engraving by Simon Gribelin for the 1714 edition of Shaftesbury's *Characteristicks*. Oxford, Bodleian Library

THE

THEATRES.

A

POETICAL DISSECTION.

. By Sir NICHOLAS NIPCLOSE, Baronet.

Suppoſe JOB living 'midſt the Critic train,
Our Theatres would ope his angry vein.

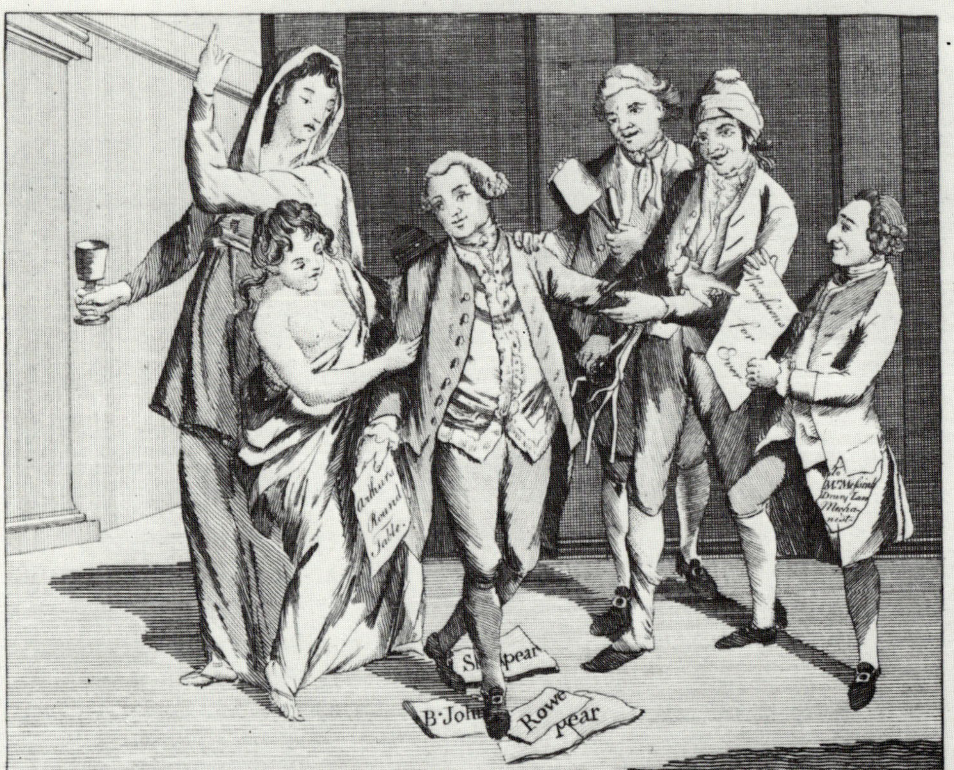

Behold the Muses ROSCIUS sue in Vain,
Taylors & Carpenters usurp their Reign.

LONDON:

PRINTED FOR JOHN BELL, IN THE STRAND, AND C. ETHERINGTON, AT YORK.

M.DCC.LXXII.

74. *Garrick between the Muses and the Tailors and Carpenters.* Engraved title-page to *The Theatres. A Poetical Dissection*, by Sir Nicholas Nipclose, Baronet, 1772. London, British Library

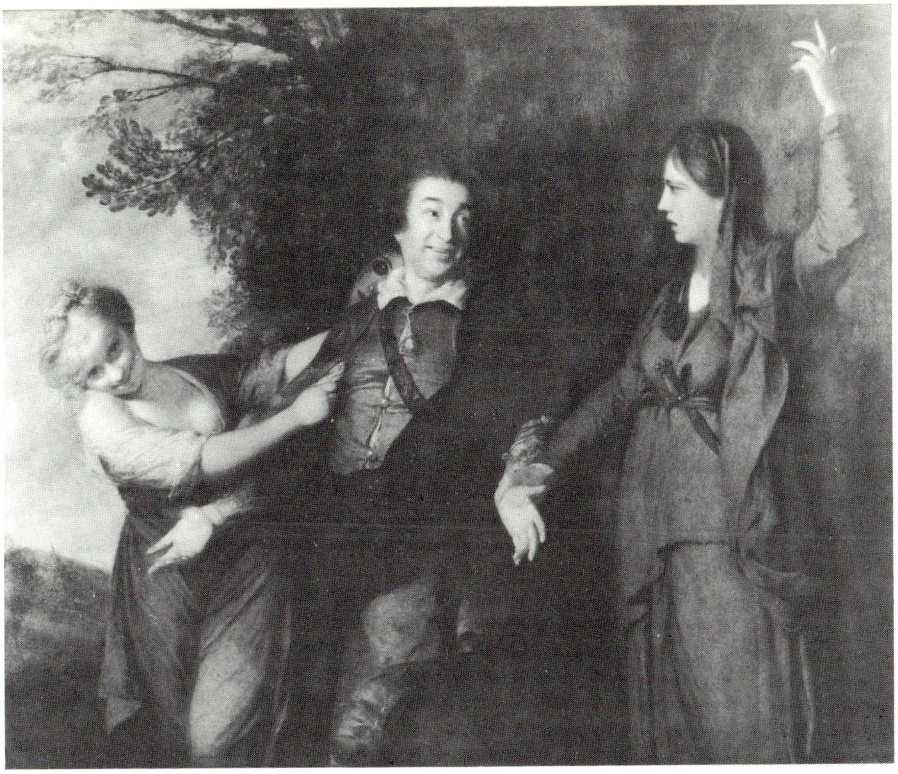

75. Reynolds: *Garrick between Tragedy and Comedy*. Private Collection

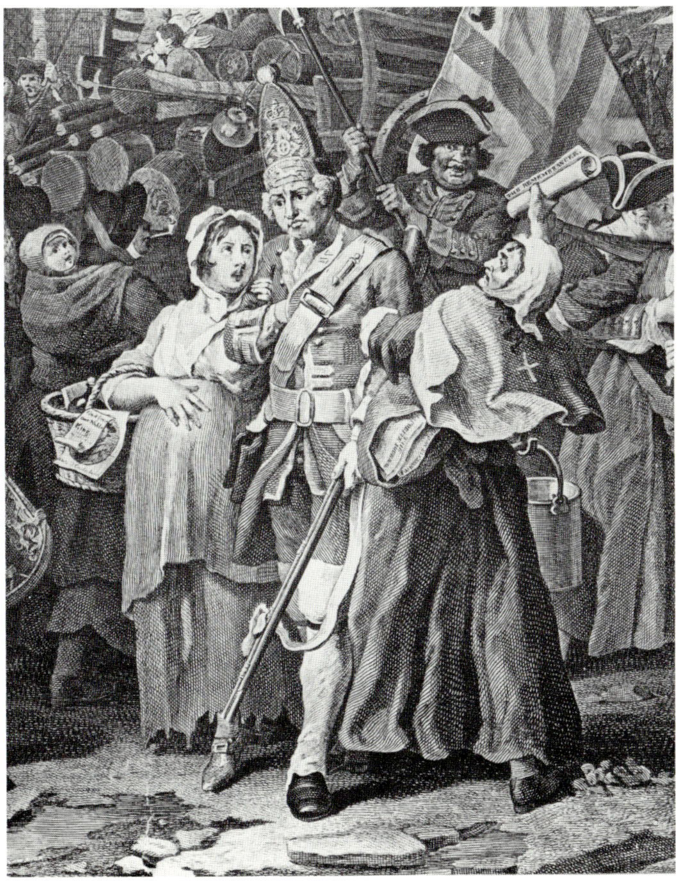

76. Hogarth: *The March to Finchley*, detail. Engraving.
Oxford, Ashmolean Museum

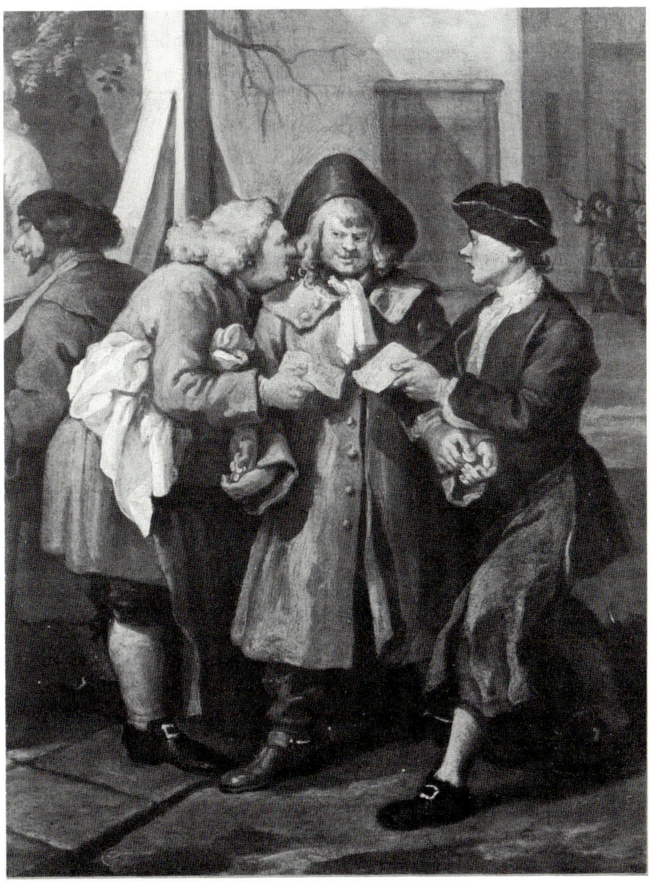

77. Hogarth: *The Election, II. Canvassing for Votes*, detail.
London, Sir John Soane's Museum

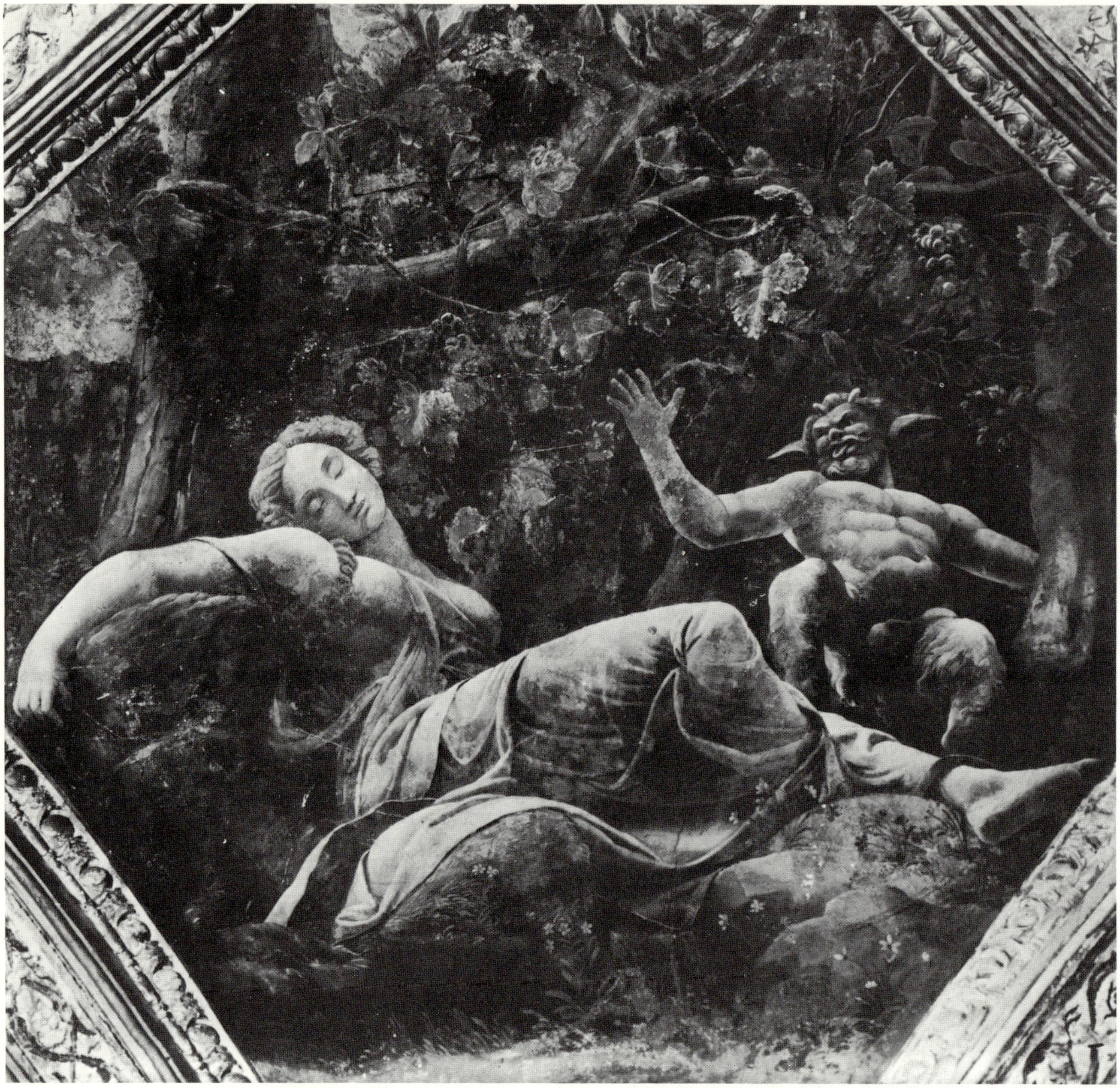

78. After Giulio Romano: *Psyche sleeping*, from the ceiling of the Sala di Psiche. Mantua, Palazzo del Te

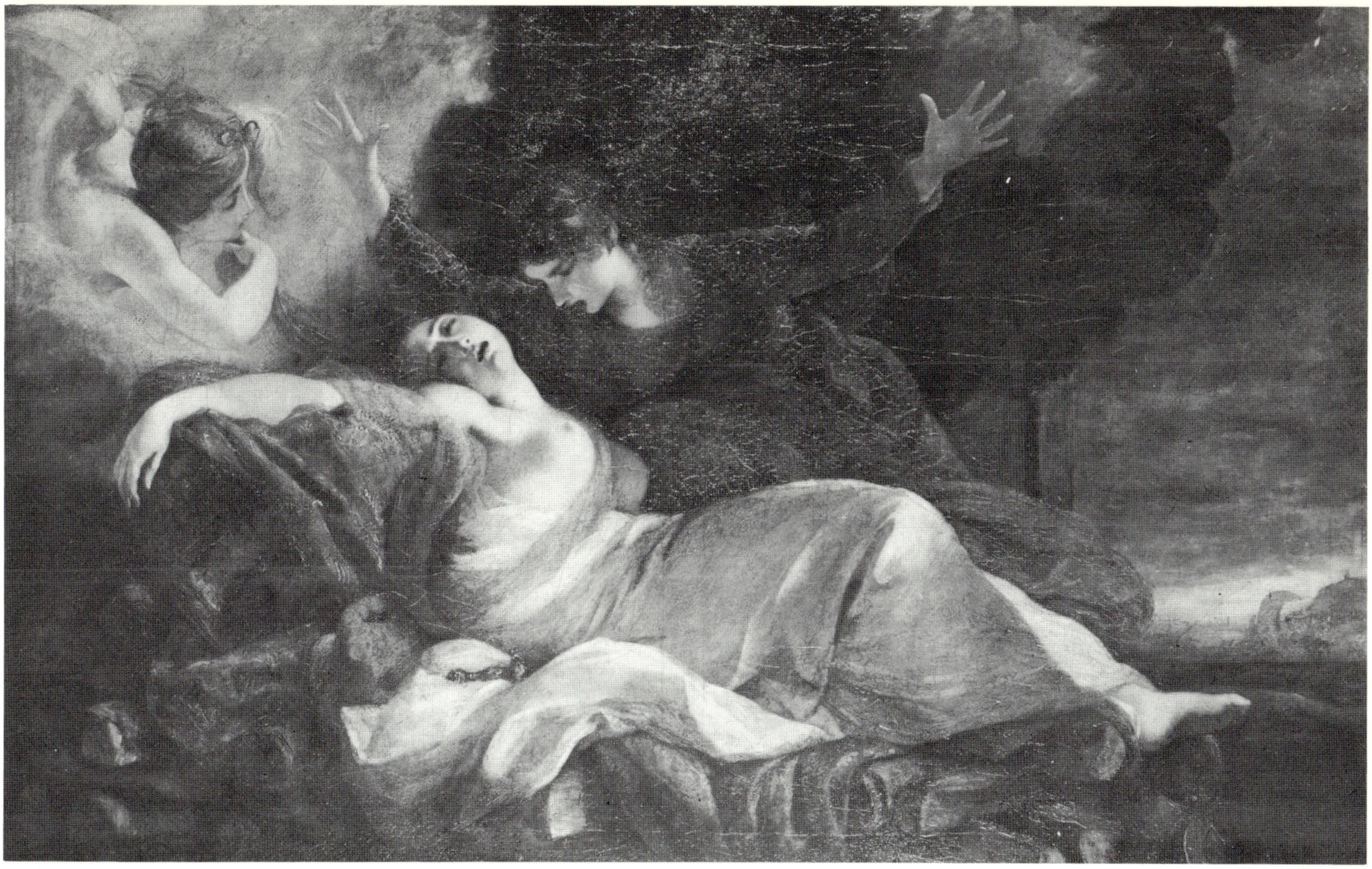

79. Reynolds: *The Death of Dido*. Buckingham Palace. By gracious permission of Her Majesty the Queen

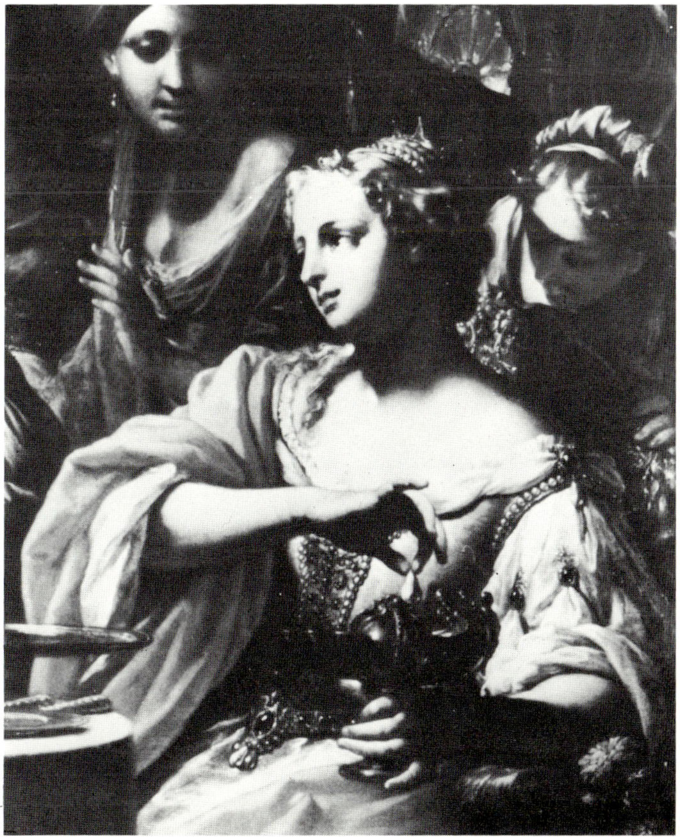

80. Francesco Trevisani: *The Banquet of Antony and Cleopatra*, detail of Cleopatra. Rome, Galleria Spada

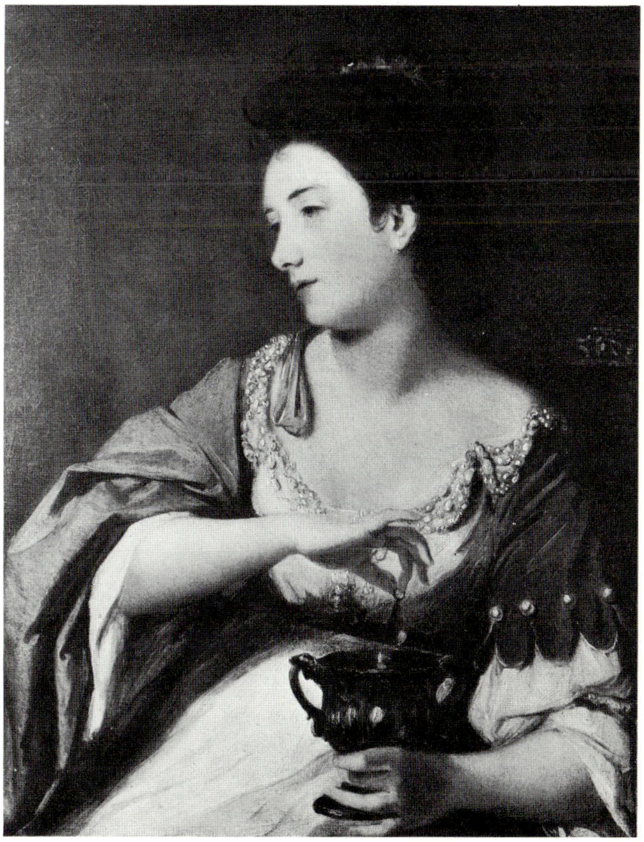

81. Reynolds: *Kitty Fisher as Cleopatra*. Kenwood, The Iveagh Bequest

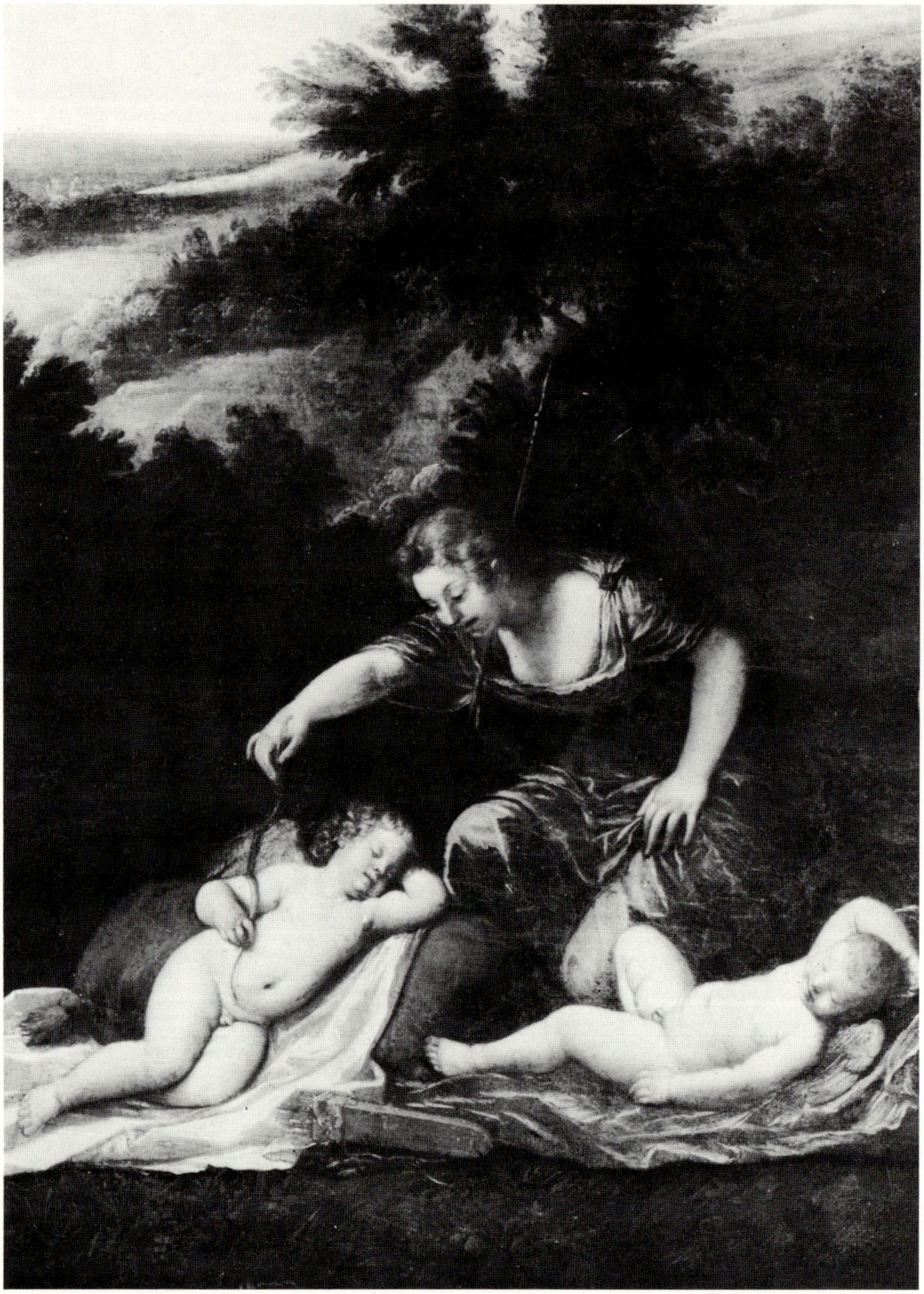

82. Francesco Albani: *Love disarmed*, detail of Diana. Paris, Musée du Louvre

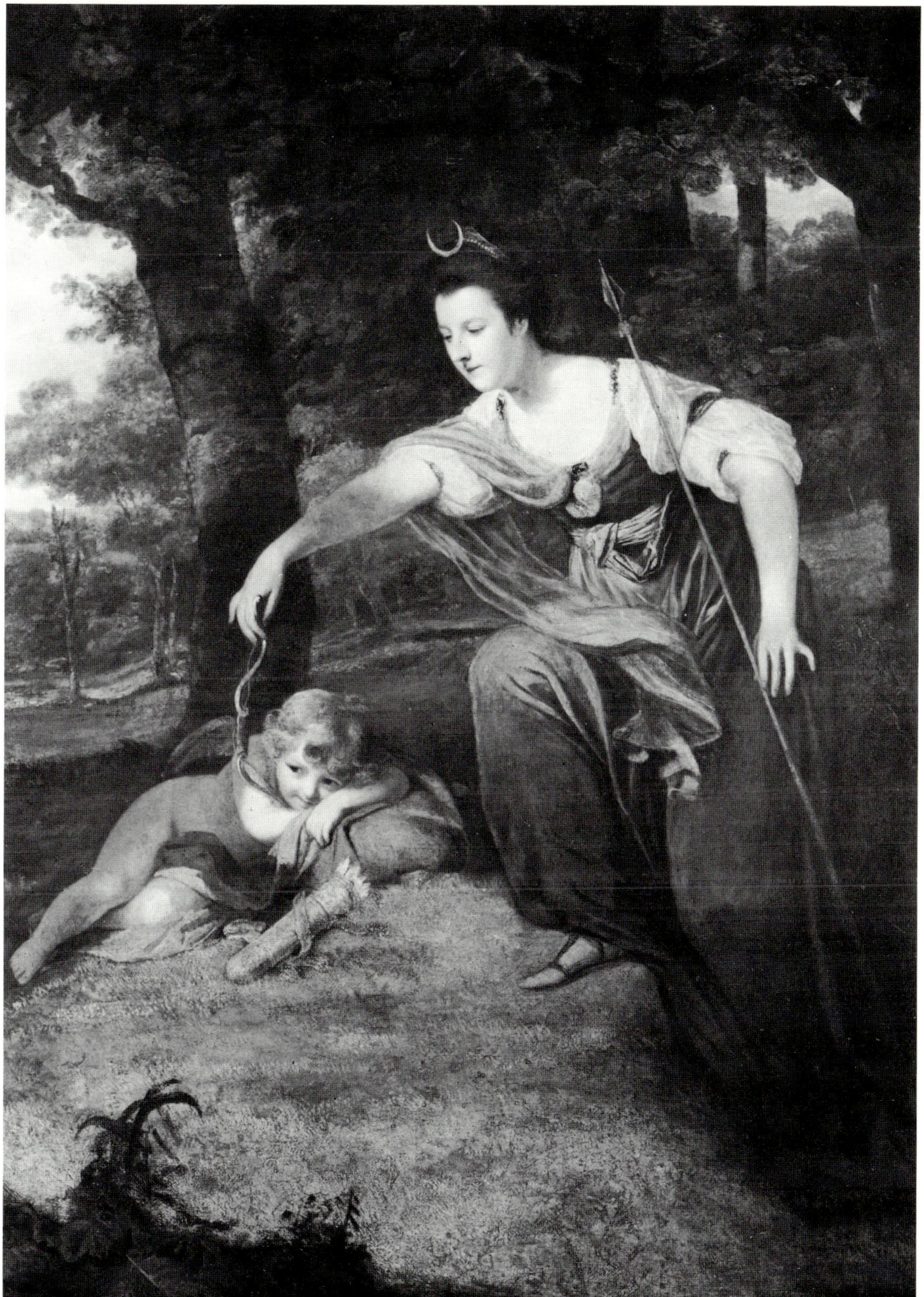

83. Reynolds: *Elizabeth, Duchess of Manchester, as Diana with her son George, Viscount Mandeville, as Cupid.* Wimpole Hall, The National Trust

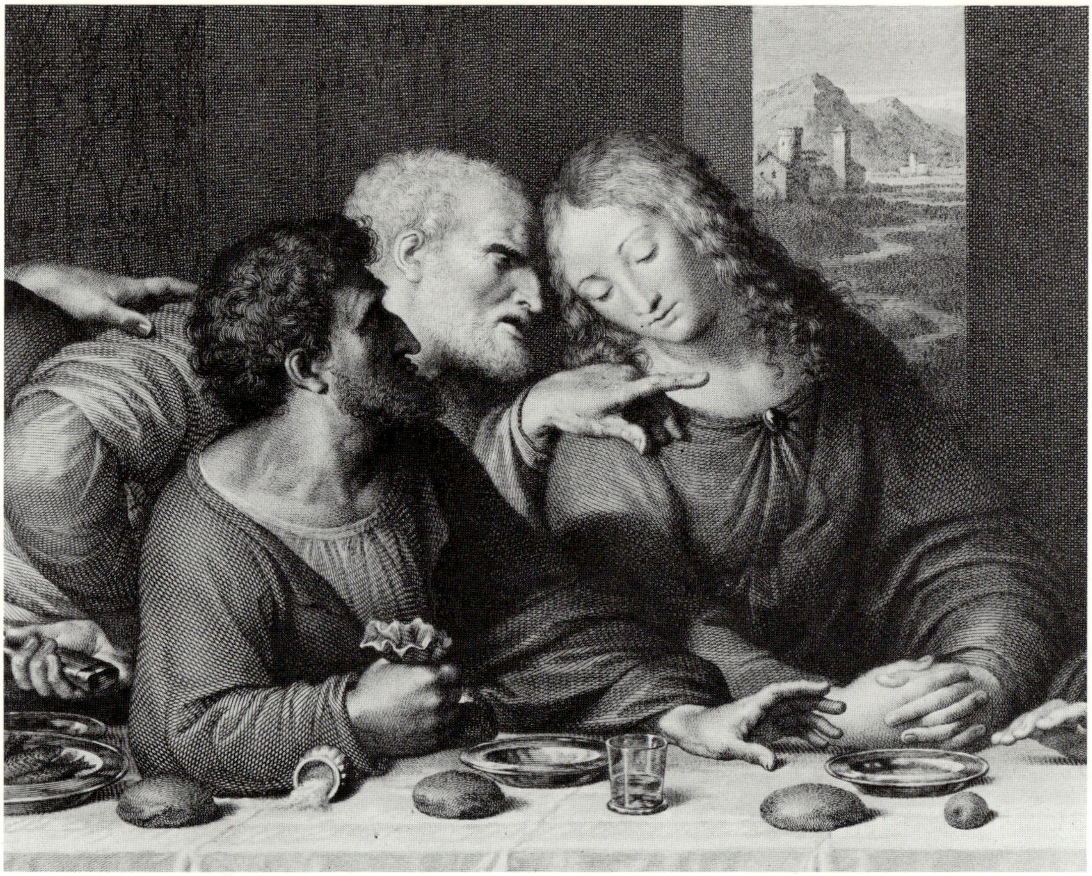

84. Leonardo da Vinci: *The Last Supper*, detail of Judas with St. Peter and St. John. Engraving by Raphael Morghen. London, British Museum

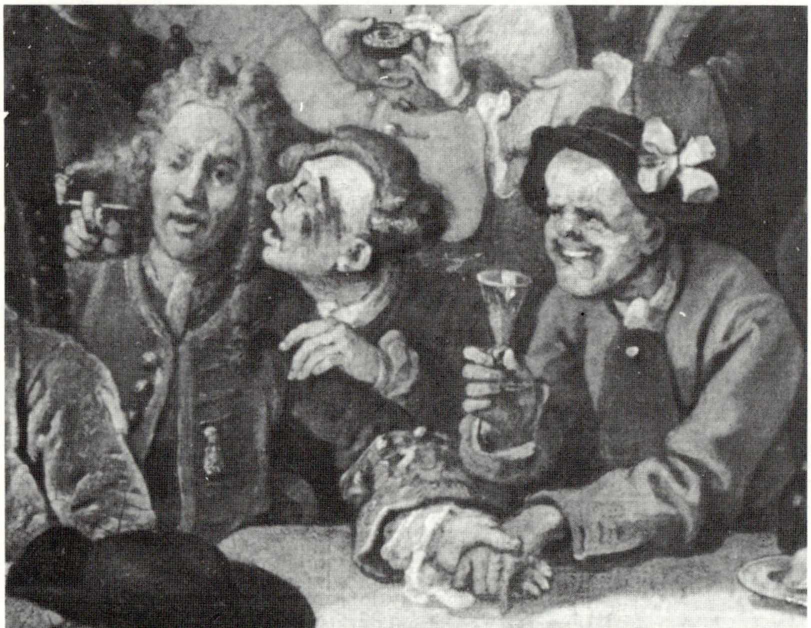

85. Hogarth: *The Election*, *1. An Election Entertainment*, detail. London, Sir John Soane's Museum

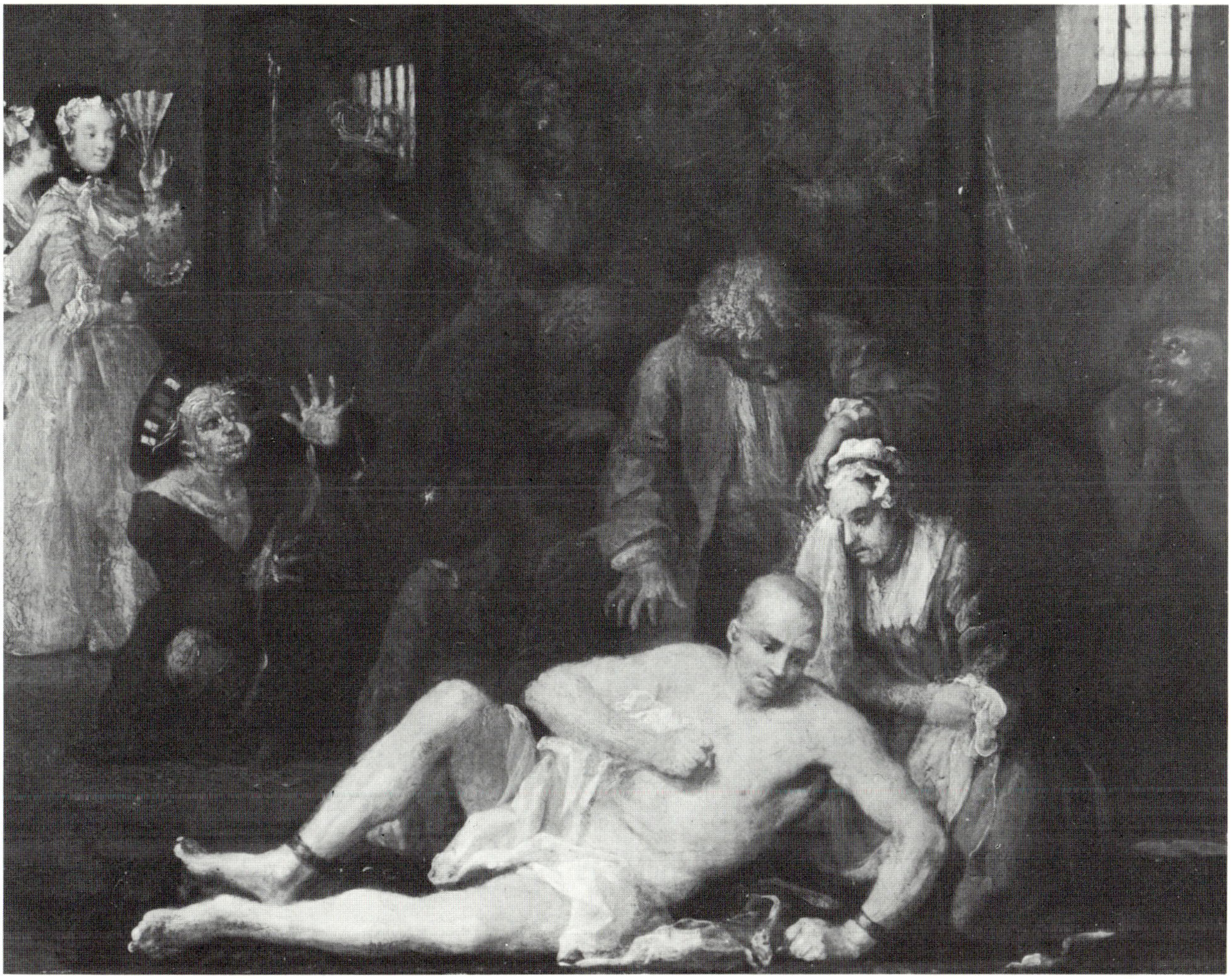

86. Hogarth: *A Rake's Progress*, VIII. *The Madhouse*, detail of the Rake chained. London, Sir John Soane's Museum

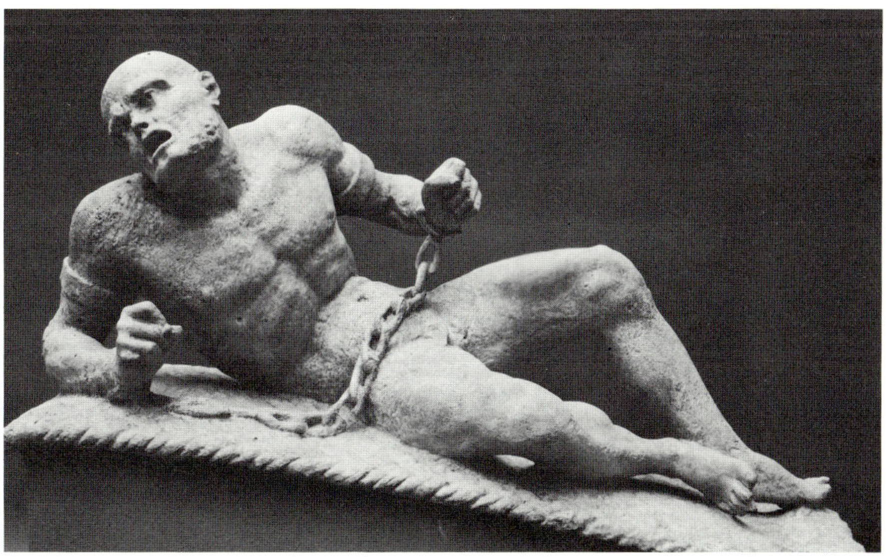

87. Caius Gabriel Cibber: *Raving Madness*, from the gate to old Bedlam Hospital, Moorfields. London, Victoria and Albert Museum

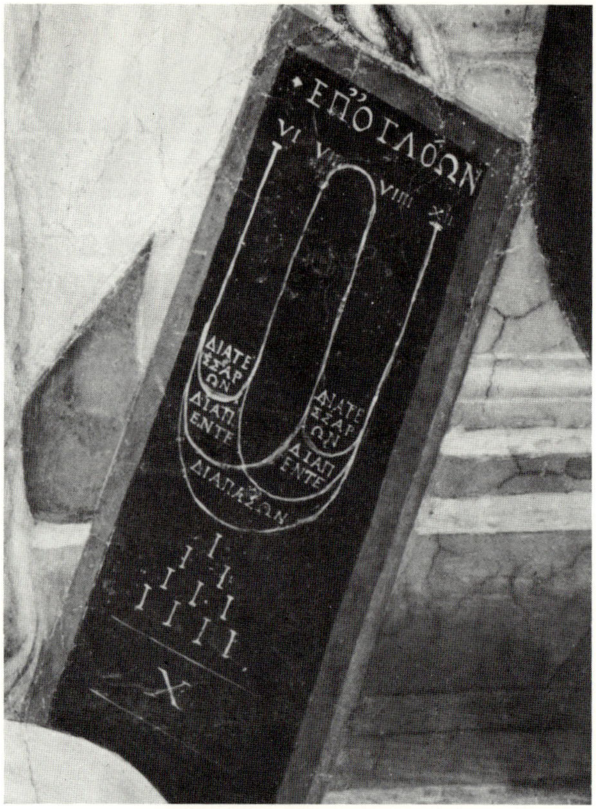

88. Raphael: 'Pythagorean' musical tablet, detail from *The School of Athens*. Palazzo Vaticano, Stanza della Segnatura

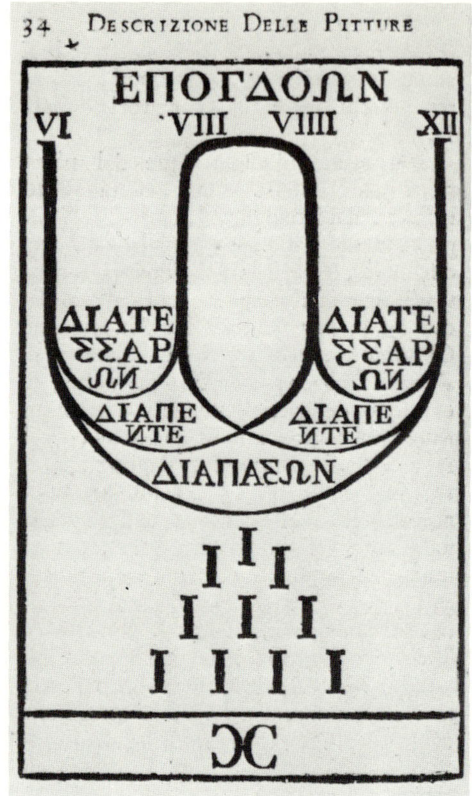

89. Diagram of Raphael's 'Pythagorean' tablet, from Bellori's *Descrizione delle immagini* . . ., 1751. Woodcut. Oxford, Bodleian Library

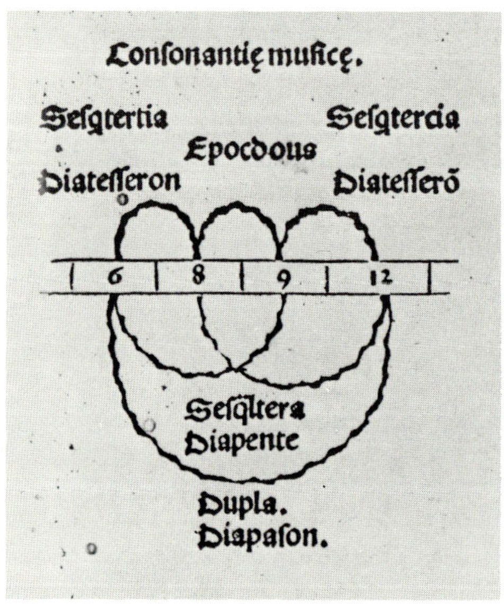

90. *Maxima armonia*, diagram at the end of Boethius' *Arithmetica*, 1488. Woodcut. Oxford, Bodleian Library

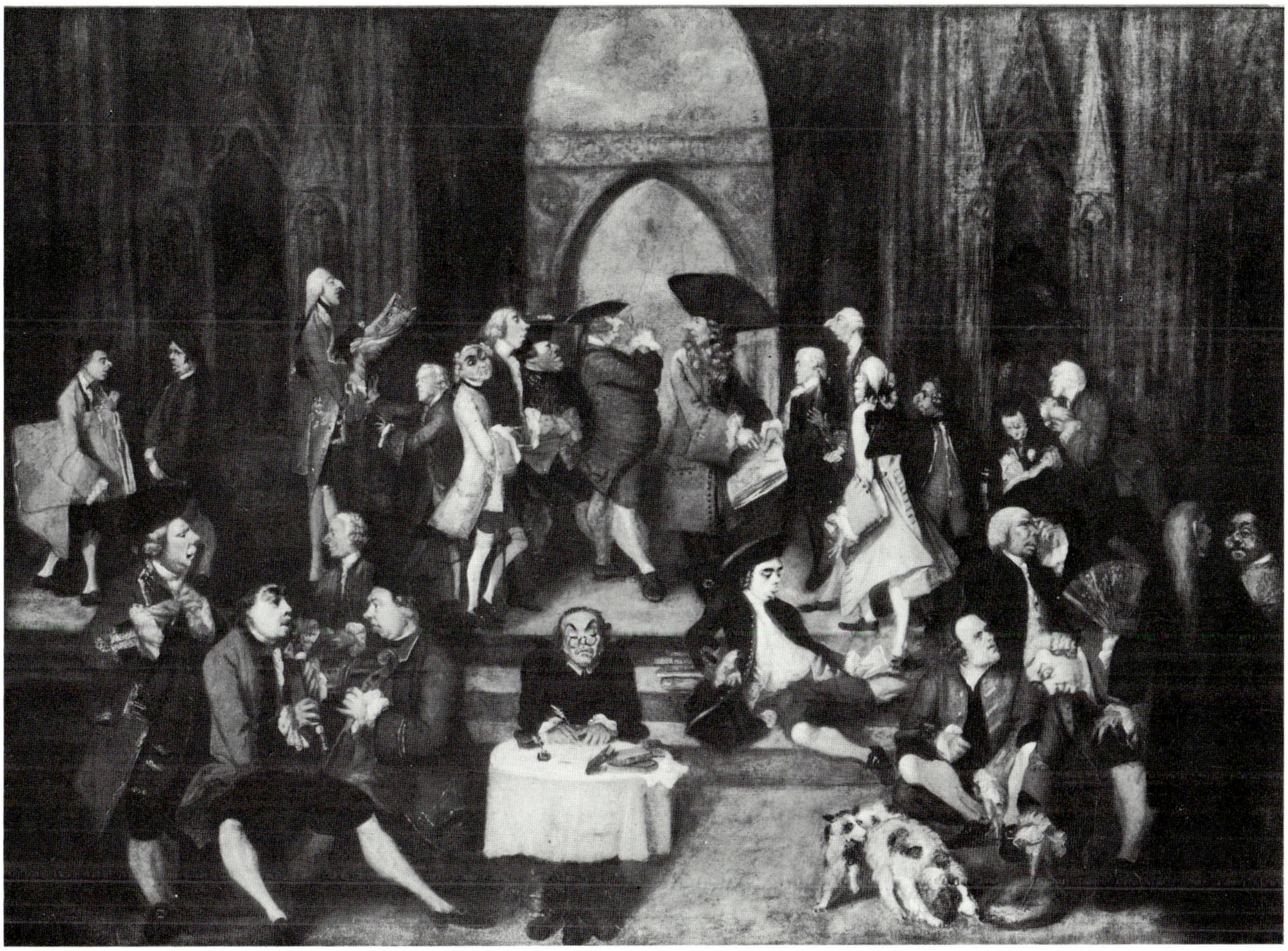

91. Reynolds: *Parody of Raphael's 'School of Athens'*. Caricatures of English Visitors to Rome. Dublin, The National Gallery of Ireland.

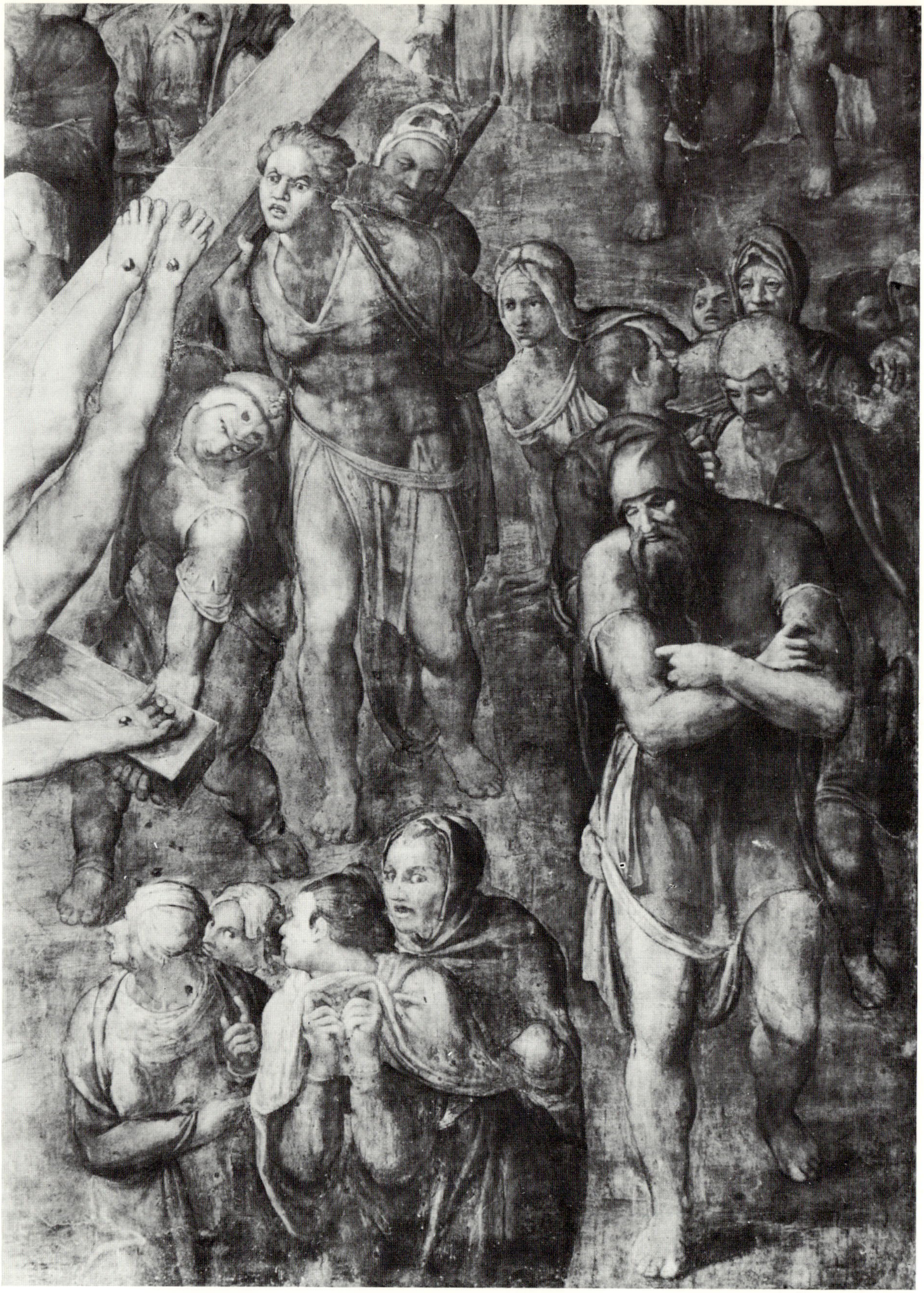

92. Michelangelo: Detail from *The Crucifixion of St. Peter*. Palazzo Vaticano, Cappella Paolina

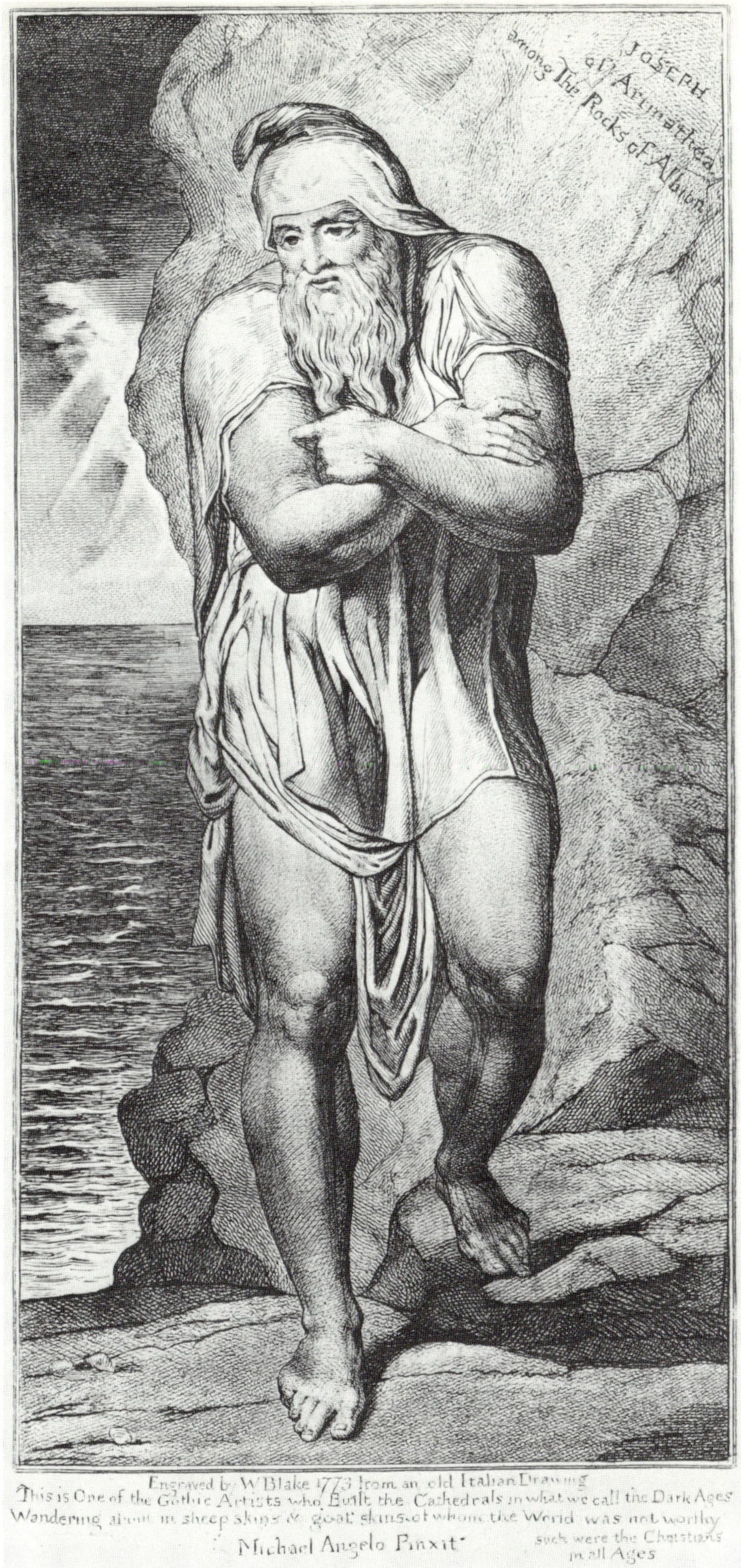

93. Blake: *Joseph of Arimathea among The Rocks of Albion.*
Engraving of 1773 from an old Italian drawing after Michelangelo. Second State.
London, British Museum

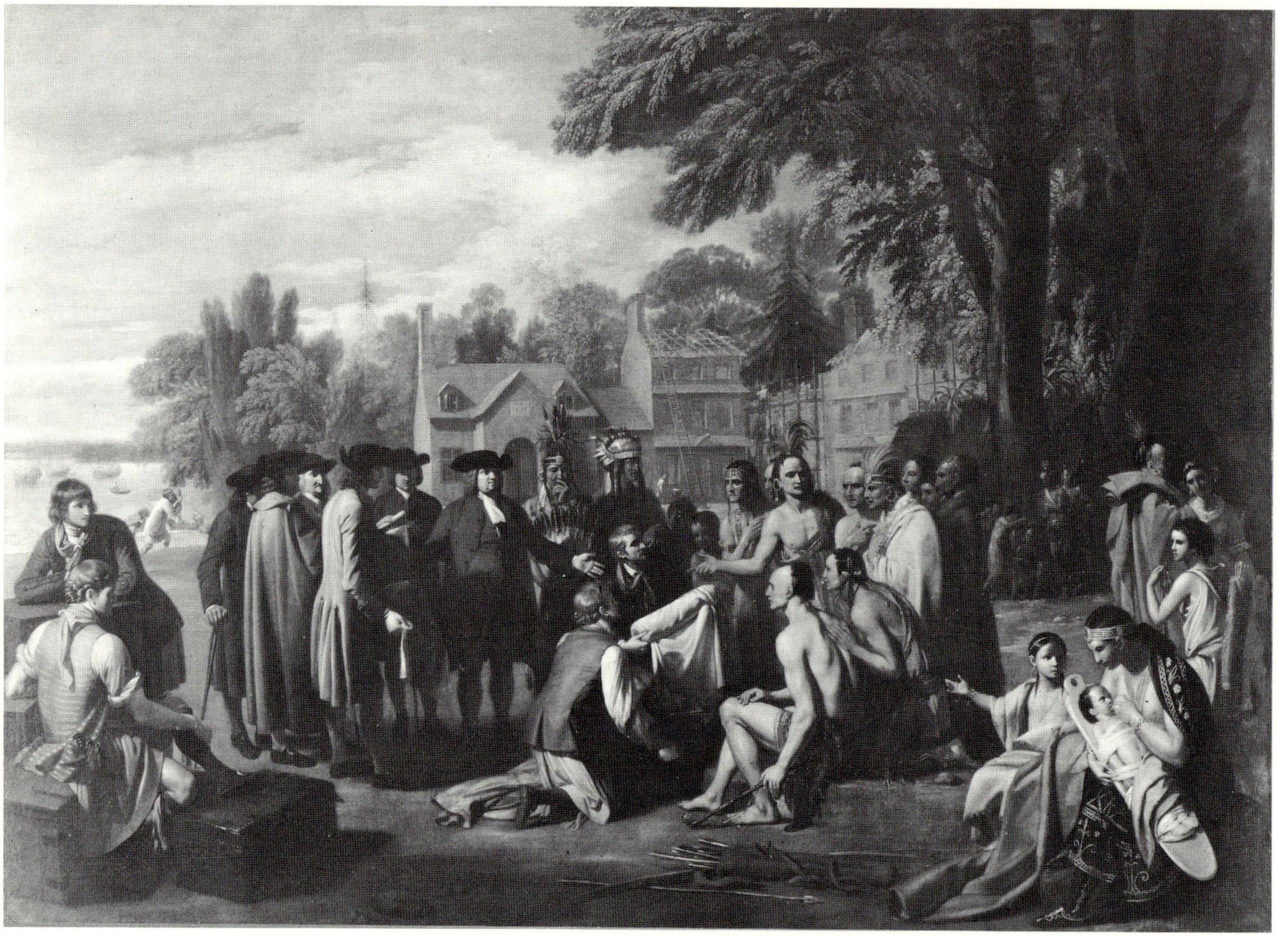

94. Benjamin West: *William Penn's Treaty with the Indians*. Philadelphia, The Pennsylvania Academy of the Fine Arts, Joseph and Sarah Harrison Collection

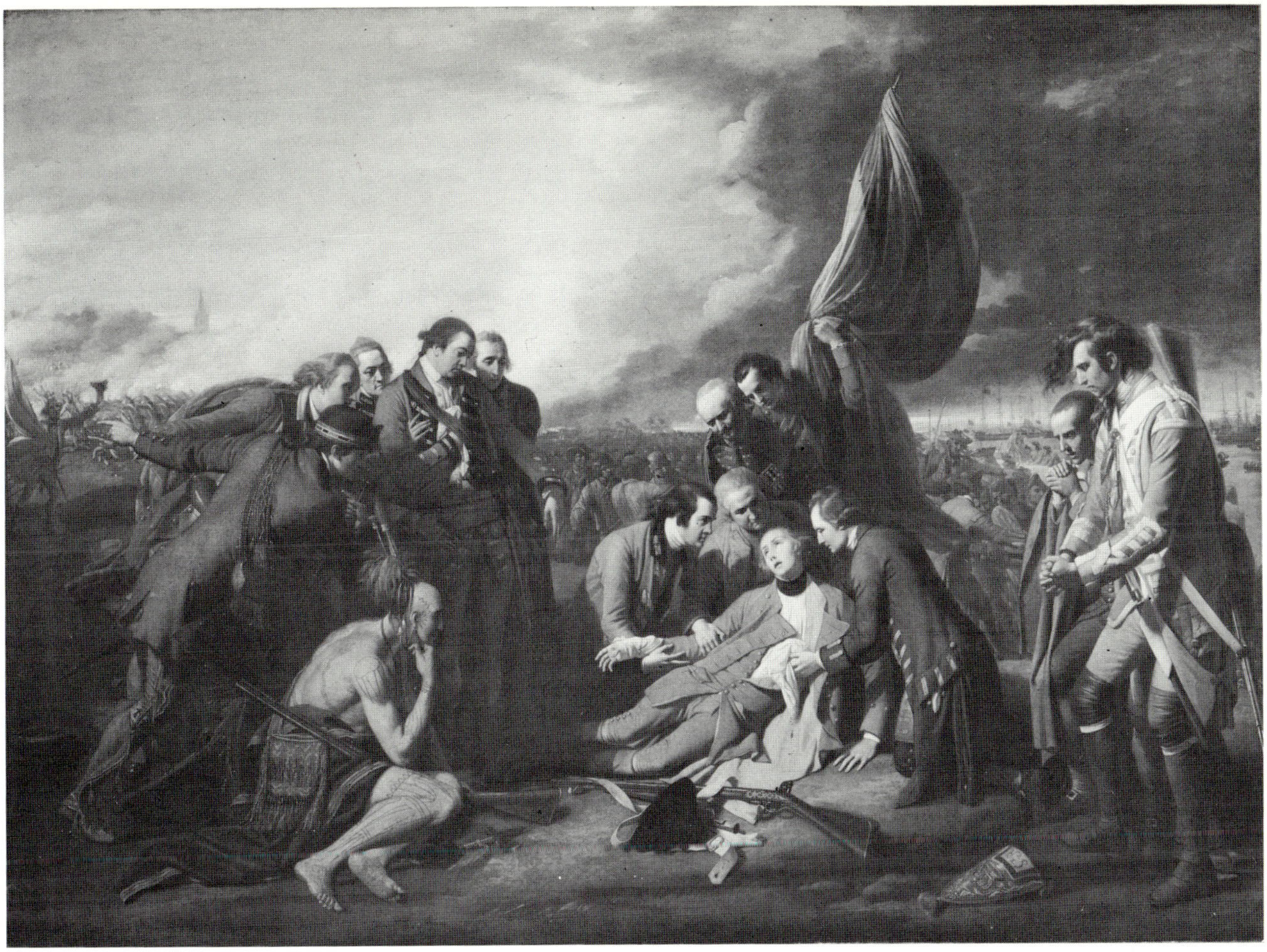

95. Benjamin West: *The Death of General Wolfe*. Ottawa, The National Gallery of Canada. Gift of the Duke of Westminster

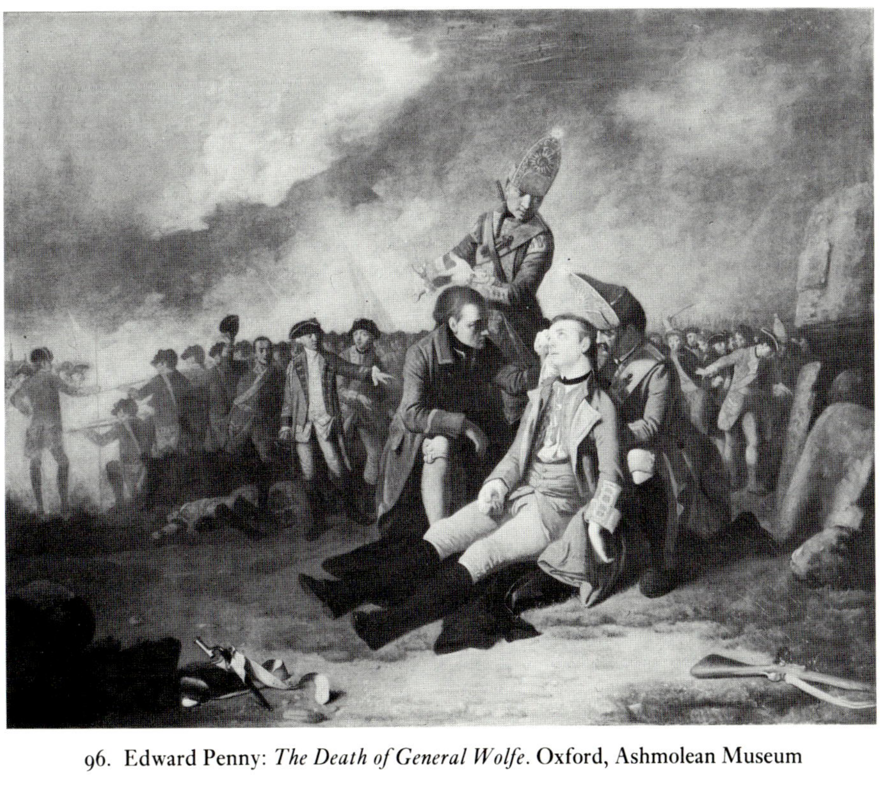

96. Edward Penny: *The Death of General Wolfe*. Oxford, Ashmolean Museum

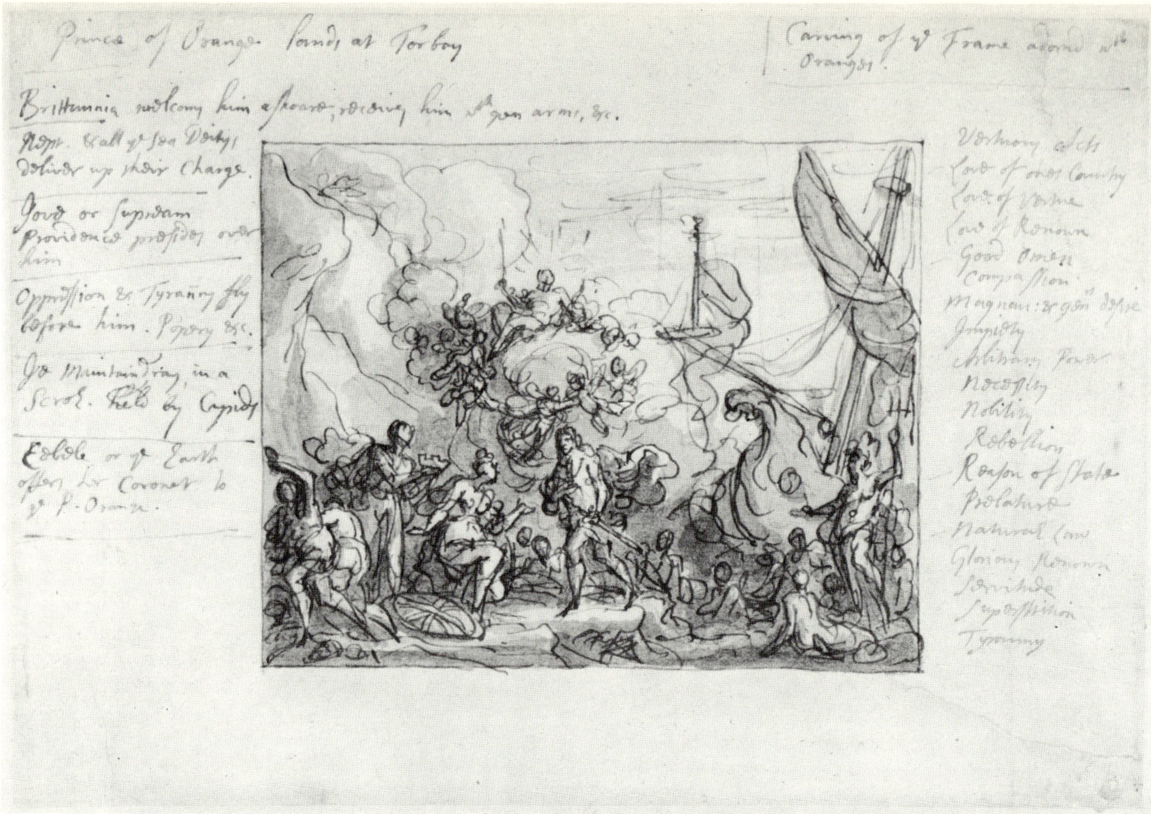

97. Sir James Thornhill: *Landing of the Prince of Orange at Torbay*. Drawing. London, British Museum

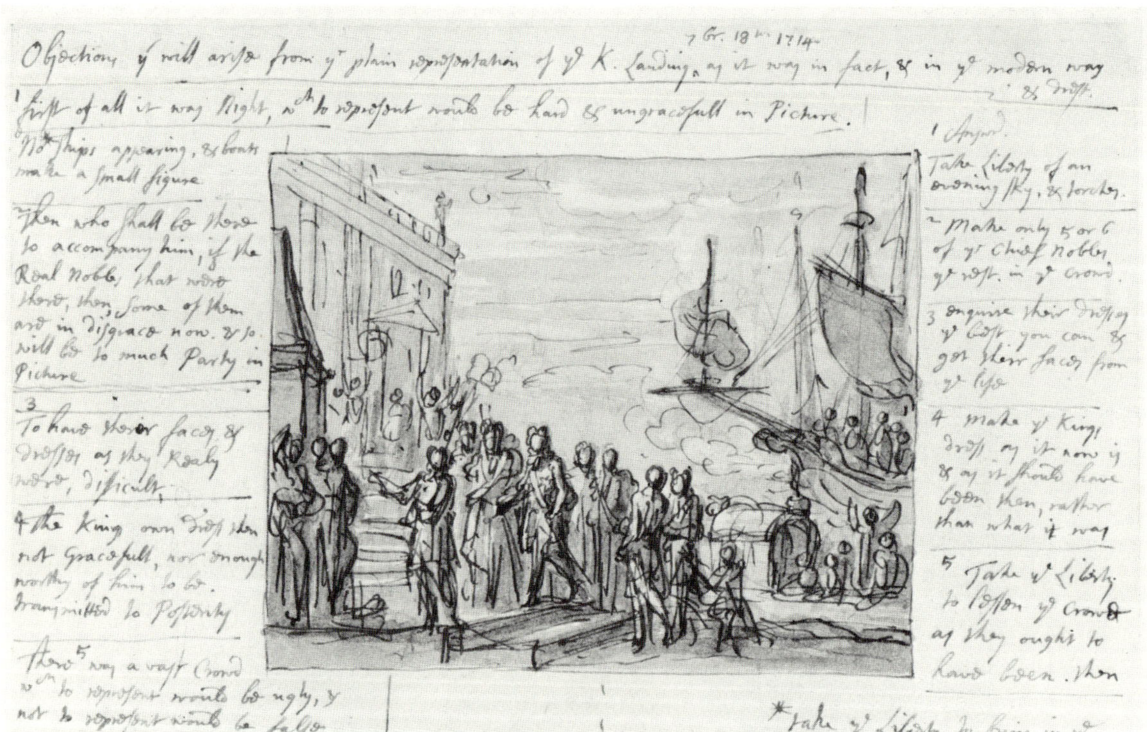

98. Sir James Thornhill: *George I landing at Greenwich*. Drawing. London, British Museum

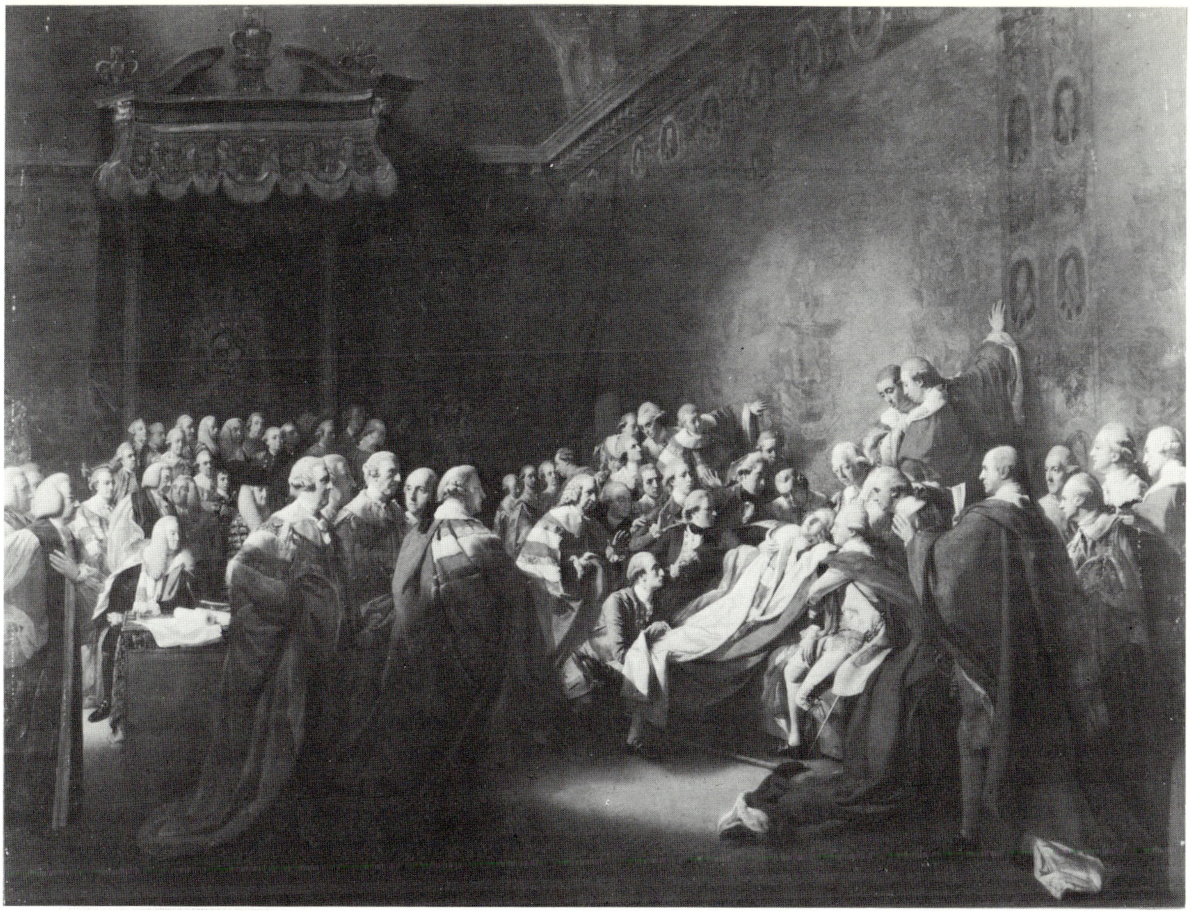

99. John Singleton Copley: *The Death of the Earl of Chatham*. London, Tate Gallery

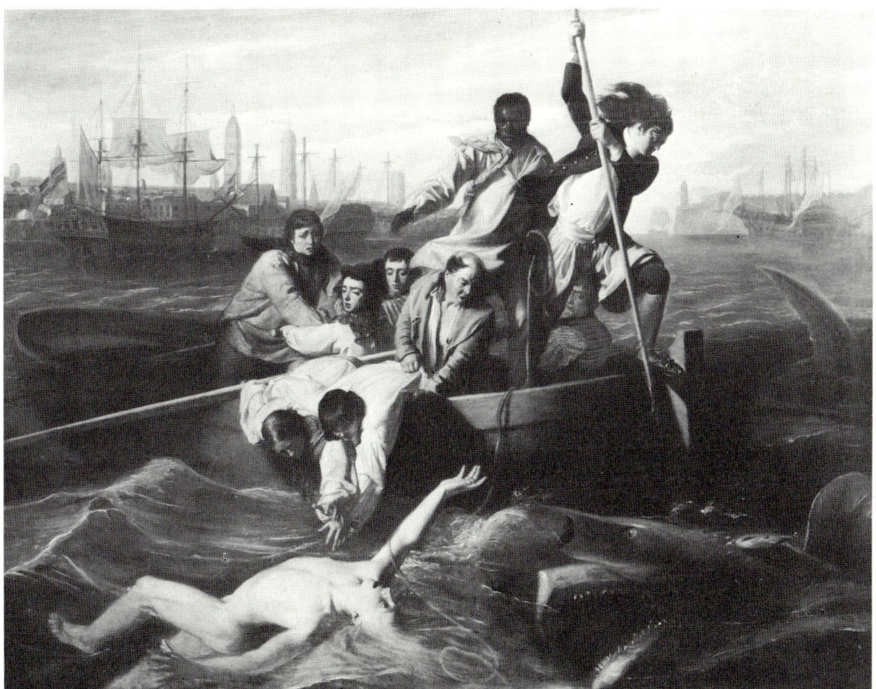

100. John Singleton Copley: *Brook Watson and the Shark*. Washington, DC, National Gallery of Art, Ferdinand Lammot Belin Fund

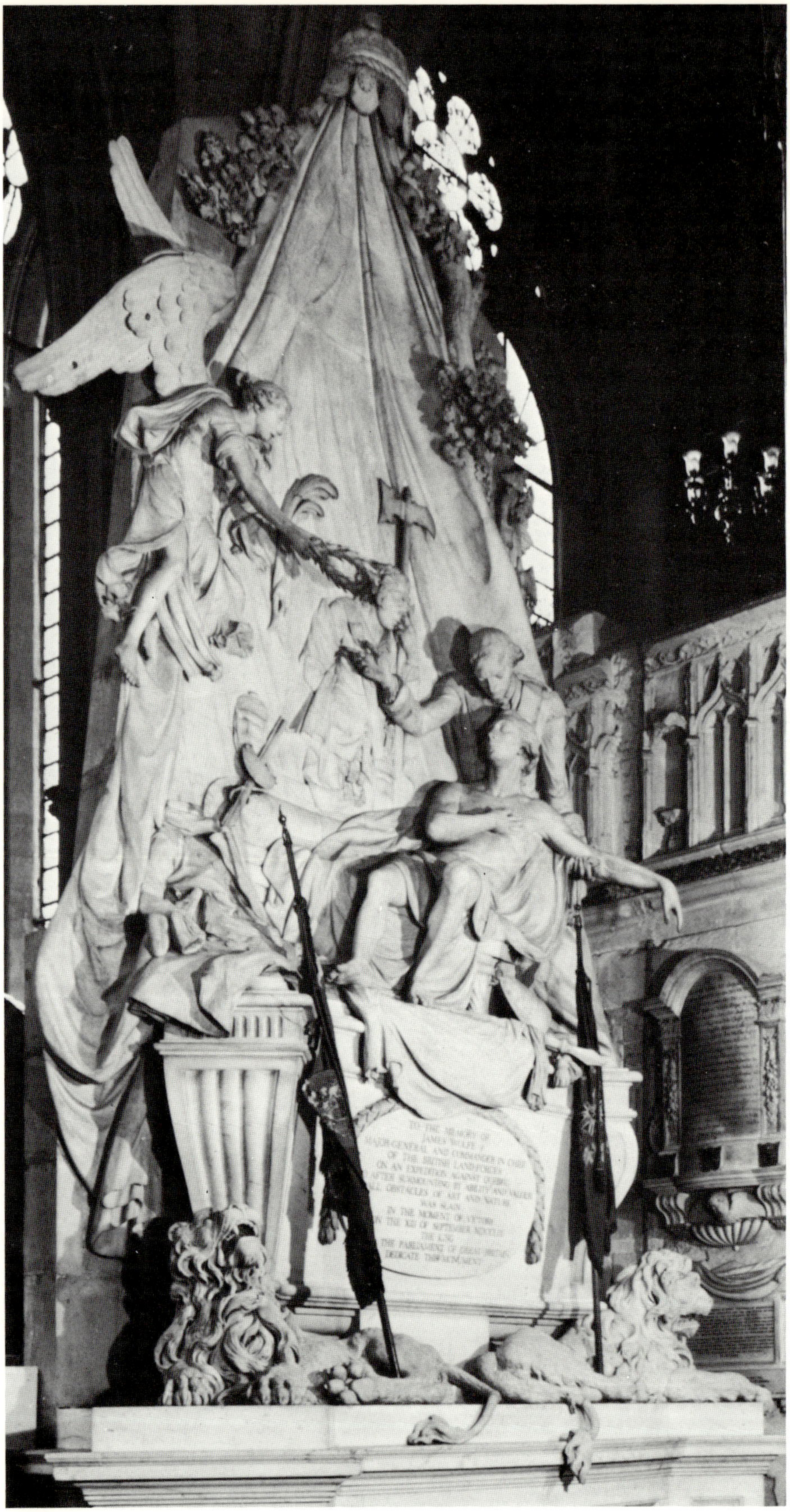

101. Joseph Wilton: Monument to General Wolfe. London, Westminster Abbey

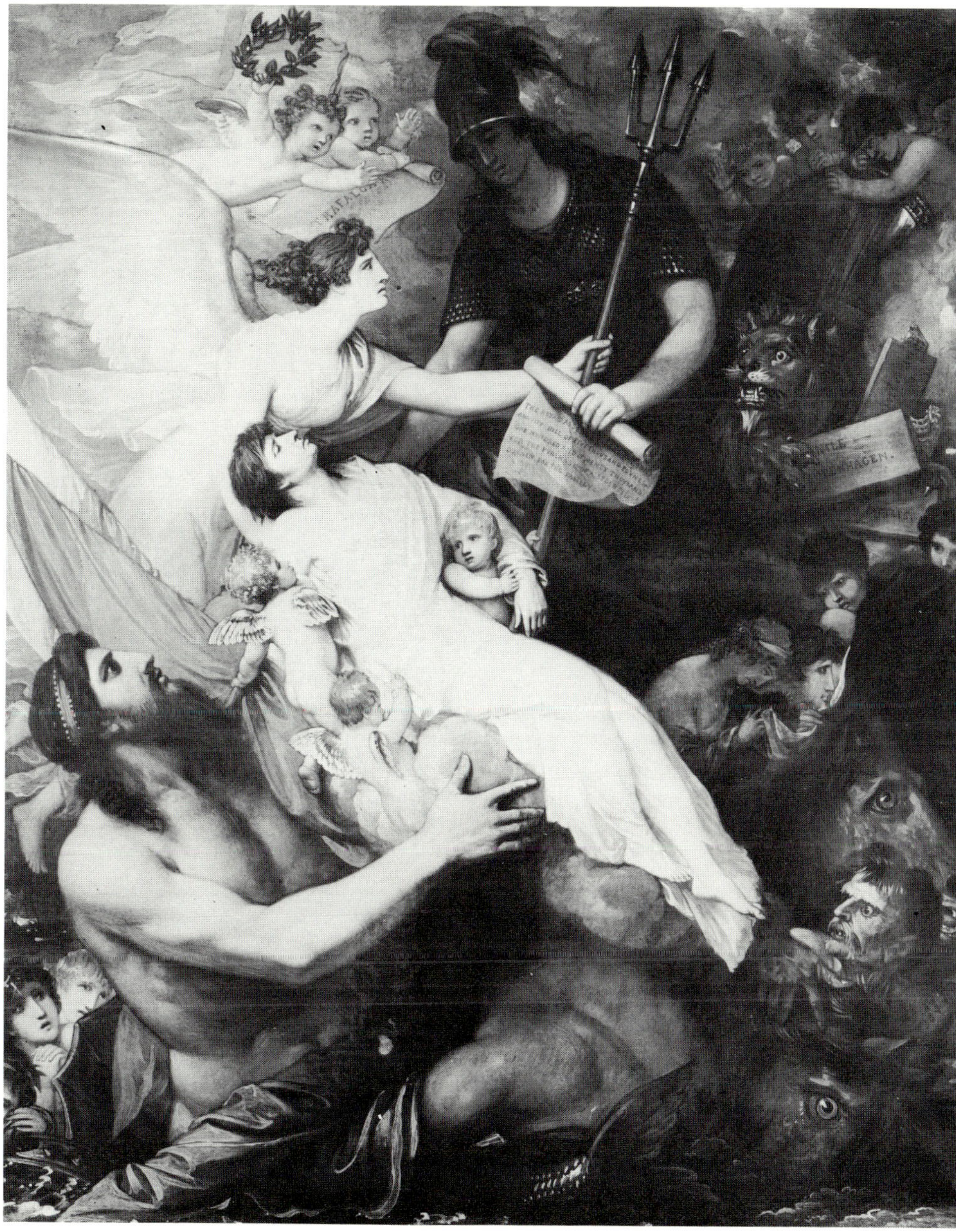

102. Benjamin West: *The Apotheosis of Nelson*.
London, The National Maritime Museum – Greenwich Hospital Collection

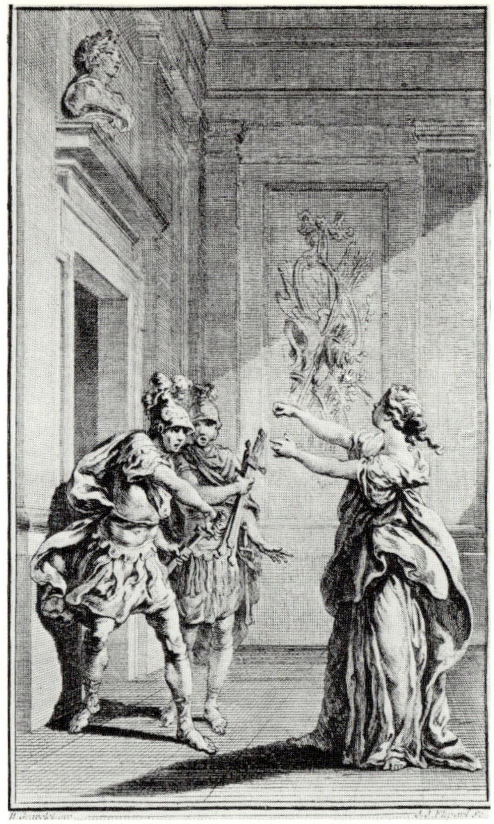

103. Hubert-François Gravelot:
The Death of Camilla. Engraving by J. J. Flipart.
Frontispiece to Corneille's *Horace*, in *Théâtre de P. Corneille, avec des commentaires . . .*, 1764.
Oxford, Taylor Institution

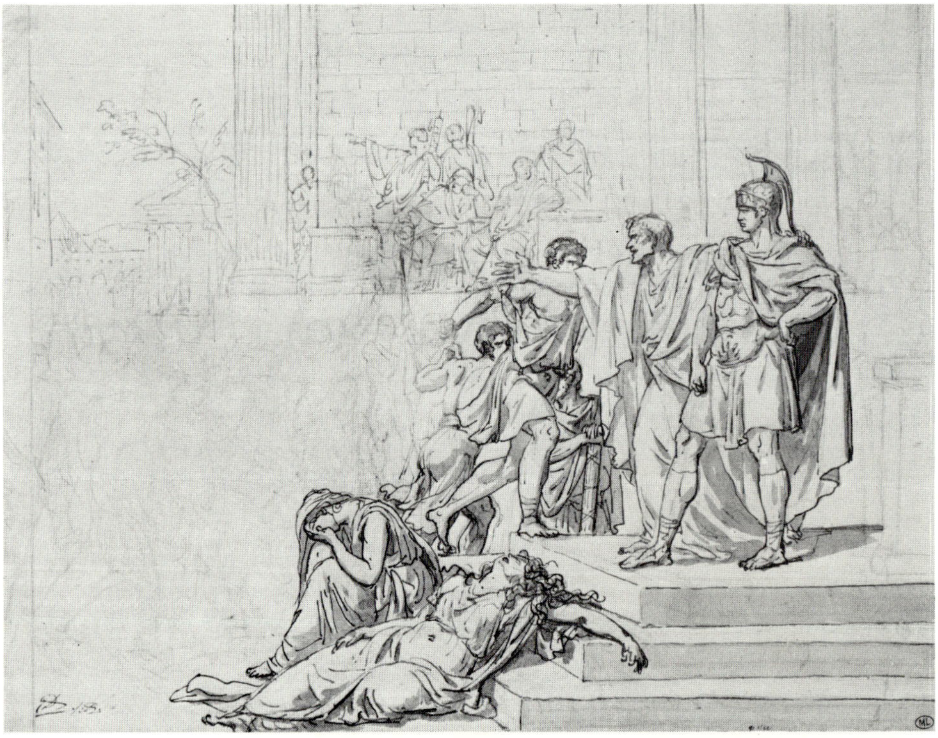

104. David: *Horatius defending his Son*. Drawing. Paris, Musée du Louvre

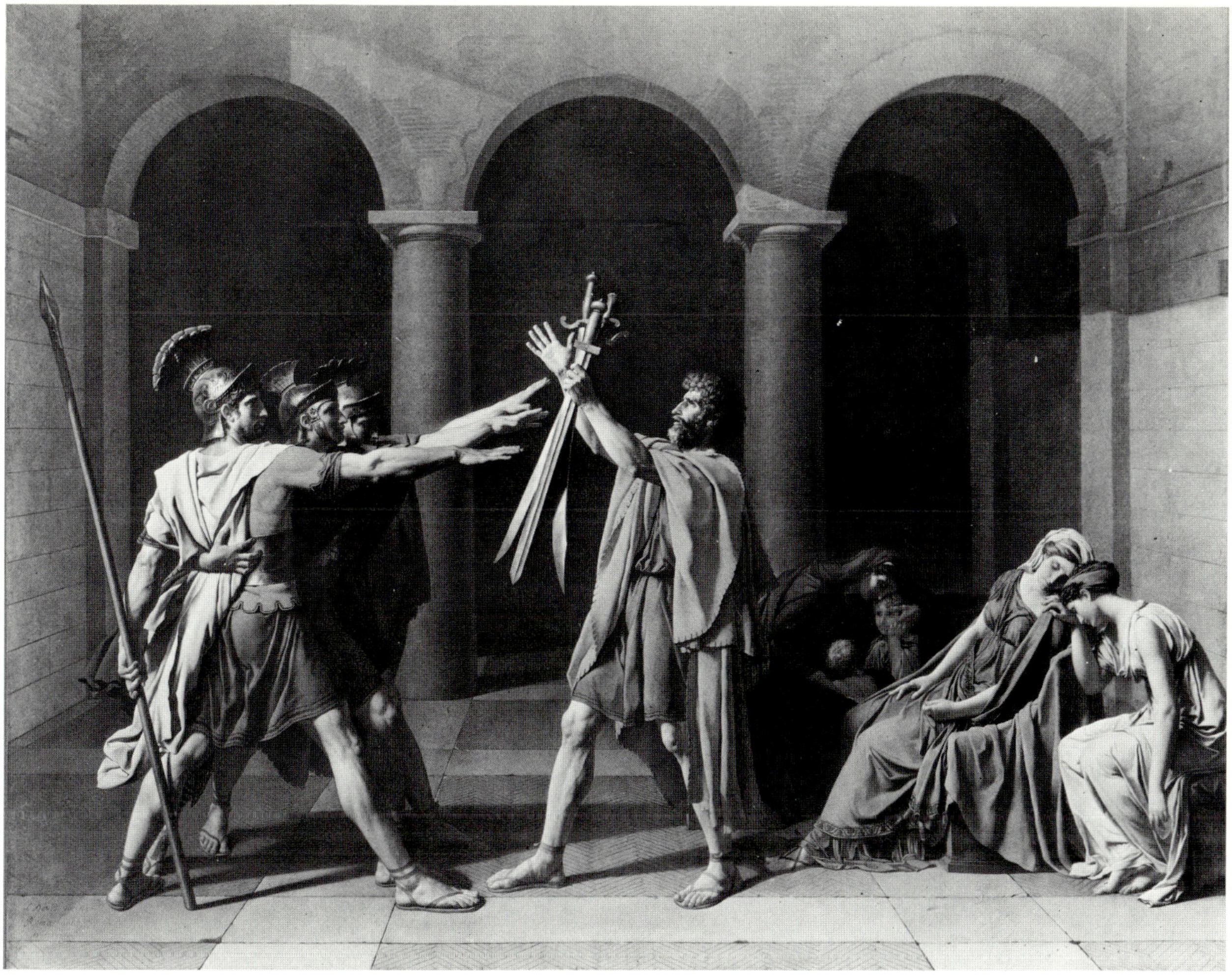

105. David: *The Oath of the Horatii*. Paris, Musée du Louvre

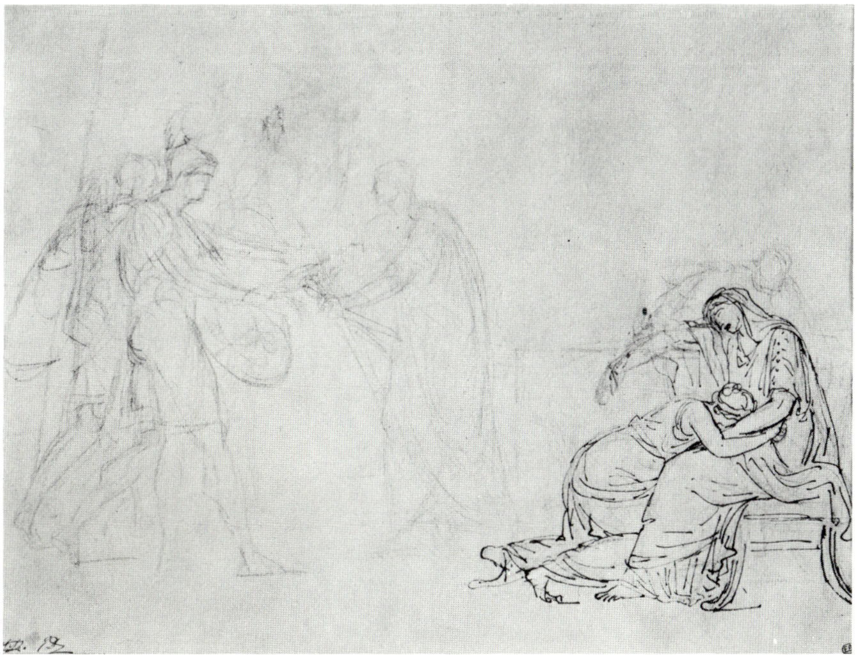

106. David: *Horatius preparing his Sons for battle*. Drawing.
Paris, École nationale supérieure des Beaux-Arts

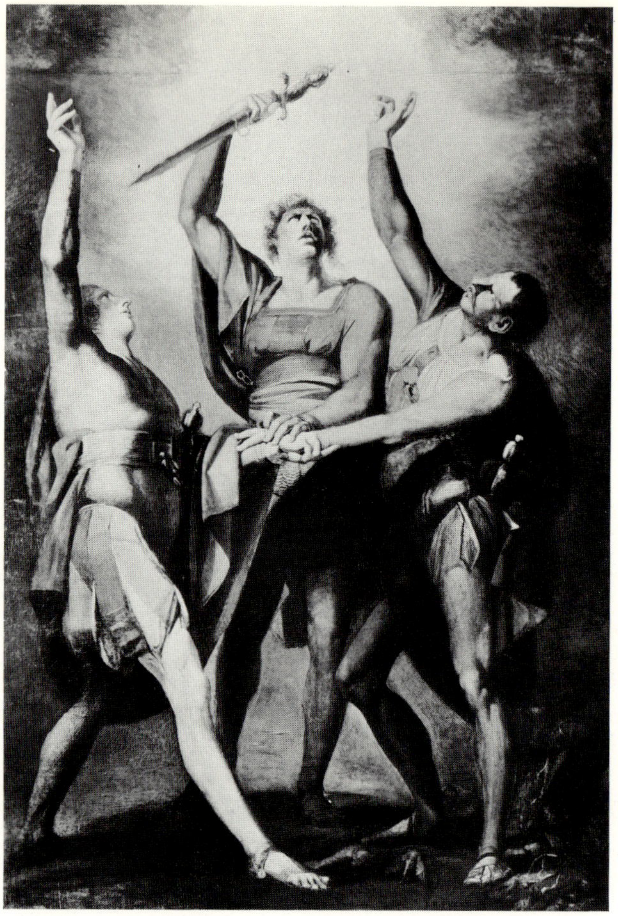

107. Fuseli: *The Oath on the Rütli.* Zurich, Rathaus

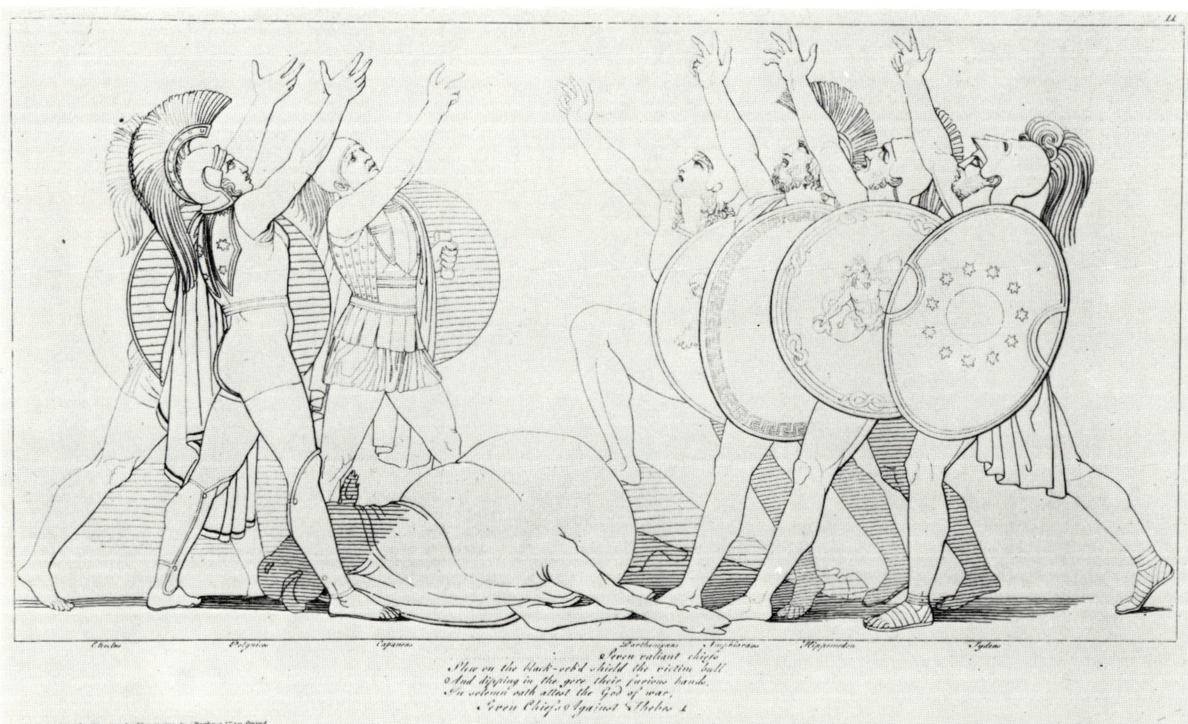

108. Flaxman: *Seven Chiefs against Thebes*, I.
From *Compositions from the Tragedies of Æschylus designed by Iohn Flaxman, engraved by Thomas Piroli*, 1795.
London, British Library

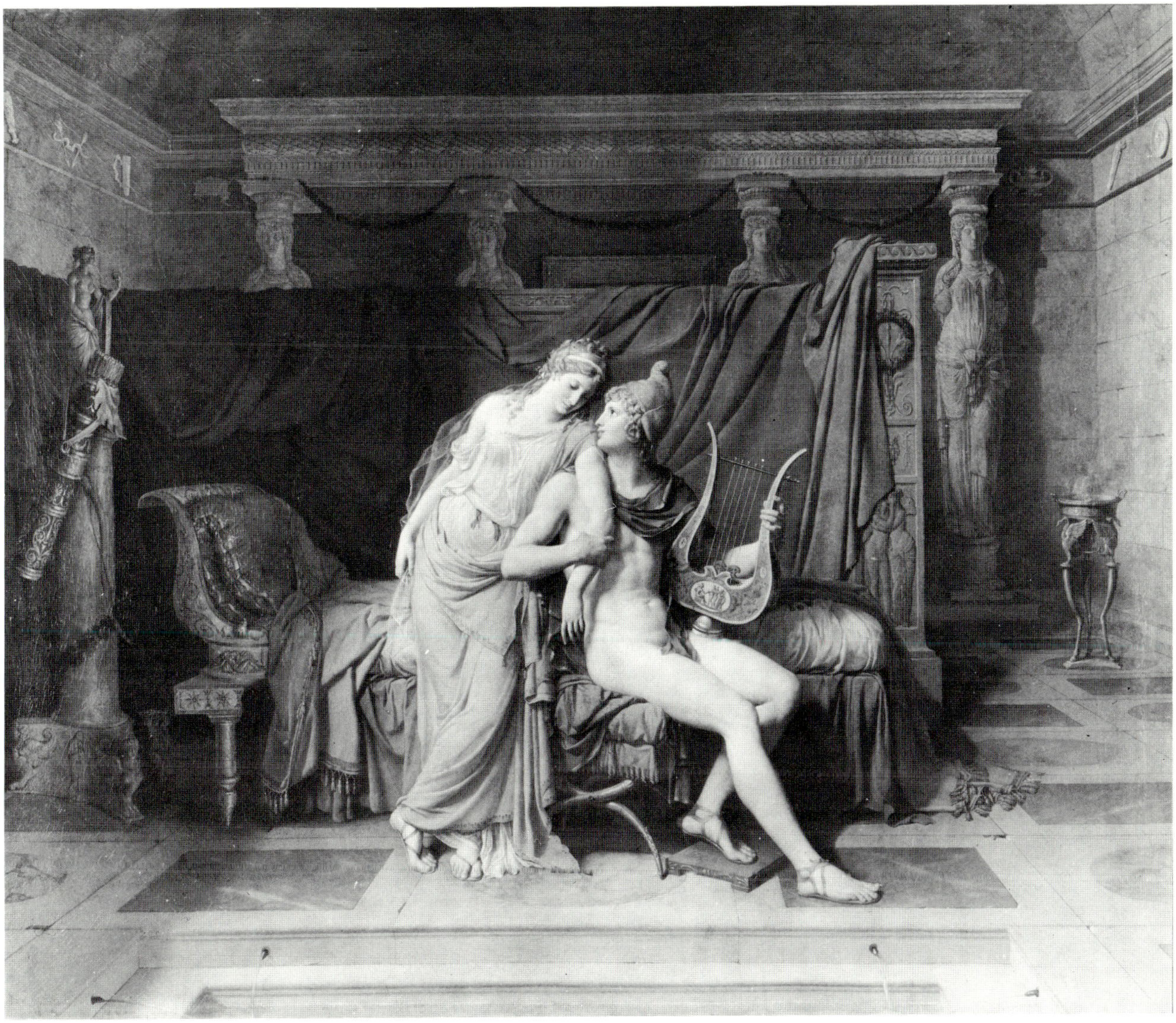

109. David: *Paris and Helen*. Paris, Musée du Louvre

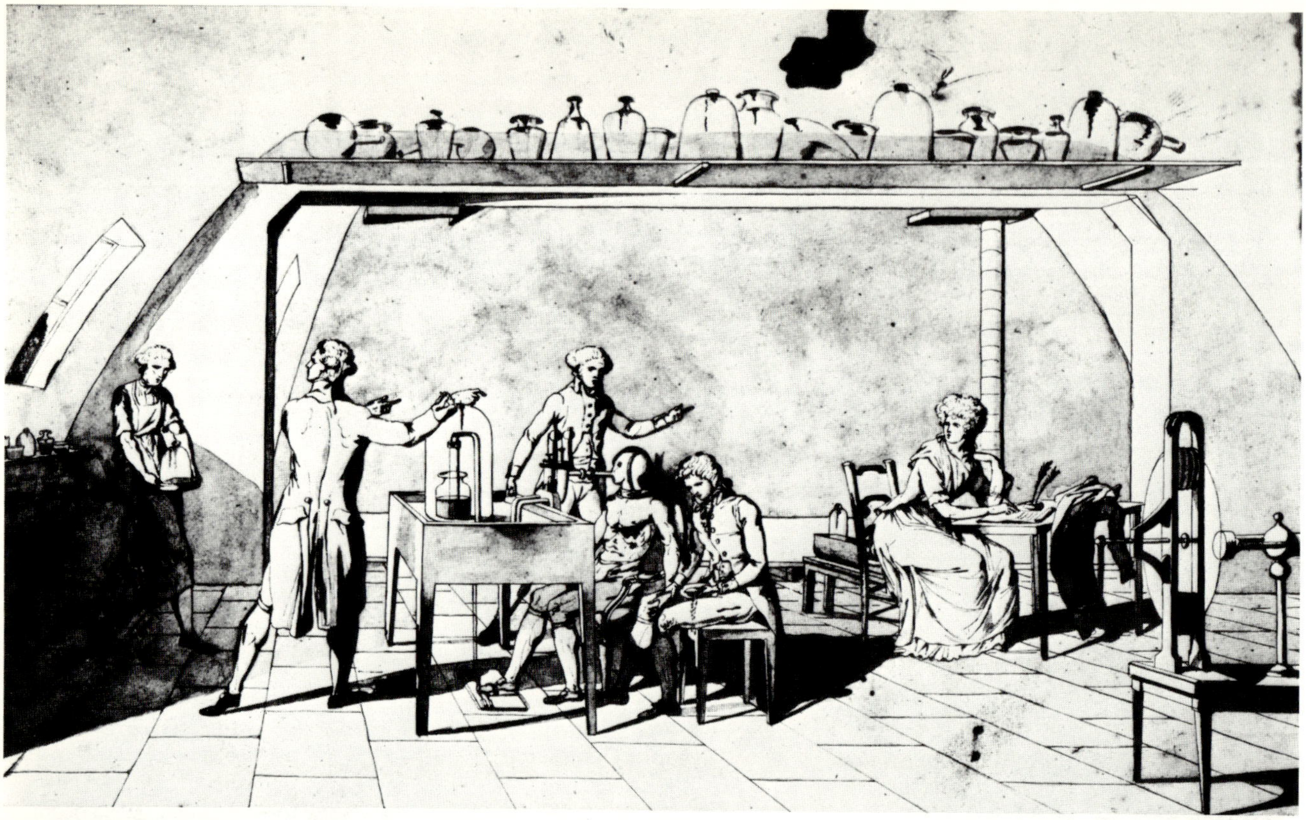

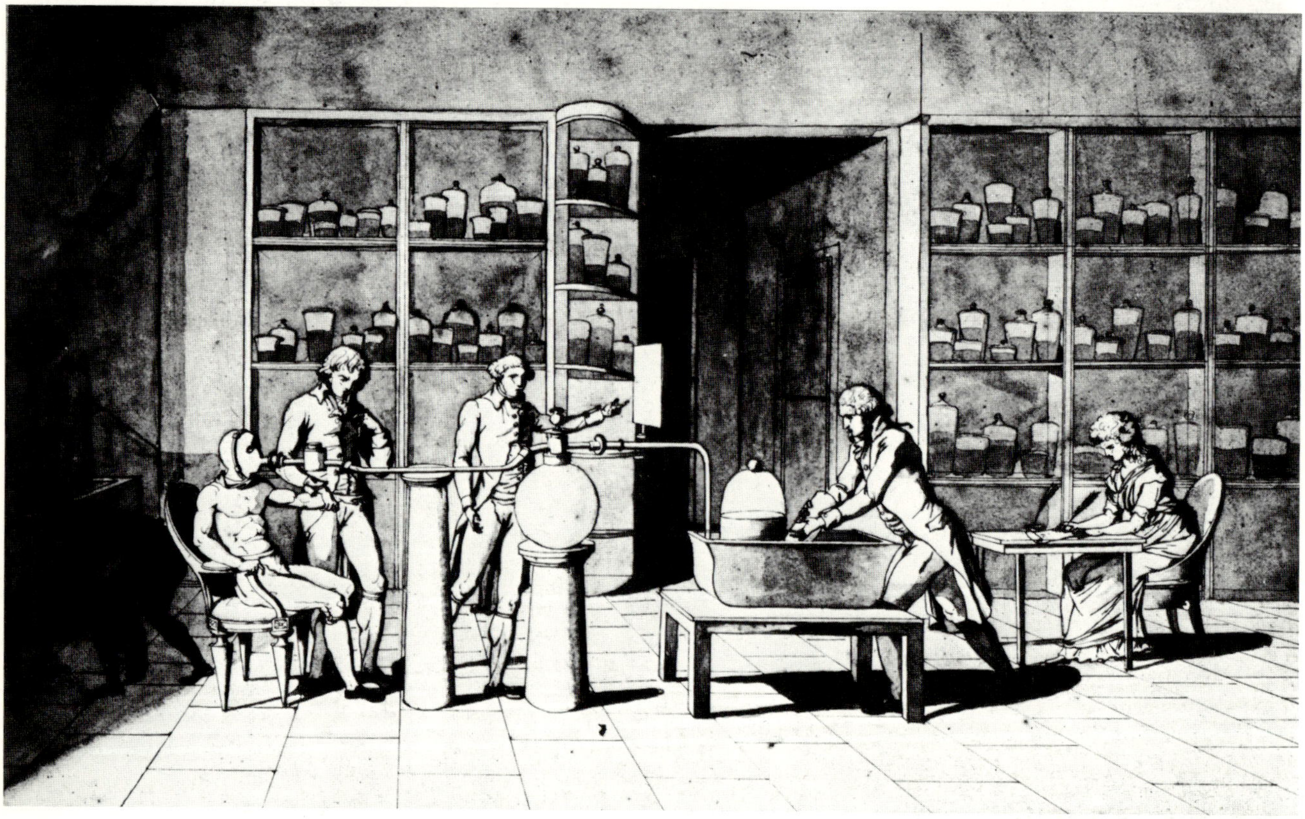

110–11. Madame Lavoisier: *Lavoisier in his Laboratory conducting Experiments on Respiration*.
He is comparing the quantity of oxygen consumed by his masked assistant, Armand Seguin, during exercise (above) and at rest (below).
Drawing. Private Collection

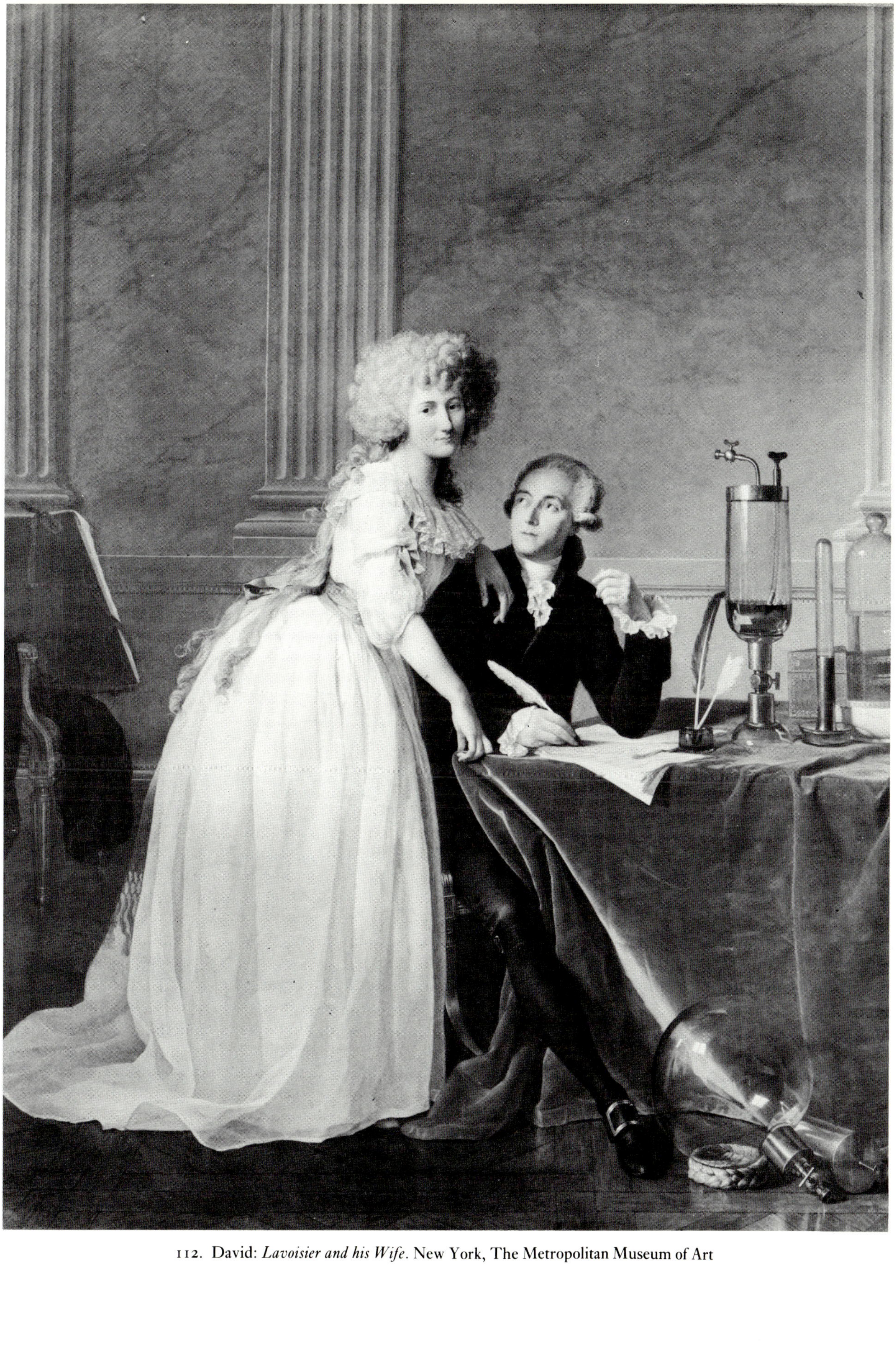

112. David: *Lavoisier and his Wife*. New York, The Metropolitan Museum of Art

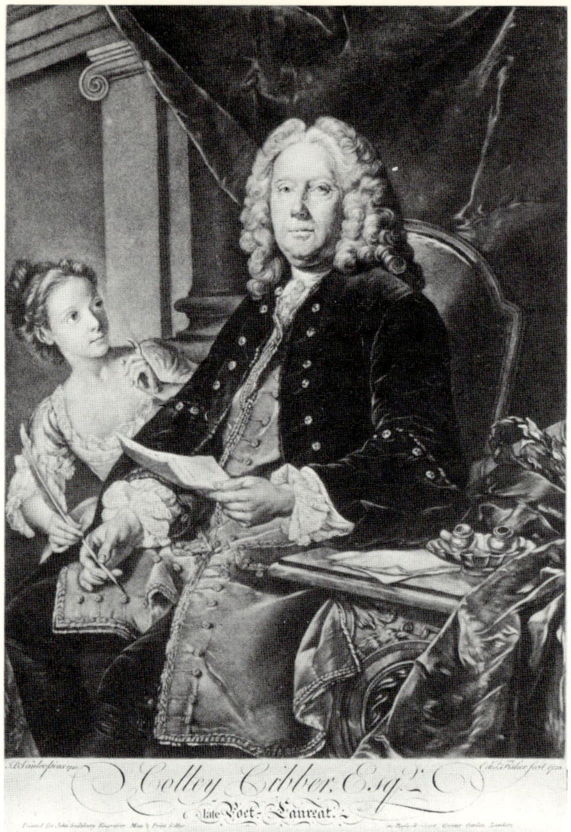

113. J.-B. Van Loo: *Colley Cibber with his Muse.*
Mezzotint by Edward Fisher. London, British Museum

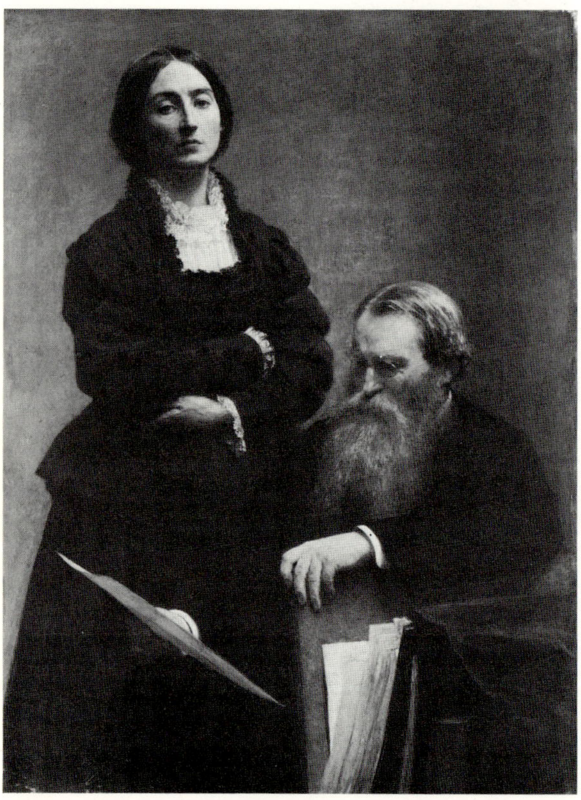

114. Fantin-Latour: *Mr and Mrs Edwin Edwards.*
London, Tate Gallery

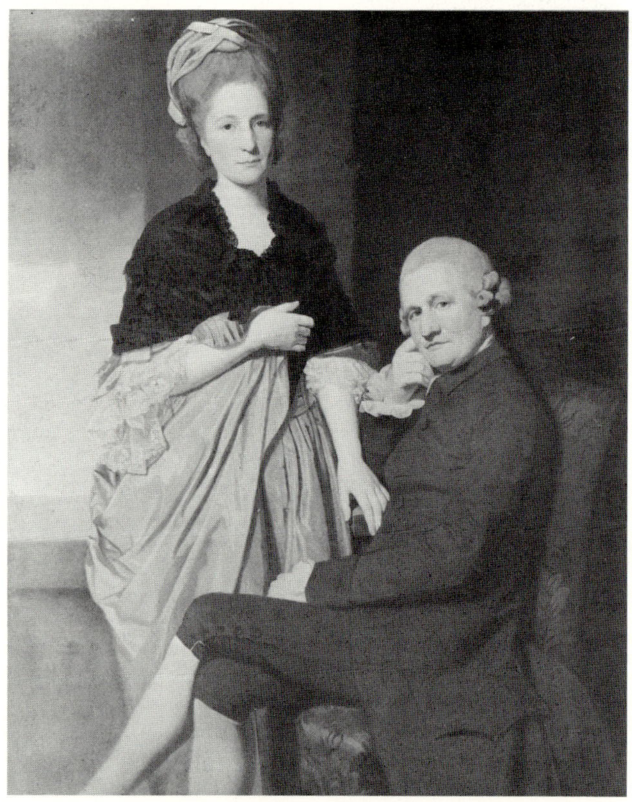

115. Romney: *Mr and Mrs William Lindow.*
London, Tate Gallery

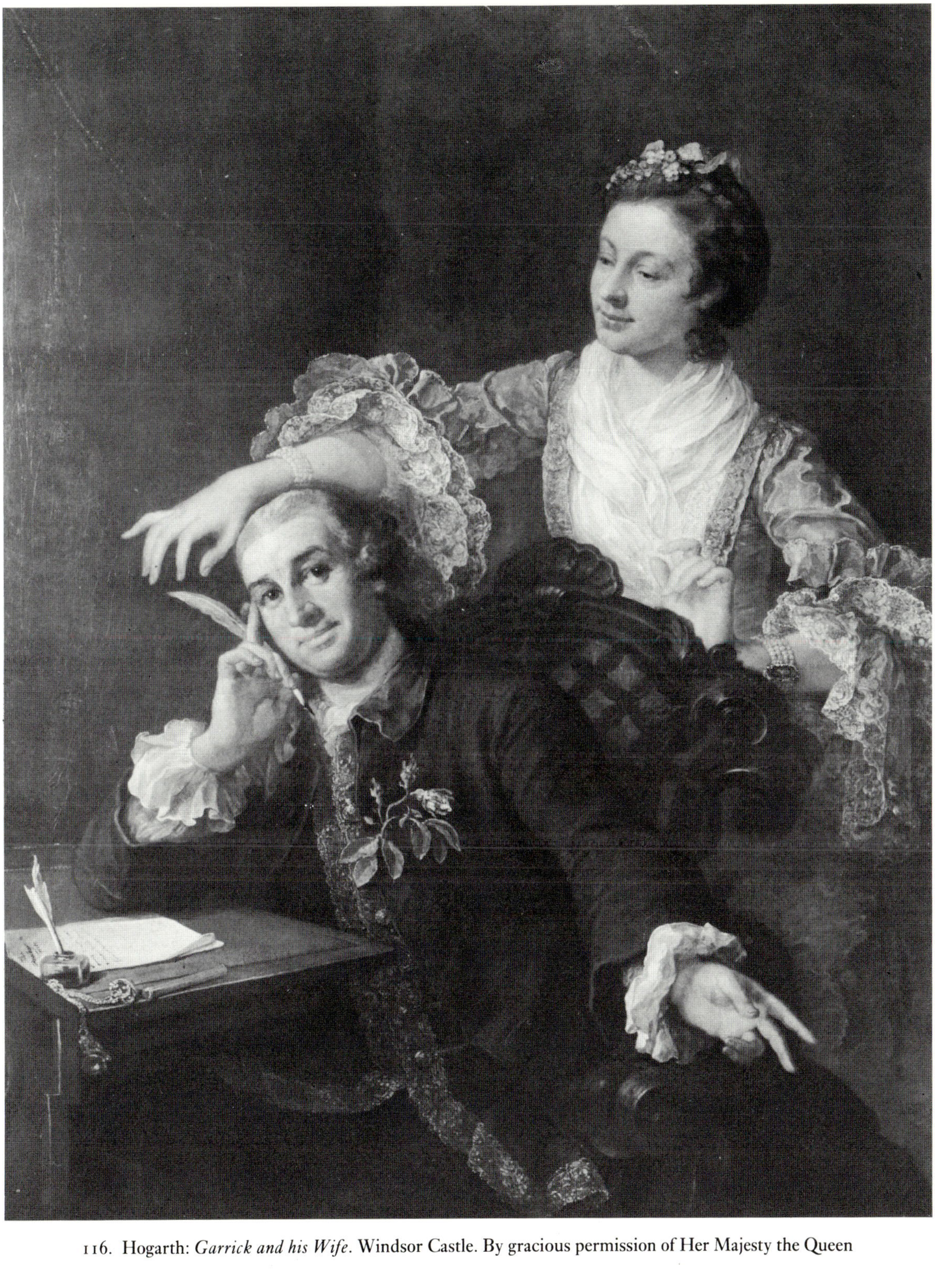

116. Hogarth: *Garrick and his Wife*. Windsor Castle. By gracious permission of Her Majesty the Queen

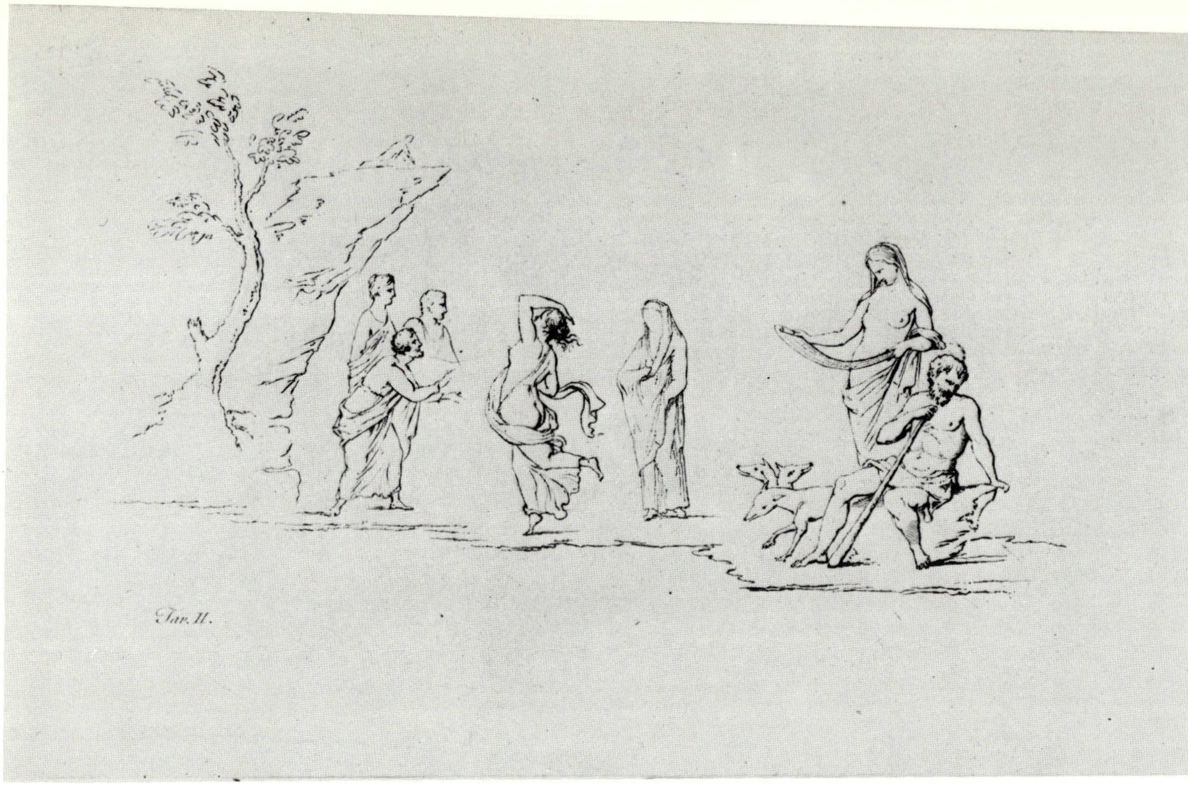

117. Bas-relief from the so-called Tomb of the Dancer near Cumae: *The Arrival in the Beyond*. Engraving from Andrea de Jorio's *Scheletri cumani dilucidati*, 1810. Oxford, Bodleian Library

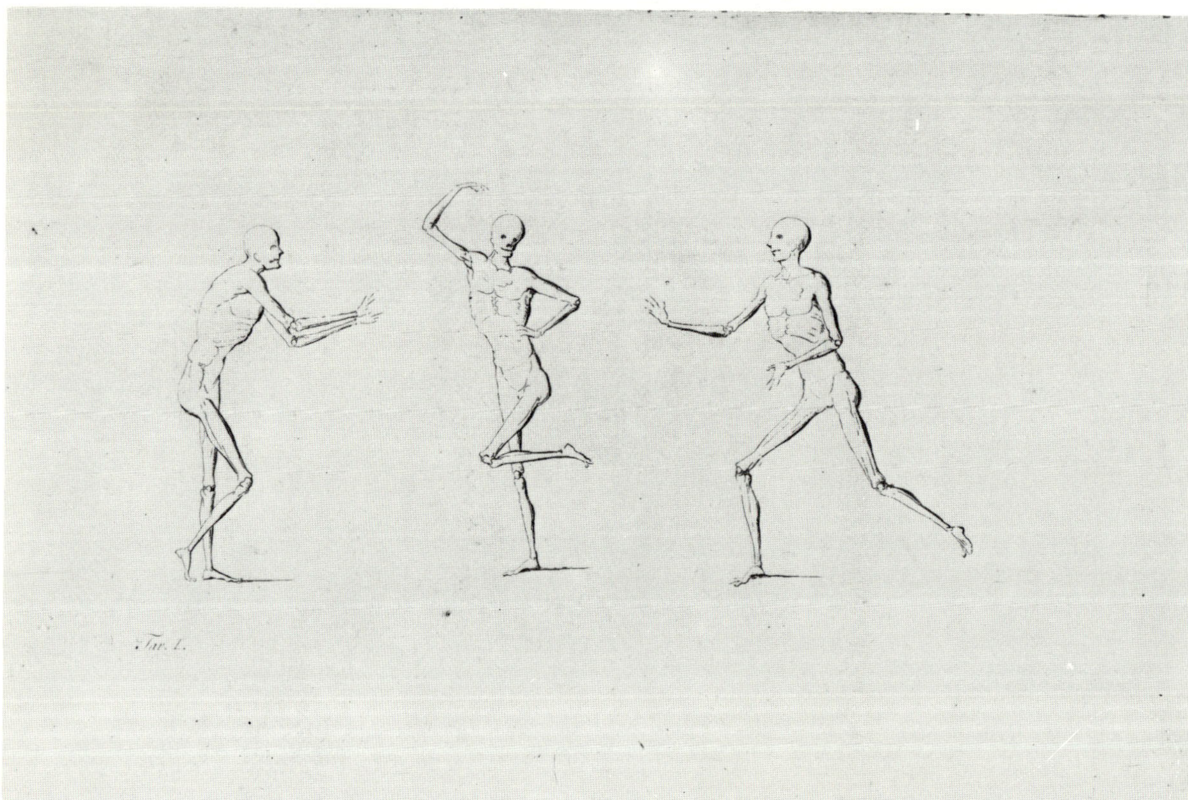

118. Bas-relief from the Tomb of the Dancer: *Skeletons dancing*. Engraving from Jorio's *Scheletri*. Oxford, Bodleian Library

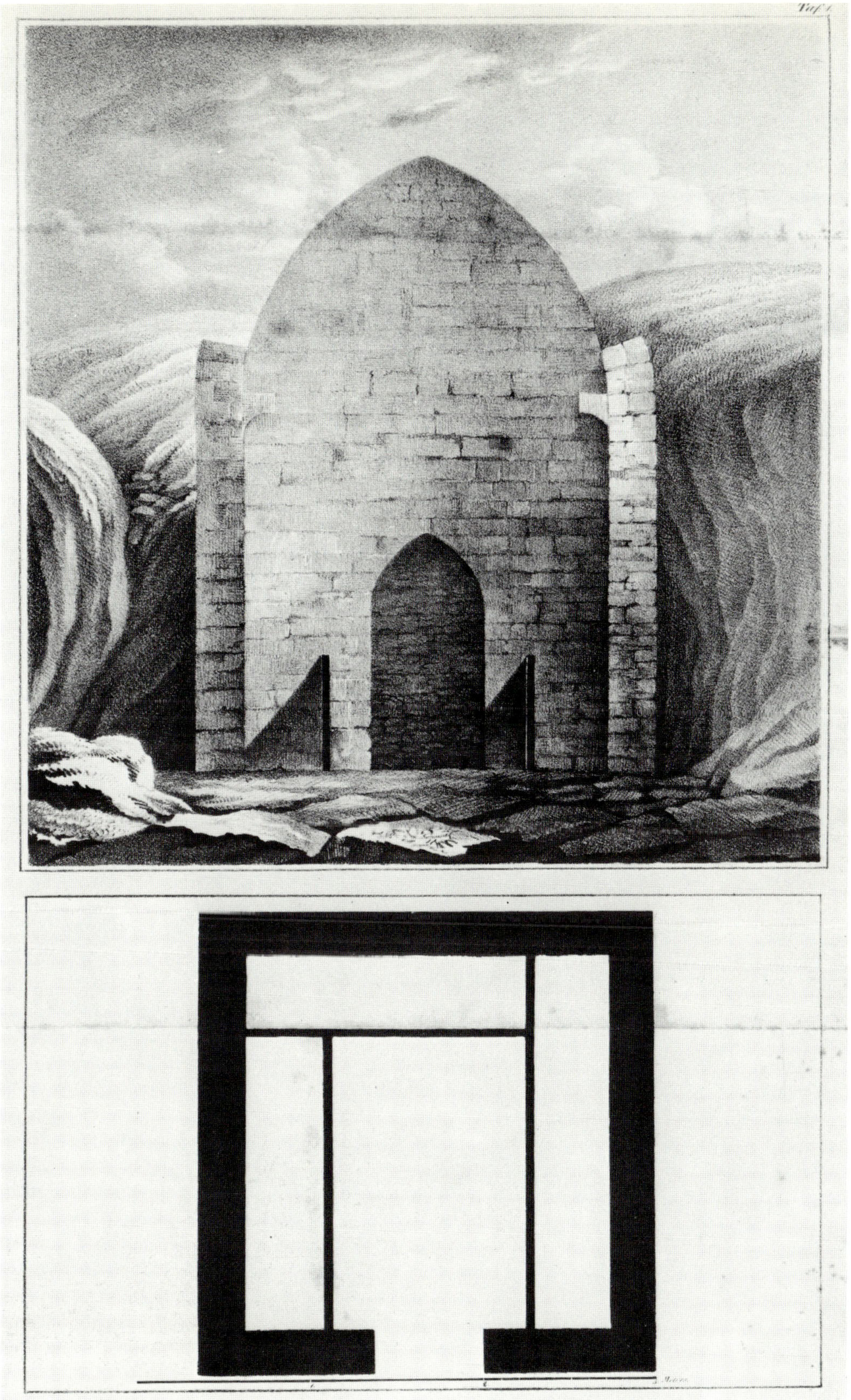

119. The Tomb of the Dancer, exterior view (location now unknown). Engraving from J. Fr. M. von Olfers's *Über ein merkwürdiges Grab bei Kumae . . .*, 1830. Oxford, Bodleian Library

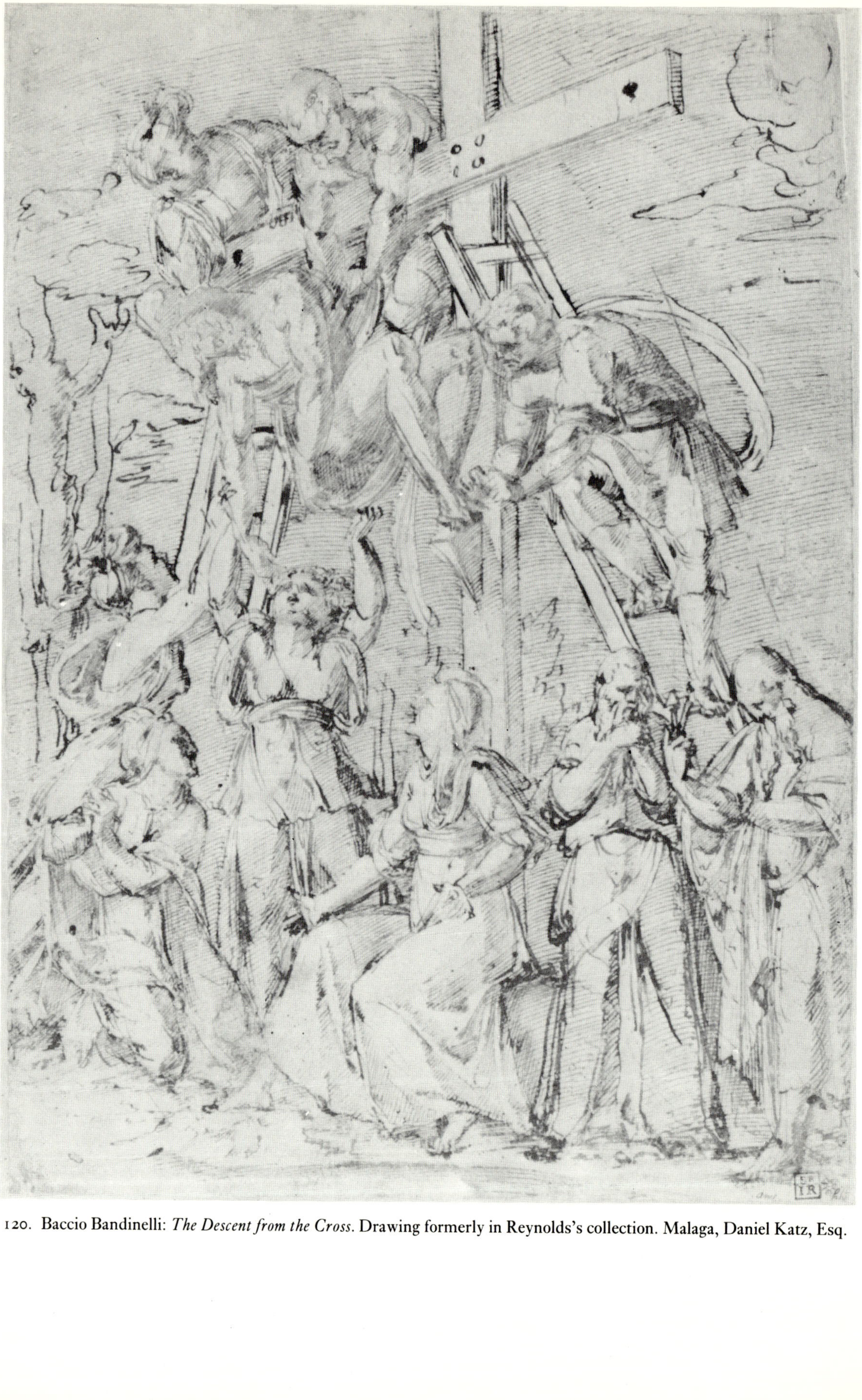

120. Baccio Bandinelli: *The Descent from the Cross*. Drawing formerly in Reynolds's collection. Malaga, Daniel Katz, Esq.

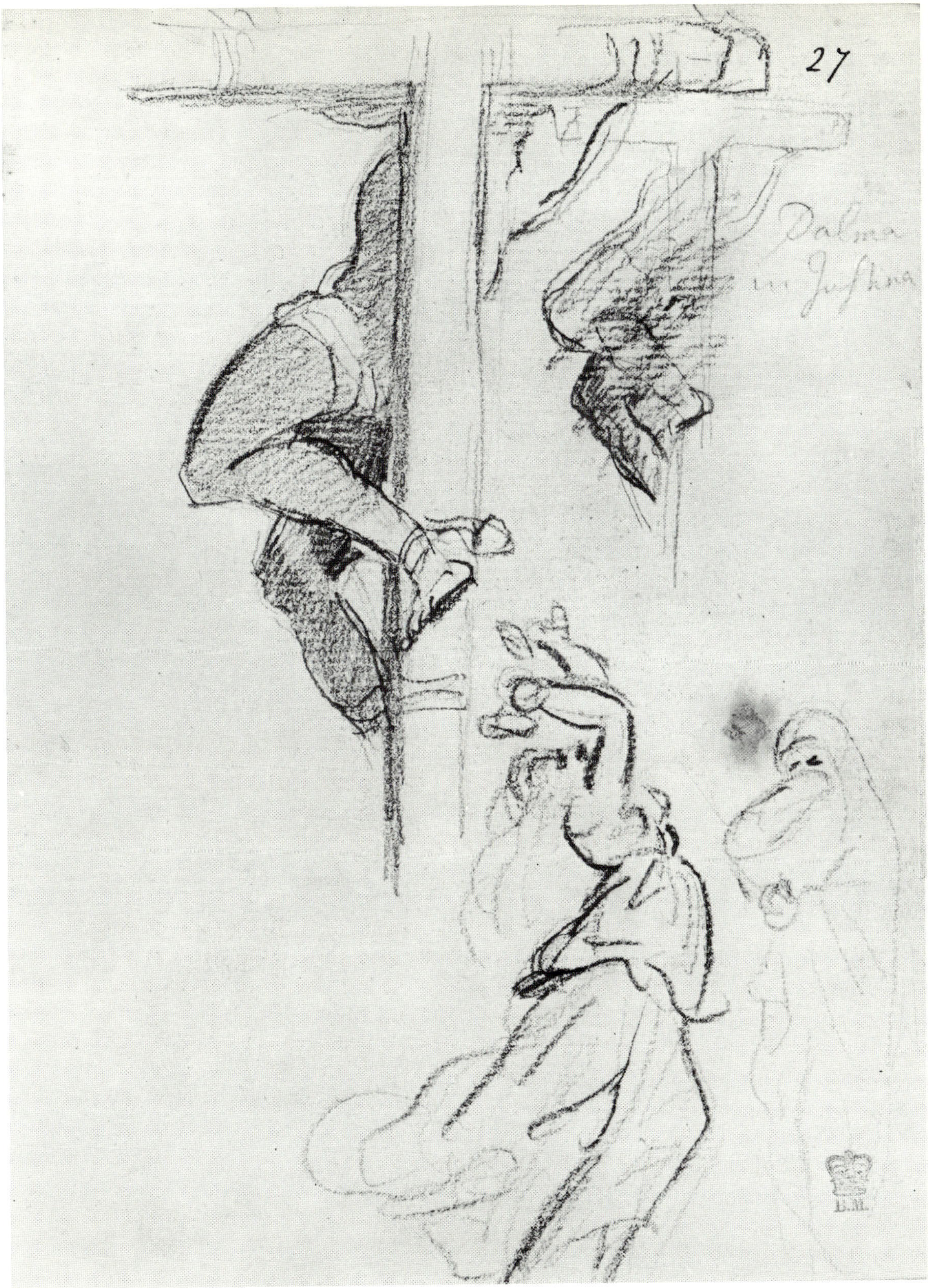

121. Reynolds: Pencil drawing inscribed *Palma | in Justina*, from an Italian Sketch-book. London, British Museum

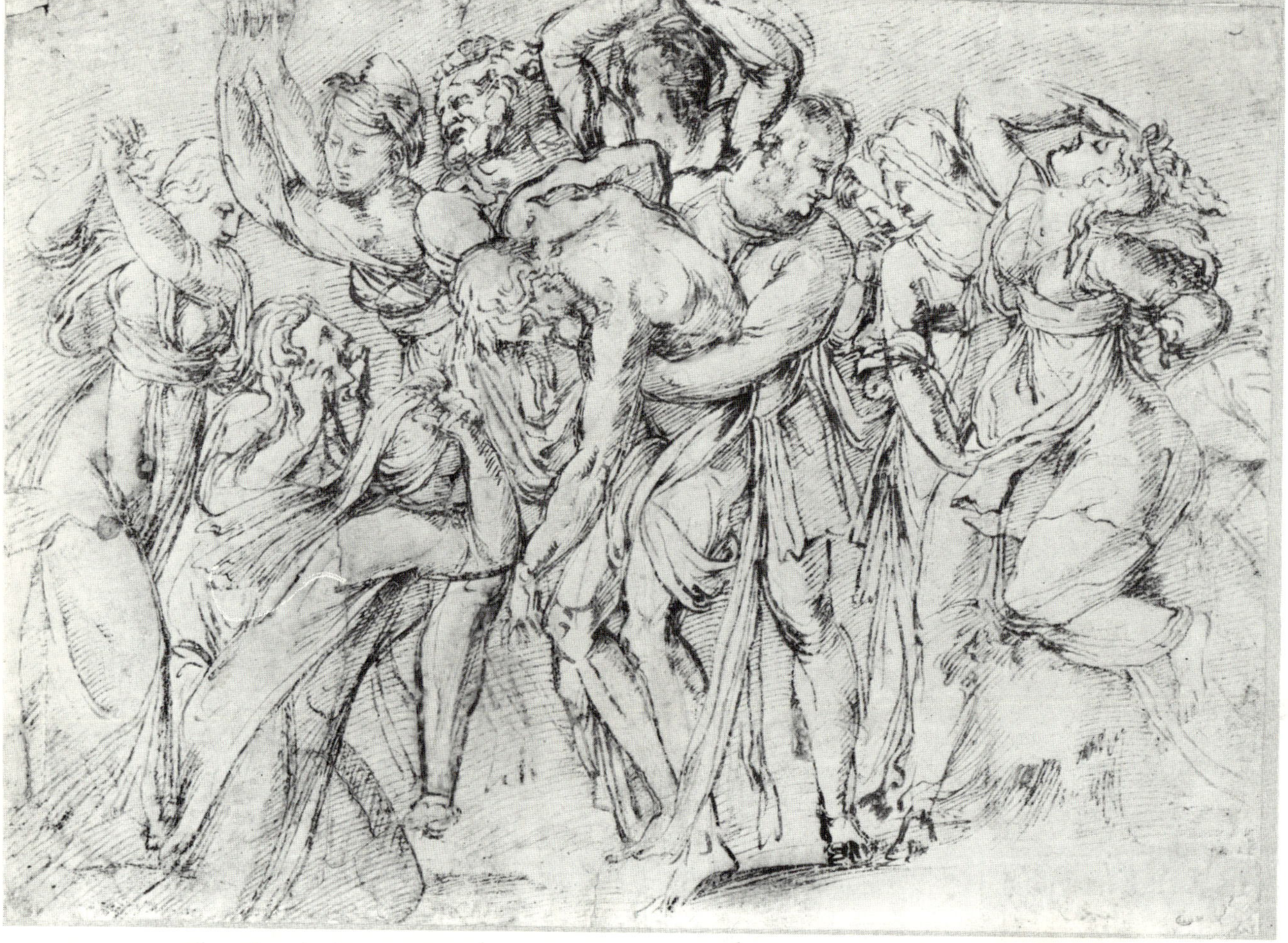

122. Baccio Bandinelli: *The Descent from the Cross*. Drawing. Paris, École nationale supérieure des Beaux-Arts

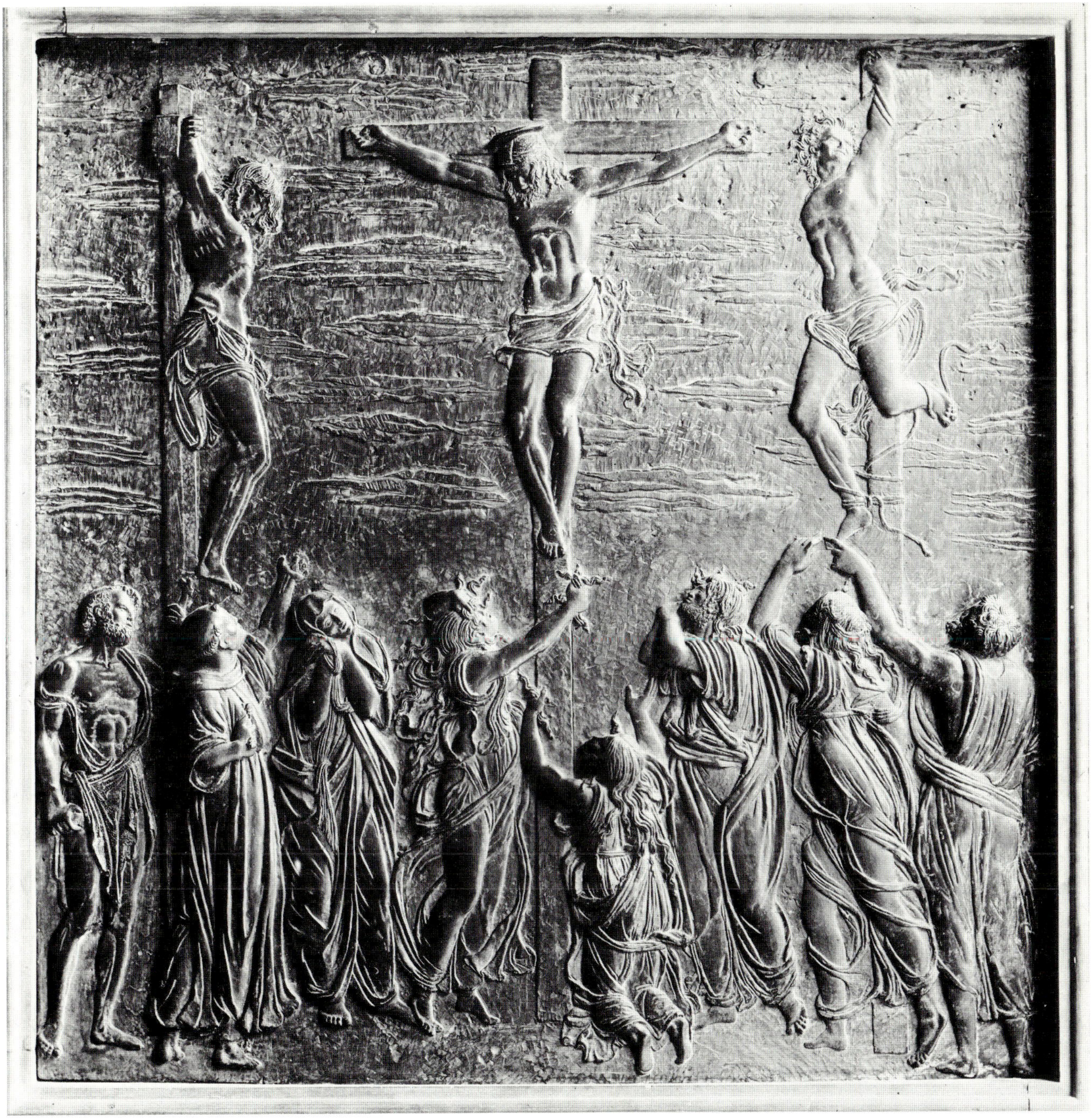

123. Bertoldo di Giovanni: *The Crucifixion*. Florence, Museo nazionale del Bargello

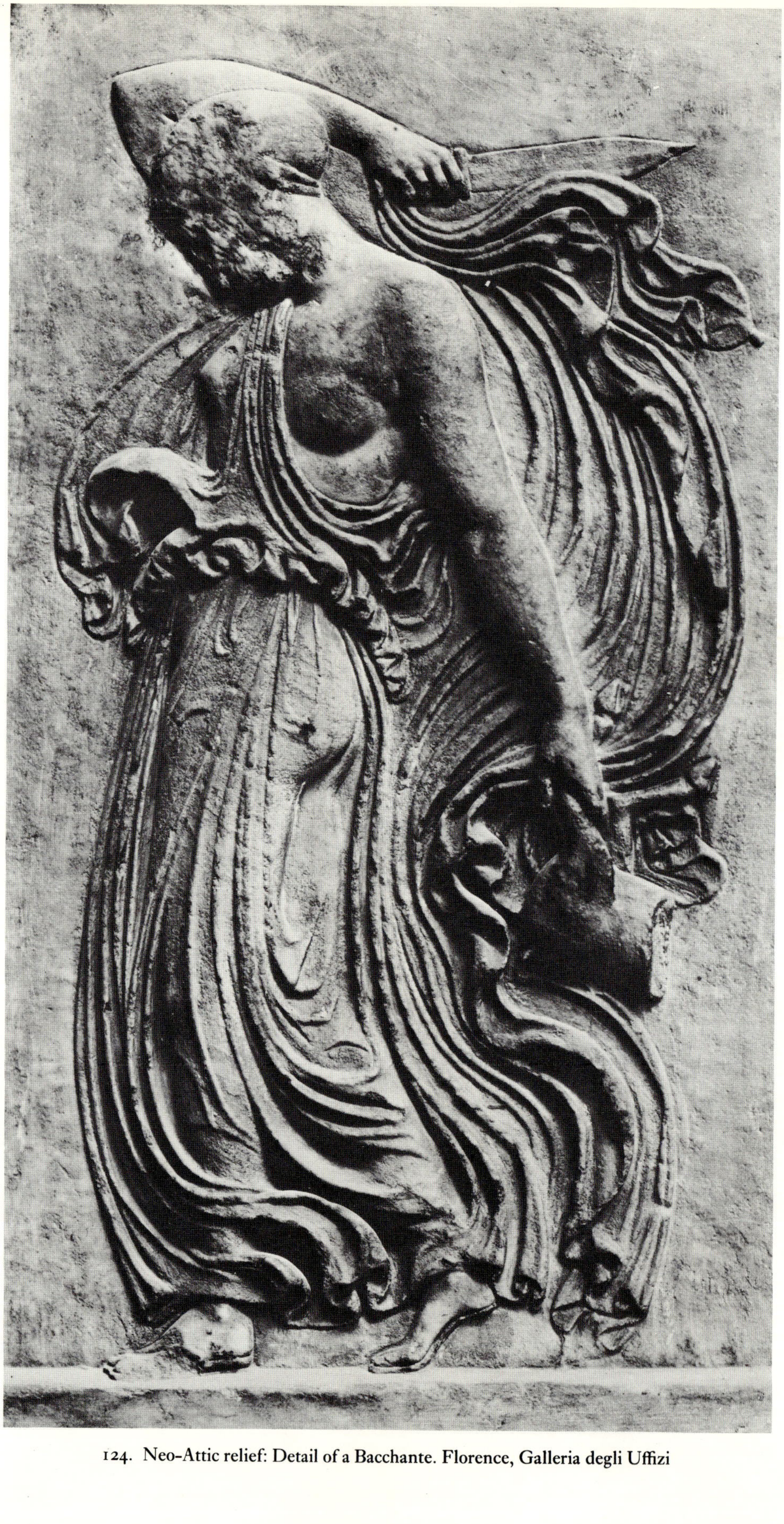

124. Neo-Attic relief: Detail of a Bacchante. Florence, Galleria degli Uffizi